NATURALISTIC DECISION MAKING
AND MACROCOGNITION

T0264694

Naturalistic Decision Making and Macrocognition

Edited by
JAN MAARTEN SCHRAAGEN
TNO Human Factors, The Netherlands

LAURA G. MILITELLO
University of Dayton Research Institute, US

TOM ORMEROD
Lancaster University, UK

RAANAN LIPSHITZ
The University of Haifa, Israel

CRC Press
Taylor & Francis Group
Boca Raton London New York

CRC Press is an imprint of the
Taylor & Francis Group, an **informa** business

CRC Press
Taylor & Francis Group
6000 Broken Sound Parkway NW, Suite 300
Boca Raton, FL 33487-2742

First issued in paperback 2017

© 2008 by Jan Maarten Schraagen, Laura G. Militello, Tom Ormerod and Raanan Lipshitz
CRC Press is an imprint of Taylor & Francis Group, an Informa business

No claim to original U.S. Government works

Version Date: 20160226

ISBN 13: 978-1-138-07270-1 (pbk)
ISBN 13: 978-0-7546-7020-9 (hbk)

Visit the Taylor & Francis Web site at
http://www.taylorandfrancis.com

and the CRC Press Web site at
http://www.crcpress.com

Contents

PART III MICRO-MACRO RELATIONSHIPS

PART IV ALTERNATIVE APPROACHES

List of Figures

List of Tables

Notes on Contributors

Shilo Anders is a doctoral candidate in Cognitive Systems Engineering at Ohio State University. She is currently working with Drs Woods and Patterson on problems in information analysis, data overload, predictive analysis, and adaptive capacity in both military and healthcare domains. Prior to joining Ohio State University, she worked in the Human Factors Group at the University of Dayton Research Institute where she collaborated with Ms Militello applying methods of cognitive task analysis in both military and healthcare domains. Ms Anders also participated in the Institute for Collaborative Innovation at Ohio State University where she was a member of an interdisciplinary team investigating advancements in intelligence analysis. She received her MA in Experimental Psychology and Human Factors from the University of Dayton in 2004.

Steven Asch MD, MPH is the Deputy Associate Chief of Staff for Health Services Research at the VA Greater Los Angeles Healthcare System, a health policy analyst at RAND and an Associate Professor of Medicine at the David Geffen School of Medicine at UCLA. His research applies quality measurement systems to ensure that patients get needed medical care, particularly in the areas of communicable disease and public health. Dr Asch has led several national projects developing broad-based quality measurement tools for veterans, Medicare beneficiaries, and the community. He also directs a multi-site center for quality improvement in HIV-positive patient treatment as part of the Quality Enhancement Research Initiative (QUERI). His educational efforts are focused on training physician fellows in health services research, and he serves as the director of the VA Los Angeles Health Services Research Fellowship and Associate Director of the UCLA RWJ Clinical Scholars Program. Dr Asch is a practicing internist and palliative care physician and the author of more than a hundred peer-reviewed articles.

Adrian Banks is a Lecturer in Psychology in the Department of Psychology, the University of Surrey, UK. The goal of his research is to understand how people use their knowledge to reason and make decisions in naturalistic situations. Of particular concern is the role of prior beliefs, causal cognition, expertise, and distributed cognition. These studies involve experiments and computer simulation, in particular the ACT-R cognitive architecture.

Emma C. Barrett has worked for over a decade with United Kingdom (UK) law enforcement agencies and government departments and is currently employed as a Behavioural Science Adviser with the UK government. In addition to various work-related projects, Emma is also carrying out research on the development of investigator expertise, in conjunction with several UK police forces. The focus of this research, part of a PhD program at the University of Birmingham, UK, is the cognitive mechanisms underlying investigative situation assessment, the process by which investigators make sense of information available during complex criminal investigations. Her other research interests include interview strategies for informants and suspects, interpersonal persuasion and deception.

Craig Bonaceto is a Senior Information Systems Engineer in MITRE's Command and Control Center, where he applies Cognitive Systems Engineering, Cognitive Task Analysis, and Software Engineering methods to improve the design of systems in air traffic control, military command and control, and intelligence analysis. He performs research on methods to model and understand the cognitive and decision-making demands of complex work domains as a basis for the design of decision support concepts and work teams. Mr Bonaceto also performs research on human "mental models" to improve Human-System Integration. He holds a BS degree in Computer Science and Psychology from Rensselaer Polytechnic Institute.

Kevin Burns is a Principal Scientist at the MITRE Corporation (1999–present). His interest is computational modeling of cognitive processing, including strategic decisions and visual perception, to improve the design of decision support systems. Before MITRE was a twenty-year career in the commercial nuclear power industry, performing computational modeling of reactor physics and thermal hydraulics, as well as probabilistic risk assessment and human reliability analysis. Kevin holds engineering degrees from the Massachusetts Institute of Technology where he also studied cognitive science and media arts.

Margaret T. Crichton (MA(Hons), MSc, PhD Psychology) is founder of People Factor Consultants, and is a Chartered Psychologist. She held a post as Research Fellow at the University of Aberdeen, from 1998 until March 2005. Her main interests are developing and presenting training in human factor skills such as decision making (especially under stress), situation awareness, and communication in high-hazard organizations, for example, offshore oil and gas drilling teams, UK nuclear power production, emergency services (Police, Fire and Rescue Services), and the Scottish Executive. She has also been involved in developing computer-based training for decision-making skills improvement, as well as observing and monitoring emergency exercises. Her publications include *Safety at the Sharp End: A Guide to Non-technical Skills* by R. Flin, P. O'Connor and M. Crichton (Aldershot: Ashgate, 2008), and "Incident Command Decision Making" in *Incident Command: Tales from the Hot Seat* (Flin and Arbuthnot, Ashgate, 2002), as well as other articles published in academic and industry publications.

Stephen Deutsch is a Division Scientist at BBN Technologies where he has been actively involved in human performance modeling for more than twenty years. Steve led the development of the original Operator Modeling Architecture (OMAR) suite of software tools for human performance modeling. He made important contributions to the instantiation of the psychological framework that forms the basis for several OMAR-based human performance models. Steve was then principal investigator leading the evolution of the human performance modeling framework as the Distributed Operator Model Architecture (D-OMAR). Steve was the principal investigator for a series of NASA Ames Research Center research efforts that developed models of error in human performance on commercial aircraft flight decks. He has led related research that examined decision making on the flight deck. Steve has also developed human performance models to portray flight crew situation awareness and air traffic controller workload within the complex information flow environment of flight crews and controllers. Steve was principal investigator leading BBN's effort as the moderator for the Agent-based Modeling and Behavior Representation (AMBR) experiments that compared and contrasted the performance of four human performance modeling systems with human subjects on selected tasks in the air traffic control domain.

Steven Estes is a Senior Human Factors Engineer with the MITRE Corporation's Center for Advanced Aviation System Design. His research interests included computational cognitive modeling, working memory load, and aviation psychology. His experience includes work in the areas of HCI, runway safety, air traffic flow management, en route and terminal air traffic control, and commercial flight decks.

John M. Flach Due to some serendipitous events I ended up in graduate school at The Ohio State University working with Rich Jagacinski in 1978. Six years later, I finally completed the PhD and joined the faculty at the University of Illinois where I had joint appointments in Mechanical and Industrial Engineering (50 percent), Psychology (25 percent); and the Institute of Aviation (25 percent). I soon learned that I wasn't able to satisfy three masters. In 1990, I failed to get tenure. Fortunately, I found a home at Wright State University where supportive colleagues helped me over the tenure hurdle and up the academic ladder to the rank of Professor. Due to another twist of fate, I became Chair of the Psychology Department in 2004. In addition to chair responsibilities, I teach graduate and undergraduate courses in the areas of cognitive psychology and cognitive systems engineering. My research stems from my failures at sports, ineptitude with music, and uneasiness with technology. I came to psychology seeking reasons why things that seemed to come so easily to others were difficult for me. Despite over twenty years of research, however, I remain a clumsy athlete, an inept musician, and am always the last to learn how to use new technologies.

Rhona Flin (BSc, PhD Psychology) is Professor of Applied Psychology in the School of Psychology at the University of Aberdeen. She is a Chartered Psychologist, a Fellow of the British Psychological Society and a Fellow of the Royal Society of Edinburgh. She directs a team of psychologists working with high-risk industries on research and consultancy projects concerned with the management of safety and emergency response. The group has worked on projects relating to aviation safety, leadership and safety climate in shipping and in the offshore oil and gas industry, team skills and emergency management in the nuclear industry. She is currently studying surgeons', anesthetists' and scrub nurses' cognitive and social skills and is leading the Scottish Patient Safety Research Network. Her latest book (with O'Connor and Crichton) is *Safety at the Sharp End: A Guide to Non-Technical Skills* (Aldershot: Ashgate, 2008).

Julie Gore is a Senior Lecturer in Organizational Behaviour at the School of Management, the University of Surrey, UK. She completed her PhD in Applied Psychology (Oxford Brookes University) and is a Chartered Psychologist. Her current research interests focus upon the application of behavioral science to management research in particular: cognitive psychology; qualitative methodologies including cognitive task analysis and a long-term interest in NDM (having first joined the NDM community in Dayton, Ohio in 1994). Publications in NDM include Gore et al. (2006). "Naturalistic decision making: reviewing pragmatic science." in Raanan Lipshitz, Gary Klein, and John Carroll (eds), *Special Issue: Naturalistic Decision Making in Organizations. Organization Studies*, 27 (7): 925–42; and (2004) "Recruitment and selection in hotels: experiencing cognitive task analysis." In H. Montgomery, R. Lipshitz, and B. Brehmer (eds), *How Professionals Make Decisions*, Mahwah, NJ: Lawrence Erlbaum Associates.

Simon Henderson is a Principal Consultant in QinetiQ's Command and Intelligence Systems group in the UK, where he has worked since 1987. His work has involved the development of a wide range of data collection and knowledge-elicitation tools and techniques; experimental assessment of teamwork and decision making; the enhancement of team-based planning and decision-making processes, and the assessment and enhancement of organizational learning practices. Simon has worked across a variety of domains, including all military services and ranks, police armed-response units, software engineers, surgical teams, space crews, oil installation managers, project managers, and operational analysts. He has been the UK representative on a number of NATO, TTCP and other international panels on decision making and teamwork, and has published a wide range of papers, book chapters, and journal articles.

Robert R. Hoffman is a Senior Research Scientist at the Institute for Human and Machine Cognition in Pensacola, FL. He is a Fellow of the American Psychological Society and a Fulbright Scholar. He received his BA, MA, and PhD in experimental psychology at the University of Cincinnati. He was a Postdoctoral Associate at the Center for Research on Human Learning at the University of Minnesota, and then joined the faculty of the Institute for Advanced Psychological Studies at Adelphi University. He has been recognized internationally in psychology, remote sensing, weather forecasting, and artificial intelligence, for his research on human factors in remote sensing, terrain analysis, and weather forecasting, for his work in the psychology of expertise, the methodology of knowledge elicitation, and for work on human factors in the design of workstation systems and knowledge-based systems. Hoffman's major current project involves the methodologies of human-centered computing and the theory of complex cognitive systems. His latest books are; *Working Minds: A Practitioner's Handbook of Cognitive Task Analysis*; *Expertise Out of Context*; *Minding the Weather: How Expert Forecasters Think*; and *The Cambridge Handbook of Expertise and Expert Performance*.

Erik Hollnagel is Industrial Safety Chair at École des Mines de Paris (France) and Professor of Cognitive Systems Engineering at Linköping University (Sweden). Since 1971, he has worked at universities, research centers, and industries in several countries with problems from several domains, such as nuclear power production, aerospace and aviation, software engineering, healthcare, and land-based traffic. His professional interests include industrial safety, accident analysis, cognitive systems engineering, cognitive ergonomics and intelligent human-machine systems. He has published widely including 12 books, the most recent titles being *Resilience Engineering: Concepts and Precepts* (Aldershot: Ashgate, 2006), *Joint Cognitive Systems: Patterns in Cognitive Systems Engineering* (London: Taylor and Francis, 2005), and *Barriers and Accident Prevention* (Aldershot: Ashgate, 2004). Erik Hollnagel is joint Editor-in-Chief with Pietro C. Cacciabue of the international journal, *Cognition, Technology and Work*.

David W. Jones is the Department Head for Environmental and Information Systems at the Applied Physics Laboratory, University of Washington (APL-UW). In his own research, he studies the application of cognitive engineering solutions for decision tasks in the meteorological and oceanographic domains. Recently, he has participated in several multidisciplinary teams that have studied the impact of web-based applications on military planning and decision making. Prior to his position at APL-UW, he served as a United States Navy officer for twenty-one years, specializing in operational meteorology and oceanography. His last position in the Navy was as the Director of Operations at Fleet Numerical Meteorology and Oceanography Center, Monterey, CA.

Dr. Susan Joslyn is a cognitive psychologist who conducts research that investigates cognition in applied settings. Her work focuses on decision making in several different domains including air traffic control, 911 dispatching and weather forecasting. Her work on individual differences in multi-tasking abilities won the American Psychological Association's Division of Experimental Psychology 1999 New Investigator Award. She is currently involved in a project studying how people understand and use weather forecast uncertainty information to make decisions. She is also a lecturer at the University of Washington, teaching courses on general cognition, human memory, human performance, working memory and applied issues.

Gary Klein (PhD) is Chief Scientist of Klein Associates, a group he formed in 1978 to better understand how to improve decision making in individuals and teams. The Klein Associates Division is now part of Applied Research Associates, Inc. Dr Klein is one of the founders of the field of Naturalistic Decision Making. His work on recognitional decision making has influenced the design of new systems and interfaces, and decision training programs. He has extended his work on decision making to describe problem detection, option generation, sensemaking, planning, and replanning. In order to perform research on decision making in field settings, Dr Klein and his colleagues have developed new methods of Cognitive Task Analysis. Dr Klein received his PhD in experimental psychology from the University of Pittsburgh in 1969. He was an Assistant Professor of Psychology at Oakland

University (1970–74) and worked as a research psychologist for the US Air Force (1974–78). He is a fellow of the American Psychological Association. He has written more than 70 papers and has authored three books: *Sources of Power: How People Make Decisions* (1998, MIT Press); *The Power of Intuition* (2004, A Currency Book/ Doubleday); and *Working Minds: A Practitioner's Guide to Cognitive Task Analysis* (Crandall, Klein and Hoffman, 2006, MIT Press).

Helen Altman Klein is Professor of Psychology and member of the Graduate Faculty in the Human Factors/Industrial and Organizational Psychology Program at Wright State University. She has held visiting positions at Western Australia Institute of Technology and The Hebrew University in Israel. Dr Klein directs the Applied Psychology Laboratory that undertakes theory-based empirical and applied research in macrocognition and naturalistic decision making. She developed the Cultural Lens Model and pioneered the inclusion of national differences in culture into the study of naturalistic decision making. Dr Klein's research group has investigated domains including multinational civil aviation safety, multinational peacekeeping collaborations, foreign command decision-process modeling, and asymmetrical operations planning. They have extended the Cultural Lens Model to Middle Eastern groups and are developing simple measures for predicting complex cognition. The Laboratory also investigates health care systems problems including medication instructions, health care delivery to foreign nationals, and Type 2 diabetes self-management. Dr Klein has presented her research to a wide variety of professionals in the United States and abroad, has reviewed for a number of journals, and has served as a consultant to many governmental and private-sector organizations.

Kristina Lauche is Associate Professor of Design Theory and Methodology at the Faculty of Industrial Design engineering at Delft University of Technology, the Netherlands. She trained as a work psychologist at the Free University of Berlin and received her PhD from the University of Potsdam in 2001. Dr Lauche has worked as a researcher and lecturer at the universities in Munich (1995–97), ETH Zurich (1997–2001) and Aberdeen (2001–2005). Her research centers on team interaction in complex environments and the design and use of complex systems. She has worked in high-risk industries (oil and gas, police, fire service) as well as manufacturing and consumer goods. Dr Lauche currently supervises PhD projects on organizational change, team mental models and strategic decision making in new product development.

Mei-Hua Lin is a Doctoral Candidate in the Human Factors/Industrial and Organizational Psychology Program at Wright State University. She received her Bachelor of Science degree in Psychology at University of Iowa and her Master of Science degree in Industrial and Organizational Psychology at Wright State University. Her master's thesis, "The Role of Analytic-Holistic Thinking on Sensemaking," examined national differences in Analytic-Holistic Thinking and its influence on information recalled during sensemaking. She has presented her work on topics including cross-cultural cognition, performance measures, group processes, and multinational teamwork at the Society for Industrial and Organizational

Psychology (SIOP) Conference and the 7th Naturalistic Decision Making (NDM) Conference. She has also published in the area of human development. Ms Lin is currently involved in data management and statistical analysis for a cross-cultural research effort that is measuring the cognition of over a thousand participants in seven nations from the regions of North America, South Asia, East Asia, and Southeast Asia.

Jasper Lindenberg is currently a researcher and program manager at TNO, the Netherlands organization of applied scientific research. Jasper received his MSc in Cognitive Artificial Intelligence from Utrecht University in 1998. His current and past research includes: adaptive automation, human system integration, cognitive support systems, intelligent interfaces, mental load assessment and performance prediction of human-agent systems. Jasper has applied his research and expertise in various domains. He has worked on system optimization and concept development for the Royal Netherlands Navy, the development and evaluation of various electronic consumer services, the specification of astronaut support systems, improving knowledge sharing and error prevention for the police, designing support for mobile services and realizing effective alarm handling for the maritime sector.

Claire McAndrew is a PhD researcher in Organizational Behaviour at the School of Management, the University of Surrey, UK. She holds an MSc in Occupational and Organisational Psychology attained from the School of Human Sciences at the University of Surrey. Her research in NDM is essentially underwritten by a concern for cross-disciplinary dialogue as a means of methodological advance. Particular areas of interest lie in the application of NDM to the fields of Managerial Cognition and Strategy As Practice to understand both decision making and the nature of micro-strategizing in organizations. Research to date has involved the combinatorial use of Applied Cognitive Task Analysis, cognitive mapping methods and cognitive modeling architectures such as ACT-R and Convince Me. These support her drive for the closing of the quantitative versus qualitative and macrocognitive versus microcognitive divides that have occurred within the Social Sciences generally. Her PhD extends this interdisciplinary focus to the investment arena, exploring NDM within fund managers' and day traders' expertise. She has presented her work within the NDM community at the 7th and 8th International Conferences on Naturalistic Decision Making and through a number of other research forums.

Anna P. McHugh is a Senior Cognitive Scientist at Klein Associates Division of Applied Research Associates, specializing in uncovering the nature of human cognition and skilled performance in individuals and teams across a variety of applied domains. She has led multiple projects aimed at improving human performance through means such as team development, physical space redesign, and decision-support technology. She has recently led several efforts to uncover cultural differences in cognition, including projects for the Air Force Research Lab to expand the Cultural Lens Model to Middle Eastern culture, and an effort to support military crowd control in the Middle East. She has also led projects for the Army Research Lab to understand differences in teamwork across cultures and to facilitate successful multinational collaboration. In addition to her work in the cultural realm, Ms McHugh has applied the NDM perspective to several efforts in the healthcare arena, in the crisis response domain, and in consumer cognition. Ms McHugh holds a MS in Industrial-Organizational and Human Factors Psychology from Wright State University, Dayton, OH and a BA in Psychology and Spanish from Denison University, Granville, OH. She is a member of the American Psychological Association (APA), the American Psychological Society (APS), and The Society for Industrial/Organizational Psychology.

Laura G. Militello is a Senior Research Psychologist in the Human Factors group at the University of Dayton Research Institute. She has extensive experience conducting cognitive task analysis across a broad range of domains including critical care nursing, air campaign planning, weapons directing, and consumer decision making. Ms Militello contributed to the development of a set of applied cognitive task analysis methods (ACTA) for use by practitioners, She has conducted more than twenty cognitive task analysis workshops for human-factors professionals and students. Her most recent work explores the idea of expanding conative elements within cognitive task analysis. In addition, she has recently become interested in collaboration research, particularly in the context of command and control in both military and crisis management settings. She is currently examining issues of trust in collaboration via a series of lab-based studies in simulated logistics tasks. Ms Militello is also involved in an ongoing effort to examine the contributions of and interactions among technology, human agents, and organization structures in military command and control. Before joining the University of Dayton Research Institute, Ms Militello worked as a Senior User Interface Designer for Acuson Corporation, manufacturer of high-end diagnostic ultrasound equipment.

Peter Moertl is a Lead Human Factors Engineer at MITRE Corporation's Center for Advanced Aviation System Development. He leads human factors and research, engineering, and concept development for flight deck and air traffic operations for advanced aviation technologies and procedures, including ADS-B applications. He has provided human-factors research and engineering in aviation safety in cooperation with the Federal Aviation Administration and supported integration of human factors into systems engineering and safety management systems. Peter Moertl has a doctoral degree in applied cognitive psychology from the University of Oklahoma and a master's degree in psychology from the University of Graz.

Kathleen L. Mosier is a Professor and Chair of the Psychology Department at San Francisco State University. She is also an elected Fellow in the Human Factors and Ergonomics Society. Dr Mosier received her PhD in Industrial/Organizational Psychology from the University of California, Berkeley in 1990. Following this, she worked in the Human Factors Research Branch at NASA Ames Research Center for seven years before coming to SFSU. Her research has focused on decision making in applied contexts, such as aviation, medicine, and the workplace. She co-organized and hosted NDM8, the Eighth International Conference on Naturalistic Decision Making.

Mark A. Neerincx is head of the Intelligent Interface group at TNO Human Factors, and Professor of Man-Machine Interaction at the Delft University of Technology. He has extensive experience in fundamental and applied research on human–computer interaction, among other things in the domains of space, defense, security, and health. Important results are (1) a cognitive task-load model for task allocation and adaptive interfaces, (2) models of human–machine partnership for attuning assistance to the individual user and momentary usage context, (3) cognitive engineering methods and tools, and (4) a diverse set of usability "best practices." Dr Neerincx has published an extensive set of papers for peer-reviewed journals and international conferences, and has been involved in the organization of conferences, workshops, and tutorials to discuss and disseminate human-factors knowledge.

Tom Ormerod is a Professor of Cognitive Psychology and currently head of the Psychology faculty at Lancaster University, UK. He has published over a hundred refereed journal and conference outputs on problem solving (notably on insight, analogy and creative design expertise), deductive reasoning (with a recent focus upon investigative expertise), and human–systems interaction design and evaluation. His research uses mixed methods (from ethnography to experimentation) to study expertise in both laboratory and workplace settings. His PhD research on Cognitive Processes in Logic Programming led to research into the nature and support of design expertise, with more than $1 million funding from the UK Research Councils' Cognitive Engineering and PACCIT Initiatives. Recent projects have explored the design of software browsers to support the collaborative handling of digital photographs, and studies of how homicide and insurance fraud investigators differ in their use of deductive inference. He is currently working on projects that explore expertise in the identification and resolution of suspicions across a range of counter-terrorism activities.

Emily Patterson is a Research Scientist at the VA Getting at Patient Safety (GAPS) Center at the Cincinnati VAMC and at the Institute for Ergonomics at Ohio State University. At OSU, she is Associate Director of the Converging Perspectives on Data (CPoD) interdisciplinary consortium. Her research spans a wide variety of projects applying human-factors research to improve joint cognitive system performance in complex, sociotechnical settings, including healthcare, military, intelligence analysis, space shuttle mission control, emergency response, and emergency call centers. Dr Patterson's current work focuses on medical informatics for patient safety, handover communications, resilience to human error, and rigor in information analysis. She received a Career Award from the Veteran's Administration Health Services Research & Development (VA HSR&D) and the Alexander C. Williams, Jr., Design Award from the Human Factors and Ergonomics Society. She serves as an Editorial Board Member for Human Factors, Advisory Board Member for ECRI's Health Technology Forecast, and Centers Communication Advisory Group Member for the Joint Commission International Center for Patient Safety.

Jason Saleem is a Research Scientist at the VA HSR&D Center on Implementing Evidence-Based Practice, Roudebush VA Medical Center, as well as the Indiana University Center for Health Services and Outcomes Research, Regenstrief Institute.

He is also an Assistant Research Professor with the Department of Electrical and Computer Engineering, Indiana University-Purdue University at Indianapolis (IUPUI). Dr Saleem is a member of the Human Factors and Ergonomics Society (HFES) and American Medical Informatics Association (AMIA). Dr Saleem received his PhD from the Department of Industrial and Systems Engineering (ISE) at Virginia Tech in 2003, specializing in human factors. During his graduate training and postgraduate experience, Dr Saleem has been involved in the study and design of systems in complex domains such as industry, aviation, and healthcare, and has contributed original human-factors investigations to the literature in each of these areas. Dr Saleem's current research involves application of human-factors engineering to enhance clinical information systems, including electronic decision support, as well as redesign of healthcare processes for improved safety.

Jan Maarten Schraagen is the Senior Research Scientist with the Department of Human in Command at TNO Human Factors, one of the five business units of TNO Defence, Security and Safety. His research interests include task analysis, team decision making, and adaptive human–computer collaboration. He has served as Principal Investigator on projects funded by ONR and the Dutch Ministry of Defence, recently as program manager for the Navy program "Human-System Task Integration" (2004–2005). Dr Schraagen has served as chairman for NATO Research Study Group 27 on Cognitive Task Analysis (1996–98). He was first editor of *Cognitive Task Analysis* (2000, Lawrence Erlbaum Associates). He has organized the 7th International Conference on Naturalistic Decision Making (2005) in Amsterdam, having been part of the NDM community since 1994. Dr Schraagen holds a BA in Philosophy and Psychology from the University of Groningen, an MA in Psychology from the University of Groningen, and a PhD in Cognitive Psychology from the University of Amsterdam, The Netherlands. He has spent several years studying at the University of Winnipeg, Canada (1978–79), Manchester University in the UK (with Dr Graham Hitch; 1984–85), and Carnegie-Mellon University, Pittsburgh, PA (with Dr Marcel Just; 1985–86). In 2000–2001, he was a visiting exchange scientist at the Naval Air Warfare Center Training Systems Division, Orlando, FL. Dr Schraagen has worked at TNO since 1986.

Winston R. Sieck, PhD, is a Senior Scientist at Klein Associates Division of Applied Research Associates. He leads the Methods, Models, and Metrics (M3) center at Klein Associates, as well as the Culture and Cognition program. His research focuses on issues in decision making and related cognitive functions, including cultural influences on cognitive processes. He has served as Principal Investigator on projects funded by ONR, AFRL, and NSF to study issues such as uncertainty management, cross-cultural crowd control, and overconfidence. He has also led or participated in projects on tactical thinking assessment, information operations effectiveness, sensemaking, neglect of statistical decision aids, mathematical models of judgment under uncertainty, and cultural influences on judgment. He has published a number of papers on these issues in academic journals. Dr Sieck received a PhD in cognitive psychology from the University of Michigan in 2000, and an MA in statistics from the same university in 1995. He then taught statistics and was a research fellow at the University of Michigan (2000–2001), and then worked as a post-doctoral researcher in quantitative psychology at Ohio State University (2001–2003).

Jennifer L. Smith is a Research Scientist at Klein Associates Division of Applied Research Associates. Her interests include studying multinational teams and differences in teamwork across cultures. Ms Smith is currently working on a project sponsored by the Air Force to determine the influence of culture on crowd dynamics and crowd assessment. The goal of this project is to understand how to enhance warfighters' abilities to assess and manage crowds of varied cultures. Ms Smith holds a BA degree in Psychology from Kenyon College, Gambier, OH and is currently pursuing her doctorate in Social Psychology at Loyola University, Chicago, IL.

Kip Smith is a Professor of Cognitive Ergonomics and Cognitive Systems Engineering at Linköping University with appointments in the departments of Computer and Information Science (IDA) and Management and Engineering (IEI). His research and consulting focus on the quantitative assessment of the behavior of individuals and groups in dynamic environments, for example, distributed command and control, surface transportation, emergency management, and commercial aviation. Recent projects include assessing the impact of remote command and control on soldiers' trust in their leader, developing methods for detecting constraints on collaborative action, empirical studies of the effects of cultural differences on decision making in emergency services operations centers, and observational studies of the uses of information technology by air traffic managers at four Air Route Traffic Control Centers. He received his MS in Geophysics from the University of New Mexico in 1977 and his PhD in Management from the University of Minnesota in 1996. He is the author of twenty journal articles and more than a hundred conference papers and holds one patent. Intemperate immersion in hydrothermal systems is his primary avocation.

James J. Staszewski is a cognitive psychologist in the Carnegie Mellon University (CMU) Psychology Department. He holds a BA from Holy Cross College and a PhD from Cornell University. Herbert Simon advised his dissertation work, which was performed at CMU. His basic research has focused on understanding the mechanisms underlying expert performance and its development. Longitudinal laboratory studies have shown how individuals with normal abilities achieve extraordinary performance on high-demand tasks. Theoretical analyses have identified the content and organization of acquired knowledge underlying expertise and have shown how experts relax information-processing constraints imposed by working memory to develop and exploit their extensive, well-organized knowledge bases. These studies have also shown that models of expertise are useful for designing instructional programs for developing the skills of novices. Recent "use-inspired basic research" has examined this latter finding. His cognitive engineering approach has applied methods for understanding the cognitive bases of expert performance in field settings to use resulting expert models as blueprints to design instructional programs. He has used this approach to develop training programs adopted and used by the US Army to train operators of handheld landmine detection equipment, work for which he has been decorated.

Paul Taylor is Senior Lecturer in Forensic Psychology at Lancaster University, UK, and a research associate of the Police Research Laboratory (Carleton University) and the Laboratory for Bounded Rationality and the Law (Memorial University). He is a Fellow of the Royal Statistical Society and an executive board member of the International Association of Conflict Management. At Lancaster, Paul is director of the Department's postgraduate course on investigation and expertise, a proportion of which teaches naturalistic decision-making theory and research. His research focuses on sensemaking and strategic choice in law enforcement investigations. He has published over 30 journal articles and conference papers on hostage negotiation in journals such as *British Journal of Social Psychology* and *Human Communication Research*, and on police decision strategies in journals including *Applied Cognitive Psychology* and *Criminal Justice and Behavior*. Paul received the Earl Scheafer Best Research Paper award, and, more recently, the 2007 IACM best applied paper award, for this research.

Rogier Woltjer is a PhD candidate in the field of Cognitive Systems Engineering at the Department of Computer and Information Science (IDA) of Linköping University, Sweden. His research focuses on functional modeling and constraint management in the behavior of complex sociotechnical systems, in the domains of emergency/crisis management, command and control, and aviation system safety. He works particularly on how functional models that express constraints can be used in (re-) designing decision support systems, tasks, and working and training environments, and in understanding incidents and accidents. He received an MSc in Artificial Intelligence from the Free University of Amsterdam, The Netherlands, and the LicEng. in Quality and Human-Systems Engineering from Linköping University.

Laura A. Zimmerman, PhD, is a Senior Scientist at Klein Associates, a Division of Applied Research Associates. Her research interests include domain-specific cognitive skills and expertise development, and decision-centered training. The focus of her work has been on vignette development, knowledge elicitation, and web-based decision skills training. As a member of Klein Associates, she has worked on projects developing effective training for complex, multi-agency emergency responder and military command environments, with focus on developing cognitively relevant training scenarios within contextually relevant computerized and live environments. Prior to joining Klein Associates, Dr Zimmerman conducted research on the decision processes of experienced and novice police officers and developed a decision skills training program for patrol officers. Dr Zimmerman holds a BA in Psychology from San Francisco State University, an MA in Experimental Psychology from University of Texas at El Paso, and a PhD from University of Texas at El Paso in Experimental Psychology with a focus on Legal Psychology.

Foreword

Naturalistic decision making (NDM) is all about how experts make decisions in the real world. It is about how people *actually* make decisions—not about how they *should* make decisions. It is also about *real-life* tasks as opposed to *laboratory* tasks. Real-life tasks are frequently characterized by uncertainty, time pressure, risk, and multiple and changing goals. They involve multiple individuals and experienced decision makers often working in high-stakes organizational settings rather than the single inexperienced college student that so often serves as participant in laboratory experiments. In the NDM approach, it is the analysis of knowledge and skills underlying novice and expert performance that provides the basis for identifying leverage points for improving performance and specifying requirements for training and decision aids.

More recently, NDM has expanded to include the analysis of macrocognition. Similarly focused on the behavior of experts, it concentrates on developing a description of a wide range of cognitive functions. This focus is somewhat broader than historical NDM research and includes processes such as attention management, mental simulation, common ground maintenance, mental model development, uncertainty management, and course of action generation. These processes support the primary macrocognitive functions: decision making, sensemaking, planning, adaptation/replanning, problem detection, and coordination. Some of these, such as problem detection, emerge in field settings, are rarely considered in controlled laboratory-based experiments, and would be unlikely to emerge in typical laboratory studies of cognition. Macrocognition is what NDM is really about, after all.

It should come as no surprise, then, that the present volume links the concepts of naturalistic decision making and macrocognition in its title. The linkage also reflects the broadening in scope that was clearly apparent at the Seventh International Conference on Naturalistic Decision Making, on which this volume is largely based. The history of NDM goes back to 1989 when the first conference was held in Dayton, Ohio. Subsequent conferences were held in 1994 (Dayton, OH), 1996 (Aberdeen, Scotland), 1998 (Warrenton, VA), 2000 (Stockholm, Sweden), and 2003 (Pensacola Beach, FL). The seventh in this series of conferences was held in 2005 in Amsterdam, The Netherlands. Five themes were emphasized in this conference: decision making and training, adaptive decision support, cognitive ethnography, crime and investigation, and medical decision making. In sessions, the NDM framework was applied to new and diverse domains, such as landmine detection, judgments in crime situations, and space exploration. A panel session on macrocognition proved essential to the genesis of this book.

The editors would like to thank the following sponsors for making the seventh NDM conference possible: the European Office of Aerospace Research and Development, the Office of Naval Research Global, the Army Research Lab European Research Office, the Human Effectiveness Directorate Air Force Research Laboratory, the Air Force Office of Scientific Research, the Office of Naval Research, and the Human Research and Engineering Directorate Army Research Laboratory. In particular, we would like to thank Valerie Martindale, PhD of EOARD London and Mike Strub, PhD, then at ARL, for their indispensable help in obtaining funds. Jan Maarten Schraagen, the primary conference organizer, would like to thank the Dutch Ministry of Defence for providing funds under program V206 Human System Task Integration, and TNO Human Factors for its support, encouragement and guidance.

We would also like to express our gratitude to Guy Loft and Emily Jarvis at Ashgate for believing in this work and working with us so enthusiastically and so diligently.

Jan Maarten Schraagen,
Laura G. Militello,
Tom Ormerod,
and Raanan Lipshitz

PART I
Orientation

Chapter 1

The Macrocognition Framework of Naturalistic Decision Making

Jan Maarten Schraagen, Gary Klein, and Robert R. Hoffman

Introduction

Naturalistic Decision Making (NDM), as a community of practice, has the goal of understanding cognitive work, especially as it is performed in complex sociotechnical contexts. In recent years, the concept of "macrocognition" has emerged as a new and potential umbrella term to designate the broader paradigm that underlies NDM. In addition, the notion of macrocognition presents challenges and opportunities for both theory and empirical methodology. The present volume is a contribution to this literature, the seventh volume in the NDM series.

In this chapter we accomplish a number of things. First, we chart the history of NDM as a community of practice and then describe its stance concerning cognitive work and research methodology. Next, we chart the history of the concept of macrocognition and then show how NDM converges with it philosophically. Finally, we use these contexts to overview the chapters in this volume.

Emergence of the NDM Community of Practice

NDM as a community of practice began with the first conference in 1989 in Dayton, Ohio. That first conference was kept small—only about 30 researchers were invited, based on their interests and activities. Many had been funded by the Basic Research Unit of the Army Research Institute. Judith Orasanu, who was then working in this unit, provided ARI funding to organize the 1989 meeting. The goal of the meeting was simply to assess whether these researchers did in fact have a common, and perhaps distinctive set of goals and methods, and whether those were in any way coherent—even though many of them were studying different domains for different reasons.

The 1989 meeting was intended as a workshop to allow sharing of recent results and interests, but it sparked demand for follow-on gatherings. The NDM community has met every two to three years since then, alternating between North American and European venues. Seven such meetings have been held to date. Thus far, each of the NDM meetings has generated a book describing the research and the ideas of the conference participants (Hoffman, 2007; Klein et al., 1993; Zsambok and Klein, 1997; Flin et al., 1997; Salas and Klein, 2001; Montgomery, Lipshitz and Brehmer, 2005).

How the NDM researchers have managed to maintain their community of practice is perhaps somewhat mysterious. There is no formal society, no officers, no dues, and no newsletters. At the end of each conference, all the attendees who are interested in helping with the next conference gather together to select a host and site. There are always several volunteers to organize the next conference. The community is sustained by common interests and by a desire to find out what the other researchers have been up to. There is of course a great deal of behind-the-scenes work focused on securing sponsorships that really make the meetings possible. Supporters have included the Office of Naval Research, the Army Research Institute, the Army Research Laboratory, the Human Effectiveness Directorate of the US Air Force, the National Aeronautics and Space Administration (NASA), the US Navy, and TNO Human Factors.

In addition to the NDM meetings, many NDM researchers gather every year as part of the Cognitive Ergonomics and Decision Making Technical Group within the Human Factors and Ergonomics Society, and at meetings on Situation Awareness.

The Paradigm of Naturalistic Decision Making

In the 1980s, a number of researchers adopted a concept of decision making that seemed quite different from the standard "option generation and comparison" framework. Lipshitz (1993) tabulated nine different models of decision making that had been proposed by this emerging community of researchers over that decade. Two of the most widely cited models were Rasmussen's (1983, 1988) Skills/Rules/Knowledge account along with the "decision ladder," and Klein's (1989) Recognition-Primed Decision (RPD) model. The concept of decision making had often been defined in terms of a gamble: given two or more options, with certain information about the likelihood of each option to succeed, which is selected? However, the early NDM studies found that people (domain practitioners, consumers, managers, and so on) rarely made these kinds of decisions. Some have suggested the Klein, Calderwood, and Clinton-Cirocco (1986) study of firefighters marks the beginnings of NDM. Using a structured interview method, the researchers found that fire fighters do not evaluate options. They do not conduct anything like a "utility analysis" in which a list of options is generated, and each option is evaluated. More importantly, this is a domain in which decisions could not possibly be made using utility analysis. Thus, what purchase on reality was had by "normative" models that describe how rational decisions *should* be made? The house would burn down, or worse, people would die. In many domains, decision makers often have to cope with high-stakes decisions under time pressure where more than one plausible option does exist, but the decision makers use their experience to immediately identify the typical reaction. If they cannot see any negative consequence to adopting that action, they proceed with it, not bothering to generate additional options or to systematically compare alternatives. Thus, the metaphor of a decision as a gamble didn't seem to apply very often. If the metaphor of the decision as gamble failed to describe what practitioners usually encounter and usually do, NDM would abandon the metaphor and follow the phenomena.

NDM wanted to explore how domain practitioners make decisions in the "real world," under difficult conditions, in order to help them do a better job (Orasanu and Connolly, 1993). Such a goal statement should seem straightforward and yet it triggered a surprising amount of controversy. The lead article by Lipshitz et al. (2001) in a special issue on NDM in the *Journal of Behavioral Decision Making* was accompanied by skeptical commentaries from the Judgment and Decision Making community. Some questioned whether there was anything new about NDM that had not already been embraced by Behavioral Decision Making, and others doubted that NDM had much chance of succeeding. Those criticisms are orthogonal, of course, but were voiced with almost equal vigor and sometimes by the same people.

What is it that arouses such strong feelings, pro and con, for the NDM enterprise?

Points of Contention

We see three points of contention: approach to subject-matter experts, approach to improving decision making, and approach to research. First, NDM researchers do not see domain practitioners as infallible, but nevertheless respect their dedication, skills, and knowledge. And researchers deeply appreciate any "face time" with experts. NDM researchers want to document practitioner abilities in order to make sure that the subtle skills they have are recognized, understood, and supported in training programs and in decision support systems. NDM researchers seek to understand the true work (for example, information needs and decision requirements). This stance towards the participants in research puts NDM in conflict with some other decision researchers for a number of reasons. Some argue that experts are not special in any way, or that "expertise" is a biased, elitist notion. Some researchers take a fundamental stance: The belief that people tend to follow economic (or "rational") models of costs and benefits when they make decisions. NDM also conflicts with the "heuristics and biases" approach to decision making–NDM sees the strengths in the heuristics, but it looks beyond superficial attributions of human limitations (Flach and Hoffman, 2003), and does not assume that experts are as prone to biases as the literature on heuristics and biases suggests, or even that bias is an inherent and inevitable feature of human decision making.

Second, the NDM stance on improving decision making is to help practitioners apply their expertise more effectively, and help non-experts achieve expertise faster. NDM researchers do not assume that the practitioner has to be force-fed a probability scaling task in order to avoid one or another of the dozens of biases that are believed to pervade human thought. This stance seems to conflict with the position of Behavioral Decision Making to formulate strategies and aids that can replace or fix unreliable human judgment, for example, by having them work through a Bayesian probability evaluation procedure.

Third, the NDM stance on studying decision making emphasizes cognitive field research and cognitive task analysis (Hoffman and Woods, 2000; Schraagen, Chipman, and Shalin, 2000). Today, we have a rather large palette of cognitive task analysis methods (Crandall, Klein, and Hoffman, 2006; Hoffman and Militello, 2008), including the Critical Decision Method, Concept Mapping, various forms of task and goal analysis, and various types of experimental methods such as the Macrocognitive Modeling Procedure (see Klein and Hoffman, this volume). NDM researchers sometimes use simulations, but these have to reflect key challenges of the tasks and engage practitioners in realistic dilemmas. One thing NDM research does *not* do is use artificial paradigms that can be run on college "subjects" in 50-minute sessions.

One thing NDM research usually *does* is rely on methods of structured interviewing and task retrospection. So the very nature of the investigations causes discomfort to some experimental psychologists. NDM deliberately looks for "messy" conditions such as ill-defined goals, high stakes, organizational constraints, time pressure. Such conditions are difficult to capture in the laboratory but certainly determine the types of decision strategies people use in the "real world."

The mission of NDM—to understand how people make decisions under difficult conditions, and how to help them do a better job—meant that researchers could not confine themselves to particular tried-and-true paradigms or stovepiped "fundamental mental operations." Instead, NDM researchers expanded their focus from decision making to cognitive functionalities such as sensemaking, planning, replanning, and related phenomena such as mental modeling and the formation, use, and repair of "common ground" by teams. For example, McLennan and Omodei (1996) have examined pre-decision processes that appear to be critical to success. Mica Endsley's (1995a; Endsley and Garland, 2000) work on situation awareness is central to much of the NDM research. So is David Woods' examination of resilience and disturbance management (Hollnagel and Woods, 1983, 2005) and Vicente's (1999) description of Cognitive Work Analysis methodology. The very notion that decisions are things that are "made" came into question (Hoffman and Yates, 2005).

Since the original 1989 workshop, the NDM community has expanded its mission to "understanding how people handle difficult cognitive demands of their work, and trying to help them do a better job." To have retained an exclusive focus on decision making would have lost sight of the phenomena that were being studied, and could have disenfranchised some NDM researchers, including the authors of this chapter. Thus, NDM does not seem to be just the study of decision making. Certainly no one ever saw benefit to actually limiting the scope of investigations to decision making. The focus of interest has been more directed by the real-world settings that NDM researchers explore and by the demands that these settings place on the people who are responsible for getting the job done, efficiently, effectively, and safely. As a result, some have wondered whether NDM should change its name.

As the NDM framework broadened, researchers came to realize that they were interested in the cognitive functions that were carried out in natural settings. "Naturalistic Decision Making" was evolving into "Naturalistic Cognition." The same kind of mission still applied, and the same cognitive field research and cognitive task analysis methods still applied. But it was time to recognize that the interests of the NDM community had expanded. It came to be generally understood that the designation of NDM made sense primarily in historical context—as a reminder of the initial successes in discovering how decisions are made under time pressure and uncertainty and the importance of studying decision making in real-world contexts—but no longer captured the spirit and mission of the movement.

Origins of the Concept of Macrocognition

The line of discussion that led to the term "macrocognition" began in 1985 when at a NATO-sponsored conference on intelligent decision support systems for process control, Gunnar Johanssen distinguished micro- and macro- levels in an analysis of decision-making situations:

> Decision making is required on all levels of social life and in all work situations…The macro-operational situations are characterized by the need for decision making during such tasks as goal-setting, fault management, and planning in systems operations or maintenance of man–machine systems…The micro-operations situations involve decision making as an ingredient of control processes, either manual or supervisory, in man–machine systems. [pp. 328–31]

Although this is not quite the sense of macro–micro we rely on today in NDM, it is clearly pointing in the direction of looking at the phenomenology of cognitive work (see Klein et al., 2003).

Not surprising, given that David Woods was a participant in the 1985 conference and co-editor of the resultant volume (Hollnagel, Mancini, and Woods, 1985), the notion of macrocognition, and the distinction with microcognition was manifest in Woods and Roth's (1986) discussion of a hierarchy of decision-making situations— including organizational, macro-operational, and micro-operational levels, in reference to process control for nuclear power.

Ten years later, Pietro Cacciabue and Erik Hollnagel (1995) contrasted macrocognition with microcognition in order to present a view for human–machine systems design that would not take an information-processing approach. This alternative description is of cognitive functions that are performed in natural as opposed to laboratory settings:

> Micro-cognition is here used as a way of referring to the detailed theoretical accounts of how cognition takes place in the human mind…the focus is on "mechanisms of intelligence" per se, rather than the way the human mind works. Micro-cognition is concerned with the building of theories for specific phenomena and with correlating the details of the theories with available empirical and experimental evidence. Typical examples of micro-cognition are studies of human memory, of problem solving in confined environments (for example, the Towers of Hanoi), of learning and forgetting in specific tasks, of language understanding, and so on. Many of the problems that are investigated are "real," in the sense that they correspond to problems that one may find in real-life situations—at least by name. But when they are studied in terms of micro-cognition the emphasis is more on experimental control than on external validity…Macro-cognition refers to the study of the role of cognition in realistic tasks, that is in interacting with the environment. Macro-cognition only rarely looks at phenomena that take place exclusively within the human mind or without overt interaction. It is thus more concerned with human performance under actual working conditions than with controlled experiments. [pp. 57–8]

Cacciabue and Hollnagel argued that the forms taken by macrocognitive theories and microcognitive theories are different, with macrocognitive theories being unlike, for instance, information-processing flow diagrams or sets of procedural rules.

At this point in time the notion of macrocognition had two elements. One was what we might call the Johanssen-Woods assertion that cognitive work can only be understood through study at a number of levels or perspectives (see also Rasmussen, 1986). The other was the Cacciabue-Hollnagel assertion that the information-processing approach provides an incomplete and incorrect understanding of cognitive work.

In 2000, Klein et al. suggested the concept of macrocognition as an encompassing frame for studying the cognitive processes that emerged in complex settings. They attempted to encourage a dialog between laboratory and field researchers. Like Cacciabue and Hollnagel, Klein et al. defined macrocognition as the study of complex cognitive functions, including decision making, situation awareness, planning, problem detection, option generation, mental simulation, attention management, uncertainty management and expertise. In other words, it was dawning on people that macrocognition is what NDM is really about, after all.

Expansion of the Notion of Macrocognition

Klein et al. (2003) saw macrocognition as a broader framework for NDM, more than the Johanssen notion of levels or perspectives, and more than the mere expansion of NDM to cover phenomena other than decision making. There are explanatory models such as the Recognition-Primed Decision-making model (RPD) (Klein, 1998), decision pre-priming (McLennan and Omodei, 1996), and levels of situation awareness (Endsley, 1995a). There are emergent functional phenomena such as sensemaking (Klein, Moon, and Hoffman, 2006a, b), and problem detection (Klein et al., 2005). Macrocognition is seen as the study of cognitive phenomena found in natural settings, especially (but not limited to) cognitive work conducted in complex sociotechnical contexts. The concept of macrocognition retains the essence of NDM, but with a broader mandate. Figure 1.1 describes the key macrocognitive functions listed by Klein et al. (2003): decision making, sensemaking, planning, adaptation/replanning, problem detection, and coordination. Some of these, such as problem detection, emerge in field settings, are rarely considered in controlled laboratory-based experiments, and would be unlikely to emerge in typical laboratory studies of cognition (for example, studies of how people solve pre-formulated puzzles would be unlikely to demonstrate the phenomenon of problem-finding).

Figure 1.1 also shows supporting processes such as maintaining common ground (for example, Klein et al., 2004), developing mental models (Gentner and Stevens, 1983), uncertainty management (Lipshitz and Strauss, 1996), using leverage points (Klein and Wolf, 1998), attention management, and mental simulation (Klein and Crandall, 1995). We differentiate these from the primary macrocognitive functions. In most cases, workers and supervisors are not immediately interested in performing the processes themselves—the supporting functions are a means to achieving the primary macrocognitive functions.

The macrocognitive functions and supporting processes are performed under conditions of time pressure, uncertainty, ill-defined and shifting goals, multiple team members, organizational constraints, high stakes, and reasonable amounts

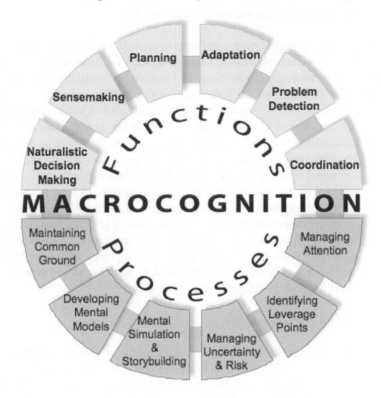

Figure 1.1   Primary and supporting macrocognitive functions

Source: Adapted from Klein et al. (2003).

of expertise—the same conditions that mark NDM. The primary macrocognitive functions are regularly found in cognitive field research. Macrocognitive functions need to be performed by individuals, by teams, by organizations, and by joint cognitive systems that coordinate people with technology. If we are going to understand cognitive work, and find ways to help practitioners and organizations, we need to learn how these functions are performed, the knowledge required, the information required, the reasoning strategies used, and the means of collaboration and cooperation.

Contrasting Microcognition and Macrocognition

We define macrocognition as *the study of cognitive adaptations to complexity.* The macrocognitive functions and processes shown in Figure 1.1 are the most salient cognitive adaptations. Of course, lying behind the conceptual definition of macrocognition is a philosophy or view, and an attendant approach to research. Macrocognition can be thought of as a cluster of "isms": naturalism, functionalism, and phenomenalism. On the other hand, microcognition seems to cluster other "isms": experimentalism, formalism, and reductionism.

Reductionism versus Phenomenology of Cognitive Work

In contrast to microcognition, which attempts to provide a reductionist, causal-chain account of behavior, macrocognition seeks to maintain a focus on the phenomena themselves. The macrocognition view is intended to spotlight something that seems apparent: In cognitive work the sorts of things that we might point to and call "mental operations"—everything from attention management during multi-tasking situations, to re-planning triggered by problem recognition, to goal and decision modification based on situation awareness—are all, always, parallel and always highly interacting. This certainly is a challenge to causal-chain explanations, including traditional information-processing flow models, hierarchies of procedural rules or conflict-free goals, and millisecond-level access to long-term memory. Decision making usually involves sensemaking and situation awareness; replanning depends on problem detection, and so on. Microcognitive analysis permits computer modeling, especially of well-specified or highly routinized tasks. However, the power of controlled investigation can be a liability in studying cognition under conditions of high stakes, context-rich tasks involving multiple participants and organizational constraints working with ill-defined and shifting goals.

Naturalism versus Experimentalism

Experimental science relies on replication. This might be contrasted with the case study method or, as Jung, Piaget and others referred to it, the "clinical method." NDM research seems to deal a lot with case studies. For instance, the Critical Decision Method (CDM) procedure of scaffolded retrospection yields, in any single study or project, any number of highly detailed case studies from which researchers identify features such as decision requirements. NDM research also seems to thrive on story-telling, using particular cases to convey key points or phenomena revealed about the nature of cognitive work. Often, such cases are rare, or, for the practitioner, tough. Thus, NDM seems to hinge on the study of unique events, situations that cannot be easily replicated. But this is not to say that the core phenomena cannot be replicated, which they are, across cases.

Laboratory science seeks to select and manipulate the key variables that define the situations in which the unique events occur. One must be able to maintain the effects of certain "key" variables. At the same time, one must cope with those "other variables" by holding them constant. Or assuming that their effects are not important and that the interactions with the "key" variables are negligible and uninteresting, and can be glossed over by multiple replications (for example, averaging over large samples). Researchers need to be able to trigger the phenomenon of interest, making it appear and disappear upon command.

But what has this approach done for (or to) cognitive psychology?

Experimental psychology has yet to really recover from the critical claim of Newell in his classic paper, "You can't play 20 questions with nature and win" (1973), or take to heart the critical claim in James Jenkins' classic paper, "Remember that old theory of memory? Well, forget it!" (1974). Newell wondered whether the paradigm of hypothesis testing in the psychology laboratory ever really solves anything, since it seems to be the endless permutation of controls and counterbalancings on topics that come and go as fads; a never-ending game of beating problems to death with pet research paradigms, until people become bored and move on, only to have a subsequent generation re-discover the phenomena and re-invent the wheels. Jenkins (1978) wondered whether theories of learning were really that, or were just models of how particular people (usually, college students) adapt to specific constraints of materials and tasks (for example, memorizing lists of pairs of words) in specific (laboratory) contexts (small quiet room, with table chairs, pencil, paper). Although we saw aftershocks of the Jenkins and Newell papers on occasion (for example, Simon, 1980), cognitive psychology never adequately accommodated, although cognitive science certainly followed Newell's suggestion of trying to escape the problem by building cognitive architectures. Leading experimental psychology journals are still populated by studies of interest to fairly small groups of people who use particular micro-paradigms. Recognizing this state of affairs, George Miller (1989) encouraged psychologists to escape the "analytic pathology" of studying isolated cognitive processes, divorced from practical applications (see Staszewski, this volume).

The study of "real-world" cognition has always been of interest to psychology (see Münsterberg, 1912). We believe that what has led to these two paths—experimental cognitive psychology versus NDM—is the fact that cognitive psychology, and to a surprising extent even applied cognitive psychology, has been situated in the traditional academy. Experimental psychology has as its foundation, programs of studies, not individual experiments, that demonstrate phenomena, control important variables, and then determine causal relations. Such work takes considerable time. One challenge for applied research is that the timeframe for effective laboratory experimentation is vastly outstripped by the rate of change in the sociotechnical workplace, including change in technology and change in cognitive work. In contrast, NDM/Macrocognition are focussed on the world outside the academy. NDM emerged in the private sector, in government, and in sponsored research (in domains such as nuclear engineering, aviation safety, nursing, and so on). NDM research is, by definition, aimed at useful application. Thus, as a community, NDM has resonated strongly to the views of Jenkins and Newell, seeing this as part of the justification for a naturalistic-functionalistic approach to the empirical investigation of cognition.

What kinds of science can be accomplished using field research paradigms? A first part of the answer is that NDM researchers do perform controlled experiments, and NDM never advocated the abandonment of the academic laboratory. Cognitive field studies can involve the control of variables through selection and manipulation (see Klein et al., 2003; Hoffman and Coffey, 2004; Hoffman and Woods, 2000). One can hold variables constant, for example, studying reasoning in a group of individuals who qualify as experts versus a group who qualify as apprentices. One can manipulate variables, for example, number of interruptions while an operator is conducting cognitive work using a prototype of a decision aid. This being said, macrocognition is definitely more akin to the category of scientific investigation traditionally referred to as naturalism. It is often taught that the scientific method begins with observing phenomena and formulating questions. Such activities generate postulates about the phenomena of interest. These need to be tested by seeing the kinds of hypotheses they entail, and looking (somehow) to see how those entailments hold up (replicate), and hold up under other circumstances, to be sure that they are not limited to specific contexts, methods, or special types of participants or conditions. Macrocognitive field research particularly emphasizes the initial steps of formulating questions and observing phenomena, as well as the final step of seeking to generalize across contexts (or domains). In this way, researchers have identified phenomena and variables and relationships that were neglected by and would not have ever been manifested in laboratory investigations characteristic of mainstream cognitive psychology.

Controlled experimentation (or we might say, Popperian hypothesis falsification) is not a privileged path to causal analysis or understanding, and it is not an absolute requirement for any form of empirical enquiry to be dubbed "science" or "good science." Darwin, we might note, tested many hypotheses. In a more recent example, Peter and Rosemary Grant (Grant, 1986) conducted a natural experiment on finches in the Galapagos. The island that they studied, Daphne Major, was hit by a drought in 1977. As a result, one of the prime food sources for the finches became scarce. The finches had to resort to a secondary food source, a type of seed that could only be eaten by finches with fairly long beaks. Approximately 70 percent of the finches on the island died. But the next generation of finches had markedly longer beaks and were larger. The researchers would have had difficulty setting up this kind of experiment, and there was no experimental group (nothing was manipulated by the researcher) or control group (nothing could be held constant). Nevertheless, the findings are a striking demonstration of natural selection in just a single generation. The story continued in 1984–85 when the unusually rainy weather resulted in more small, soft seeds and fewer large, tough ones. Now it was the finches with smaller beaks that held an advantage and produced the most offspring. Must one try to take this sort of thing into the lab before one can believe that there is a cause–effect relation?

The most often-mentioned contrast case is astronomy. Astronomers cannot manipulate stars to test theories of internal stellar processes. But they can select groups of stars by some criterion (for example, red giants) and then falsify hypotheses about stellar processes (for example, red giants should all emit a preponderance of certain wavelengths of light). Macrocognitive research has repeatedly demonstrated and studied phenomena that would be difficult or impossible to capture in the laboratory. A good case in point involves the phenomenon of perceptual learning. While long recognized within psychology (see, for example, Gibson, 1969), the phenomenon is hard to study for a number of reasons. The most salient of these is that the acquisition of a perceptual skill takes time, for example, years of practice and experience to learn to interpret aerial photographs (Hoffman and Conway, 1990). It is hard to capture perceptual learning phenomena in the laboratory (see Fahle and Piggio, 2002). (One cannot fail to notice that laboratory studies of perceptual learning involve experiments in which college students view sets of static, simple stimuli, for example, line caricatures of faces, outlines of trees, and so on.) In his study of landmine detection (see Chapter 16, this volume), James Staszewski discovered how an expert used the landmine detection device in active perceptual exploration. From an understanding of the expert's exploration strategy and the patterns perceived through the dynamic exploration, Staszewski was able to develop a training program that significantly accelerated the achievement of perceptual expertise at landmine detection. Another way of putting this is that Staszewski was able to capture a perceptual learning phenomenon in the field setting.

When queried, NDM researchers and advocates of macrocognition are more likely to regard their work as "basic science" than applied science, without giving the matter much of a second thought. The research is aimed at revealing fundamental facts and regularities about cognitive work. We started out our discussion of the macro–micro distinction by noting the "isms": Macrocognition involves naturalism, functionalism, and phenomenalism; microcognition involves experimentalism, formalism, and reductionism. We conclude here by noting the most important "ism," one that that micro- and macrocognition share: *empiricism.*

All scientific methods and strategies have appropriate uses, strengths, and limitations. The concept of macrocognition is specifically chosen to create a distinction from microcognition, but it is also intended to facilitate connections with the microcognition community, as discussed in the next section.

Bridging Macrocognition and Microcognition

Staszewski (this volume) rightly warns against presenting macrocognition and microcognition as competing or antagonistic frameworks. Klein et al. (2000) described a number of laboratory phenomena that were relevant to the macrocognitive agenda. Studies of categorization, particularly the distinction between taxonomic and thematic categories, have direct implications for the design of useful menus in human–computer interfaces. Studies of national differences in cognitive processes, such as tolerance for uncertainty and hypothetical reasoning, can inform individual and team training as well as computer interfaces. Research on polysemy (the way a single word evokes multiple related senses) shows that words are tools and their meanings evolve through changing use. The expansion of word usage suggests ways that context can be used to disambiguate meaning—a process that supports better design and use of multi-function interfaces.

Therefore, cognitive field studies identify phenomena of interest to macrocognitive models that can be tested in the laboratory; cognitive field studies also offer face validity assessment of laboratory findings and phenomena. Similarly, laboratory researchers generate findings that are of potential interest to macrocognitive investigations. We do not hold that either the laboratory or the field has a privileged position for originating discoveries (see Hoffman and Deffenbacher, 1993). The field setting obviously can serve as the test bed for gauging the practical implications of laboratory findings, and the laboratory can serve as a test bed for evaluating theories and hypotheses that emerge from field studies. One example of the synthesis of NDM and experimental psychology is in the emerging area of macrocognitive computational modeling. A number of investigators have been attempting to model macrocognitive phenomena such as Endsley's model of situation awareness (Gonzalez et al., 2006) and Klein's RPD model (Warwick and Hutton, 2007; Forsythe, 2003; Sokolowski, 2003). While these computational models have a long way to go, they are still very useful for revealing shortcomings in the macrocognitive models, limitations of computer modeling, and for stimulating us to elaborate our accounts more fully.

Another example, perhaps a surprising one, is the potential for fusion between NDM and Behavioral Decision Making (BDM). Kahneman (2003) has recently discussed the "System 1/System 2" framework. System 1 refers to forms of cognitive processing that are fast, automatic, effortless, implicit, and emotional—the kinds of processing strengthened through experience and highlighted in the NDM models of decision making discussed by Lipshitz (1993). System 2 refers to processing mechanisms that are slower, conscious, effortful, explicit, deliberate, logical, and serial. System 2 processing serves to monitor the intuitions generated from System 1. Within the RPD model, the pattern-matching process is an example of System 1, and the mental simulation serves as System 2 monitoring (see Hoffman and Militello, 2008, Ch. 9). The System 1/System 2 framework might form a bridge between NDM and BDM, a bridge that hopefully expands the horizons of both views.

Macrocognition researchers move back and forth from the field to the laboratory, using each setting as needed and as appropriate. More succinctly, *macrocognitive research involves bringing the lab into the world and bringing the world into the lab*. Thus, it encourages traditional laboratory researchers to spend more time observing the phenomena in natural settings.

Figure 1.2 is a Concept Map that summarizes some of the key ideas of macrocognition.

Putting the Notion of Macrocognition to Work

The notion of macrocognition has uses, both in the practice of research and in advancing our theoretical understandings. First, because of its phenomenological nature, macrocognition may be a practical basis for dialog between practitioners such as system designers and cognitive researchers. Macrocognitive functions concern supervisors and workers—and the sponsors of the research who see a need for new tools and technologies. One can talk with practitioners about macrocognitive functions, and gain purchase on collaborative effort.

Second, macrocognition may help us at the level of theory. For instance, macrocognition suggests an explanation of why some cognitive scientists have difficulty with the notion of "mental model" (for example, Hunt and Jossyln, 2007). It can be argued that the notion does not fit comfortably into information-processing views precisely because, as we can see in hindsight, it is actually a macrocognitive notion. The classic literature on mental models (Gentner and Stevens, 1983) points out that mental models (for example, a learner's understanding of how electricity works) involve such elementary phenomena as mental imagery and metaphorical understanding (for example, thinking of electrical circuits in terms of fluid flowing through pipes). Mental models link these to concepts and principles. Thus, a host of mental operations and elementary processes are involved all at once in mental model formation.

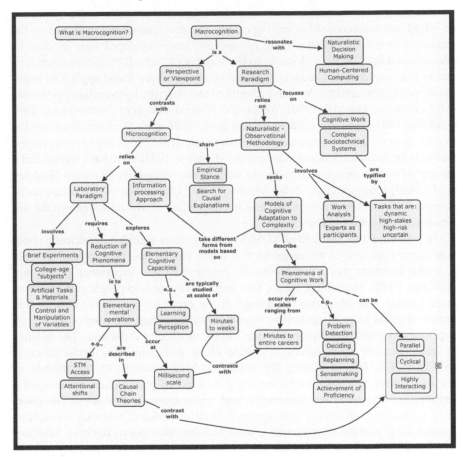

Figure 1.2   A Concept Map about macrocognition

The macro–micro distinction encourages us to make further discoveries about the core macrocognitive functions, and pursue the issue of how one can bridge between microcognitive and macrocognitive analysis. Just as the RPD model has changed the way people understand decision making and seek to support it, better models of sensemaking, problem detection, replanning and coordination might have an impact on how we view people performing complex tasks and how we seek to help them. By gaining a better understanding of macrocognitive functions we can provide an alternative to the conventional approaches to design and training and better support the needs of the practitioners—the people taking actions. Some scientists (for example, Flach, this volume) are counseling us to hold off until we are sure we have the right ontologies or metaphysical assumptions, and we look forward to further debate on all these matters. We hope to make progress by forging ahead in empirical study of macrocognitive functions, making mistakes along the way but continuing to learn how people manage to work successfully despite complexity.

Controversy is the lifeblood of scientific fields, and we believe that criticisms of NDM and macrocognition are important. But they should not overshadow the accomplishments of NDM. NDM researchers have developed new models of decision making that are a closer fit to the way people actually settle on courses of action than any previous account. These models have been found applicable many times in different settings. We have identified and described pre-decision processes (McLennan and Omodei, 1996). We have developed models of situation awareness (Endsley, 1995a) and sensemaking (Klein et al., 2006 a, b). We have described the role of common ground in team coordination (Klein et al., 2004). We have developed methods for measuring situation awareness (Endsley, 1995b). We have formulated a variety of system design approaches to support macrocognitive functions (Endsley et al., 2003; Crandall et al., 2006; Vicente, 1999). (A number of rich examples of success stories of cognitive systems engineering, many of which illustrate NDM, are recounted in Cooke and Durso, 2007.)

In gathering data, the field of NDM/Macrocognition is guided by curiosity about how experts are able to notice things that others cannot, and how experts are able to make decisions that would confound people with less experience (Klein and Hoffman, 1992). In addition, we are also seeking ways to help people gain expertise more quickly and take better advantage of the expertise they have. This applied perspective has helped to prioritize important variables that are worth studying, as opposed to variables that probably don't make much of a difference. The balance of learning more about natural cognition along with improving performance is an essential source of vitality for macrocognitive researchers. The models of macrocognitive functions are beginning to be used by practitioners for designing support systems, training programs, and organizational structures. Designers have to support uncertainty management, decision making, situation awareness, sensemaking, replanning, common ground and the other macrocognitive functions and processes. In working on these kinds of applications we have the opportunity to deepen and improve our models.

The present volume represents a contribution to this broader agenda.

Overview of the Chapters in this Volume

This volume is organized into four parts. Part I focuses on the theoretical underpinnings of the naturalistic decision making approach and its extension into macrocognition. This introductory chapter presents the case from the proponents' point of view, whereas John Flach (Chapter 2) presents a critical view of the macrocognition concept. Flach is afraid that with the proliferation of concepts involving the word "cognition," we will not come any closer to a real understanding of the phenomenon. Inventing yet another kind of cognition or another kind of research method will lead us away from a unified theory of cognition. Knowing the name of something is different from knowing something. Flach agrees that the issues raised by the construct of macrocognition are important. However, macrocognition should keep in touch with mainline cognitive science. All cognition is *macro*. It is only the paradigms of science that are *micro*. Kathleen Mosier (Chapter 3) explores several myths that persist concerning decision processes in high technology environments, including the notions that expert intuitive judgment processes are sufficient in these environments, that analysis has no place in expert decision making in dynamic environments, and that "intuitive" displays can eliminate the need for analysis. Because decision makers in hybrid naturalistic environments are required to use both analytical and intuitive processing, decision-aiding systems should recognize and support this requirement. Mosier makes the case that high-tech decision environments are significantly different from purely naturalistic domains and that this difference impacts all macrocognitive functions.

Part II of this volume ("The Heartland of Macrocognition") focuses on research best exemplifying the concept of macrocognition. Gary Klein and Robert Hoffman (Chapter 4) explore some methods for gathering data, representing, and studying mental models. Mental modeling is a supporting process, especially critical for sensemaking. This chapter is basically an invitation to naturalistic empirical inquiry: If one were conducting cognitive task analysis (CTA) on experienced domain practitioners, for purposes such as cognitive systems engineering, what sorts of methods might be used to reveal mental models? Tom Ormerod, Emma Barrett, and Paul Taylor also explore the macrocognitive concept of sensemaking, particularly in criminal contexts (Chapter 5). They focus on three relatively ill-studied domains: understanding crime reports and scenes, monitoring and decision making during ongoing hostage-taking and barricade incidents, and following up suspicious insurance claims to evaluate whether fraud has been committed. The authors claim that what these domains have in common are an infinite number of possible scenarios which prohibits a rule-based response to an immediate and perceptually-based judgment; a focus on human action and reaction rather than the state of a physical process or plant, and the deceptive nature of the investigative domain. Interestingly, experts in the investigative domain are accustomed to generating alternative hypotheses. Experts appear able to adopt different inferential stances that allow them to evaluate multiple suspicion hypotheses against potential frameworks of guilt or innocence. Crichton, Lauche, and Flin (Chapter 6) describe a particular case study of incident management training in the oil and gas industry. As it happened, key team leaders of the Incident Management Team had participated in a full-scale simulated emergency exercise shortly before an actual incident (similar to the exercise) occurred. This provided a rare opportunity for comparing the transfer from training to actual experiences. It turned out that whereas training exercises generally put a lot of emphasis on the response phase (the first days after the incident), the actual incident was much more demanding in terms of how to deal with the prolonged recovery phase. During training, there was little opportunity to experience making decisions under stress, particularly pressures of time, uncertainty, and event novelty. In contrast, these pressures and challenges were experienced for many of the decisions that arose during the actual event, particularly during the recovery phase, which involved more knowledge-based decision making. As an aside to the definition of "macrocognition," it is interesting to note that Crichton and co-workers use the term "macrocognition" to refer to an "approach" to understanding incident management skills, rather than the macrocognitive functions themselves.

Laura Zimmerman does study one particular macrocognitive function, sensemaking, in the domain of law-enforcement decision making (Chapter 7). In her study, officers were interviewed after watching a video of an actual traffic stop and analysis revealed that, compared to inexperienced officers, experienced officers provide more detailed situation assessments, make more evaluative comments, and discuss officer cognitive processing. Novice officers provide more assessments of safety issues and procedures. Anna McHugh, Jennifer Smith, and Winston Sieck address the relatively novel issue of cultural variations in mental models of collaborative decision making (Chapter 8). Mental models are viewed here as macrocognitive processes that are critical for supporting the full spectrum of macrocognitive functions, including decision making. The specific aim of the current research was to uncover the salient disconnects in mental models of collaborative decision making among people from diverse cultures. This research comes at a time when we are experiencing a strong trend toward using multinational teams to tackle highly complex problems in both commercial and governmental settings. Addressing the disconnects in mental models is important for fostering hybrid cultures in multinational teams. The researchers interviewed 61 people from a variety of nations about their beliefs on how team decision-making practices actually occur in their home culture. Their findings indicate that despite some important differences in mental models of collaborative decision making across cultural groups, the basic set of core elements comprising these mental models seems to be shared. It is the means by which these various elements are carried out that varies. The authors weave a rich tapestry of elements that need to be considered when assembling multinational teams.

Mei-Hua Lin and Helen Altman Klein also address national differences in performing macrocognitive tasks, in particular, sensemaking (Chapter 9). Theirs is a literature review on laboratory studies dealing with holistic and analytic thinking. The analytic-holistic distinction consists of four dimensions that Lin and Klein subsequently discuss in terms of cultural variation: attention, causal attribution, tolerance for contradiction, and perception of change. Each of these dimensions shows a significant amount of cultural variation, between, for instance, Americans and Japanese. It is important to at least be aware of this cultural variation when working together with people from other countries. Susan Joslyn and David Jones present a cognitive task analysis of naval weather forecasting (Chapter 10). Forecasters are thought to be creating a mental representation of the current state of the atmosphere that includes both spatial and temporal components. This has been described as a mental model. By using verbal protocol analysis, they attempt to characterize the forecaster's mental model. Only one of the four forecasters, with over twenty years of experience, appeared to consolidate a mental model of the atmosphere and check it against other information sources. The other forecasters seemed to rely on incomplete rules of thumb. Perhaps creating and maintaining a complex mental representation such as this is a skill that develops with experience and well-developed domain-specific long-term knowledge structures to support it. Moreover, Joslyn and Jones found that information-gathering and model-evaluation processes had been reduced to a routine. Forecasters relied on a few tried-and-true procedures rather than creating new solutions specific to each situation. In addition, these forecasters relied on favored models and products suggesting selection based on habit and ease of access rather than on the weather problem of the day, or on superior relative performance of a computer model in the current situation. Although this inflexible approach is unlikely to produce the optimal decision in the abstract, it provided the forecasters with adequate answers and allowed them to get the job done in a timely fashion.

We now turn to two chapters in the health care domain. Laura Militello, Emily Patterson, Jason Saleem, Shilo Anders, and Steven Asch describe a study on improving the usability and effectiveness of computerized clinical reminders used in the ambulatory care unit of a Veterans Administration (VA) hospital (Chapter 11). They address a long-standing issue in the cognitive task analysis literature: how to overcome the barrier between human factors researchers and software designers. As the clinical reminder software was already fielded, a complete redesign was not feasible. The authors chose to present their recommendations as design seeds. The goal of this approach is to communicate the intent behind the recommendation and present whatever data is available regarding its feasibility and likelihood of success. Intent information is also aimed at reducing the likelihood that design ideas will be distorted due to miscommunication in the hand off between the researchers and the implementers. Simon Henderson discusses experiential learning in surgical teams (Chapter 12). The chapter describes the development and application of a process of "Team Self-Review" for surgical theater teams, and examines how the naturalistic approach adopted throughout the study has uncovered empirical support for the concept of macrocognition. The Team Self-Review consists of a Pre-brief and a Debrief. The Pre-brief is held at the beginning of a team "session" (that is, an operating list, case, training exercise or other event) and the Debrief is held at the end of the session. The Pre-brief aims to review and clarify the plan for the session ("What do we need to know? What are we going to do? What should we do if…?") and the debrief aims to assess and review team performance in the session that has just occurred ("How well did we do? What should we do differently in the future?"). The project has gathered a great deal of press and publicity, and its impact has been cited by the Chief Executive of the British National Health Service, as an example of successful organizational learning in the NHS.

Part II of this volume is concluded by a chapter on the macrocognitive functions in the air traffic control tower (Chapter 13). The air traffic control tower (ATCT) work domain involves a full spectrum of macrocognitive functions, including naturalistic decision making, sensemaking and situation awareness, planning and re-planning, problem detection, and coordination. These macrocognitive functions require the execution of processes that include the formation and use of mental models, the management of uncertainty, and, perhaps of foremost importance in the ATCT domain, the management of attention. Peter Moertl, Craig Bonaceto, Steven Estes, and Kevin Burns discuss three methods—cognitive modeling, critical incident analysis, and coordination analysis—that allow them to make predictions about the effects of procedural changes on air traffic controller performance.

Part III of this volume focuses on the relationship between micro- and macrocognitive approaches. Claire McAndrew, Adrian Banks, and Julie Gore describe how the microcognitive focus of the ACT-R model enables the specification of micro processes that exist both within and between a number of macrocognitive processes (Chapter 14). They focus on the theoretical validation of the Recognition-Primed Decision Making model. The success of their ACT-R model as a tool for validation is most markedly characterized by the agreement between ACT-R's trace outputs to the RPD models' descriptions. A detailed examination of the differences between the two models yields recommendations for how to improve the RPD model. For instance, in ACT-R, situation familiarity is assessed after the selection of relevant cues, whereas in the RPD model the question of "Is the situation familiar?" precedes the identification of relevant cues. Also, in instances whereby the situation is not viewed as familiar, the ACT-R model is more complete than the RPD model, which is underspecified in this case. In the end, the authors consider the micro–macro distinction as artificial and unnecessary, much as Flach does in Chapter 2. Rather, they propose, like Klein et al. (2003), that "the two types of description are complementary. Each serves its own purpose, and together they might provide a broader and more comprehensive view than either by itself."

Stephen Deutsch describes a micro-approach to a macro-phenomenon: a cognitive model of a captain's planning and decision-making processes (Chapter 15). Based on the 2 July 1994 windshear accident at Charlotte/Douglas International Airport, the model fostered new insights into aircrew decision making that were sources of human error contributing to the accident. In the Charlotte windshear threat situation, there was a straightforward plan for a microburst escape to be put in place, yet a creative action selection process was employed. Had the error been grounded in a skill- or rule-based decision, the exact execution of an alternate procedure would have been expected, yet the plan and its later execution were found to be a composite that drew on elements from three procedures. Whether as a conscious or non-conscious process, the cues presented in the evolving situation led to the construction of a compelling narrative and it was that narrative that dominated the process and drove the form of the actual plan. The plan as briefed was not working and unfortunately, the belated call for a microburst escape maneuver did not lead to a successful outcome.

The final chapter in Part III investigates whether information-processing models of expert skill are useful for designing instruction in the high-stakes domain of landmine detection (Chapter 16). Landmines are a major threat to military personnel engaged in combat and peacekeeping operations and also pose a multi-faceted humanitarian threat. Eliminating mines first involves detecting their locations. The task is hazardous because it involves fallible humans operating hand-held equipment in close proximity to hidden live ordnance. It appears that there are large skill differences, with most US soldiers' performance being dangerously low, while some heavily experienced operators produce more than 90 percent detection rates. James Staszewski has designed instruction in landmine detection based on the heavily experienced operators' knowledge content and organization. Results showed that training based on expert knowledge and techniques raised aggregate detection rates to 94–7 percent and produced six-fold gains in detection of the most difficult-to-find low-metal anti-personnel targets. The results of these and other training programs developed by Staszewski demonstrate the practical utility of expert models as blueprints for instructional design. The US Army's decision to adopt and use these training programs illustrates end-users' assessment of their practical value.

Finally, Part IV of this volume presents two chapters that deal with macrocognition from a cognitive systems engineering point of view. Rogier Woltjer, Kip Smith, and Erik Hollnagel deal with the timely problem of network-based command and control (Chapter 17). This chapter describes a method for generating ecological representations of spatial and temporal resource constraints in network-based command and control, and illustrates its application in a command and control microworld. The method uses goals-means task analysis to extract the essential variables that describe the behavior of a command and control team. It juxtaposes these variables in ecological state space representations illustrating constraints and regions for opportunities for action. This chapter discusses how state space representations may be used to aid decision making and improve control in network-based command and control settings. Examples show how state space plots of experimental data can aid in the description of behavior vis-à-vis constraints. Mark Neerincx and Jasper Lindenberg present a Cognitive Engineering (CE) framework to realize an adequate human-machine collaboration (HMC) performance, in which generic human-factors knowledge and HMC solutions are refined, contextualized, and tested within the application domain, as part of an iterative HMC development process (Chapter 18). Instead of one generic method for HMC design, they propose to establish *situated* cognitive engineering methods that comprise an increasing knowledge base on the most relevant human and machine factors for the concerning operations and technical design space in the different application domains. The CE framework incorporates regular meta-engineering activities to attune the CE methods and tools to the changing situational constraints. Two case studies show the development and application of such a set of methods and tools, respectively for space stations (that is, SUITE) and naval ships (that is, SEAMATE). The evaluations in the naval and space domains improved our understanding of the relevant macrocognitive functions, such as situation assessment during naval damage control and problem detection during payload operations in a space station.

References

Cacciabue, P.C. and Hollnagel, E. (1995). "Simulation of cognition: Applications." In J.M. Hoc, P.C. Cacciabue, and E. Hollnagel (eds), *Expertise and Technology: Cognition and Human–computer Cooperation* (pp. 55–73). Hillsdale, NJ: Lawrence Erlbaum Associates.

Cooke, N. and Durso, F. (2007). *Cognitive Engineering Solutions*. Mahwah, NJ: Lawrence Erlbaum Associates.

Crandall, B., Klein, G., and Hoffman, R.R. (2006). *Working Minds: A Practitioner's Guide to Cognitive Task Analysis*. Cambridge, MA: MIT Press.

Endsley, M.R. (1995a). "Toward a theory of situation awareness in dynamic systems." *Human Factors*, 37(1): 32–64.

—— (1995b). "Measurement of situation awareness in dynamic systems." *Human Factors*, 37(1), 65–84.

Endsley, M.R., Bolte, R., and Jones, D.G. (2003). *Designing for Situation Awareness: An Approach to Human–centered Design*. London: Taylor and Francis.

Endsley, M.R. and Garland, D.J. (2000). *Situation Awareness Analysis and Measurement*. Mahwah, NJ: Lawrence Erlbaum Associates.

Fahle, M. and Piggio, T. (eds) (2002). *Perceptual Learning*. Cambridge, MA: MIT Press.

Flach, J. and Hoffman, R.R. (January–February 2003). "The limitations of limitations." *IEEE Intelligent Systems.* (pp. 94–7).

Flin, R., Salas, E., Strub, M., and Martin, L. (eds) (1997). *Decision Making Under Stress: Emerging Themes and Applications*. Aldershot, UK: Ashgate Publishing.

Forsythe, C. (2003). "RPD-based cognitive modeling: A framework for the comprehensive representation of human decision processes." Presentation at the Naturalistic Decision Making Workshop, 23–24 October, Boulder, CO.

Gay, P. (ed.) (1988). *Freud: A Life for our Time*. New York: W.W. Norton.

Gentner, D., and Stevens, A.L. (eds) (1983). *Mental Models*. Mahwah, NJ: Lawrence Erlbaum Associates.

Gibson, E.J. (1969). *Principles of Perceptual Learning and Development*. New York: Appleton Century Crofts.

Gonzalez, C., Juarez, O., Endsley, M., and Jones, D. (May 2006). "Cognitive Models of Situation Awareness: Automatic Evaluation of Situation Awareness in Graphic Interfaces." Presentation at the Conference on Behavior Representation in Modeling and Simulation (BRIMS 2006).

Grant, P.R. (1986). *Ecology and Evaluation of Darwin's Finches*. Princeton, NJ: Princeton University Press.

Hoffman, R.R. (ed.) (2007). *Expertise Out of Context: Proceedings of the Sixth International Conference on Naturalistic Decision Making*. Mahwah, NJ: Lawrence Erlbaum Associates.

Hoffman, R.R. and Coffey, J.W. (2004). "Human Centered Computing: Human Factors Design Principles Manifested in a Study of Weather Forecasters." In *Proceedings of the 48th Annual Meeting of the Human Factors and Ergonomics Society* (pp. 315–19). Santa Monica, CA: Human Factors and Ergonomics Society.

Hoffman, R.R. and Conway, J. (1990). "Psychological factors in remote sensing: A review of recent research." *Geocarto International*, 4, 3–22.

Hoffman, R.R. and Deffenbacher, K.A. (1993). "An analysis of the relations of basic and applied science." *Ecological Psychology*, 5: 315–52.

Hoffman, R.R. and Militello, L. (2008). *Perspectives on Cognitive Task Analysis*. Mahwah, NJ: Lawrence Erlbaum Associates.

Hoffman, R.R. and Woods, D.D. (2000). "Studying cognitive systems in context." *Human Factors*, 42: 1–7.

Hoffman, R.R. and Yates, J.F. (2005). "Decision(?)–Making(?)." *IEEE: Intelligent Systems*, July/August: 22–9.

Hollnagel, E., Mancini, G., and Woods, D.D. (eds) (1985). *Intelligent Decision Support in Process Environments*. Berlin: Springer-Verlag.

Hollnagel, E. and Woods, D.D. (1983). "Cognitive systems engineering. New wine in new bottles." *International Journal of Man–Machine Studies*, 18: 58–600.

——(2005). *Joint Cognitive Systems: Foundations of Cognitive Systems Engineering*. Boca Raton, FL: Taylor and Francis.

Hunt, E. and Josslyn, S. (2007). "The dynamics of the relation between applied and basic research." In R.R. Hoffman (ed.), *Expertise Out of Context: Proceedings of the Sixth International Conference on Naturalistic Decision Making*. Mahwah, NJ: Lawrence Erlbaum Associates (pp. 7–28).

Jenkins, J.J. (1974). "Remember that old theory of memory? Well, forget it!" *American Psychologist*, 29: 785–95.

—— (1978). "Four points to remember: A tetrahedral model of memory experiments." In L. Cermak and F. Craik (eds), *Levels of Processing and Human Memory*. Hillsdale, NJ: Lawrence Erlbaum Associates (pp. 429–46).

Johanssen, G. (1985). "Architecture of man–machine decision making systems." In E. Hollnagel, G. Mancini, and D.D. Woods (eds), *Intelligent Decision Support in Process Environments*. Berlin: Springer-Verlag (pp. 327–39).

Kahneman, D. (2003). "A perspective on judgment and choice: Mapping bounded rationality." *American Psychologist*, 58, 697–720.

Klein, D.E., Klein, H.A., and Klein, G. (2000). "Macrocognition: Linking cognitive psychology and cognitive ergonomics." *Proceedings of the 5th International Conference on Human Interactions with Complex Systems*. Urbana-Champaign: University of Illinois at Urbana-Champaign (pp. 173–7).

Klein, G. (1989). "Recognition-primed decision making." In W.B. Rouse (ed.), *Advances in Man–Machine System Research*, Vol. 5. Greenwich, CT: JAI Press (pp. 47–92).

Klein, G., Calderwood, R., and Clinton-Cirocco, A. (1986). "Rapid decision making on the fire ground." *Proceedings of the 30th Annual Meeting of the Human Factors Society*. Santa Monica, CA: Human Factors Society (pp. 576–80).

Klein, G. and Hoffman, R.R. (1992). "Seeing the invisible: Perceptual–cognitive aspects of expertise." In M. Rabinowitz (ed.), *Cognitive Science Foundations of Instruction*. Mahwah, NJ: Lawrence Erlbaum Associates (pp. 203–26).

Klein, G., Moon, B., and Hoffman, R.R. (2006a). "Making sense of sensemaking 1: Alternative perspectives." *IEEE Intelligent Systems*, July/August: 22–6.

—— (2006b). "Making sense of sensemaking 2: A macrocognitive model." *IEEE Intelligent Systems*, November/December: 88–92.

Klein, G., Orasanu, J., Calderwood, R., and Zsambok, C.E. (eds) (1993). *Decision Making in Action: Models and Methods*. Norwood, NJ: Ablex Publishing Corporation.

Klein, G., Ross, K.G., Moon, B.M., Klein, D.E., Hoffman, R.R., and Hollnagel, E. (2003). "Macrocognition." *IEEE Intelligent Systems*, May/June: 81–5.

Klein, G. and Wolf, S. (1998). "The role of leverage points in option generation." *IEEE Transactions on Systems, Man and Cybernetics: Applications and Reviews*, 28: 157–60.

Klein, G., Woods, D.D., Bradshaw, J.D., Hoffman, R.R., and Feltovich, P.J. (2004). "Ten challenges for making automation a 'team player' in joint human–agent activity." *IEEE: Intelligent Systems*, November/December: 91–5.

Lipshitz, R. (1993). "Converging themes in the study of decision making in realistic settings." In G.A. Klein, J. Orasanu, R. Calderwood, and C.E. Zsambok (1993). *Decision Making in Action: Models and Methods*. NJ: Ablex Publishing Corporation.

Lipshitz, R., Klein, G., Orasanu, J., and Salas, E. (2001). "Focus article: Taking stock of naturalistic decision making." *Journal of Behavioral Decision Making*, 14: 331–52.

Lipshitz, R. and Strauss, O. (1996). "How decisionmakers cope with uncertainty." *Proceedings of the Human Factors and Ergonomics Society 40th Annual Meeting*. Santa Monica CA: Human Factors and Ergonomics Society (pp. 189–92).

McLennan, J., and Omodei, M.M. (1996). "The role of pre-priming in recognition-primed decision making." *Perceptual and Motor Skills*, 82: 1059–69.

Miller, G.A. (1986). "Dismembering cognition." In S.E. Hulse and B.F. Green (eds), *One Hundred Years of Psychological Research in America: G. Stanley Hall and the Johns Hopkins Tradition*. Baltimore, MD: The Johns Hopkins University Press (pp. 277–98).

Montgomery, H., Lipshitz, R., and Brehmer, B. (eds) (2005). *How Professional Make Decisions*. Mahwah, NJ: Lawrence Erlbaum Associates.

Münsterberg, H. (1912). *Psychologie und Wirtschafstsleben*. Leipzig: Barth.

Newell, A. (1973). "You can't play 20 questions with nature and win: Projective comments on the papers of this symposium." In W.G. Chase (ed.), *Visual Information Processing*. New York: Academic Press (pp. 283–308).

Nunes, A. and Kirlik, A. (2005). "An empirical study of calibration in air traffic control expert judgment." In *Proceedings of the Human Factors and Ergonomics Society 49th Annual Meeting*. Santa Monica, CA: Human Factors and Ergonomics Society (pp. 422–6).

Orasanu, J., and Connolly, T. (1993). "The reinvention of decision making." In G.A. Klein, J. Orasanu, R. Calderwood, and C.E. Zsambok (eds), *Decision Making in Action: Models and Methods*. Norwood, NJ: Ablex (pp. 3–20).

Rasmussen, J. (1983). "Skills, rules, and knowledge; Signals, signs, and symbols, and other distinctions in human performance models." *IEEE Transactions on Systems, Man, and Cybernetics*, 13: 257–66.

—— (1986). *Information Processing and Human–Machine Interaction: An Approach to Cognitive Engineering*. New York: North–Holland.

—— (1988). "Information technology: A challenge to the Human Factors Society?" *Human Factors Society Bulletin*, 31: 1–3.

Salas, E. and Klein, G. (eds) (2001). *Linking Expertise and Naturalistic Decision Making*. Mahwah, NJ: Lawrence Erlbaum Associates.

Schraagen, J.M.C., Chipman, S.F., and Shalin, V.L. (eds) (2000). *Cognitive Task Analysis*. Mahwah, NJ: Lawrence Erlbaum Associates.

Simon, H.A. (1980). "How to win at twenty questions with nature." In R.A. Cole (ed.), *Perception and Production of Fluent Speech*. Hillsdale, NJ: Lawrence Erlbaum Associates (pp. 535–48).

Sokolowski, J.A. (2003). "Modeling the decision process of a joint task force commander." Doctoral dissertation, Department of Engineering. Norfork, VA: Old Dominion University.

Vicente, K.J. (1999). *Cognitive Work Analysis: Toward Safe, Productive, and Healthy Computer-based Work*. Mahwah, NJ: Lawrence Erlbaum Associates.

Warwick, W. and Hutton, R.J.B. (2007). "Computational and theoretical perspectives on recognition-primed decision making." In R.R. Hoffman (ed.) *Expertise Out of Context: Proceedings of the 6th International Conference on Naturalistic Decision Making*. Mahwah, NJ: Lawrence Erlbaum Associates (pp. 429–51).

Woods, D.D. and Roth, E.M. (1986). *Models of Cognitive Behavior in Nuclear Power Plant Personnel* (2 volumes). Technical Report NUREG–CR–4532, US Nuclear Regulatory Commission, Washington, DC.

Zsambok, C. and Klein, G. (eds) (1997). *Naturalistic Decision Making*. Mahwah, NJ: Lawrence Erlbaum Associates.

Chapter 2

Mind the Gap:
A Skeptical View of Macrocognition

John M. Flach

...my father would take me for walks in the woods and would tell me various interesting things...Looking at a bird he says, "Do you know what that bird is? It's a brown throated thrush; but in Portuguese it's a...in Italian a...," he says "in Chinese it's a..., in Japanese a...," et cetera. "Now," he says, "you know in all the languages you want to know what the name of that bird is and when you've finished with all that," he says, "you'll know absolutely nothing whatever about the bird. You only know about humans in different places and what they call the bird. Now," he says, "let's look at the bird." [Richard Feynman, 1999, p. 4]

I refer to this gap in scientific knowledge as an incoherence...we are not talking an ordinary area of ignorance which is being steadily eroded by advancing knowledge...No, the gap is incoherent and intractable...It is not like tunneling under a river from both sides and meeting in the middle. It is more like ships passing in the night. [Walker Percy, 1991, pp. 275–6]

Once again, I find myself playing the role of skeptic—challenging the value of a new construct (for example, Flach, 1995). My fear is that with "macrocognition" we have invented yet another name for a "bird" that we still know precious little about. Our list of names for this bird grows quickly (for example, cognition, metacognition, macrocognition, microcognition, situated cognition, distributed cognition, team cognition, cultural cognition, naturalistic decision making, ecological psychology, sensemaking, situation awareness, intuition, and so on). Perhaps, these are different birds and the different terms represent an increasingly fine discrimination of different facets of nature. However, I find myself wondering whether this list reflects a deepening understanding or an increasing incoherence? Do the different constructs represent distinct phenomena, or simply different perspectives on a single phenomenon? Does the increase in the number of constructs reflect a proportional increase in the degree of understanding? Are we closing the gap between mind and matter in order to address issues associated with *what matters*? Or are we avoiding the deeper issues, leaving the phenomenon of interest (that is, human experience) hidden by an expanding argot?

As suggested by Walker Percy (1991) in his essay "The Fateful Rift: The San Andreas Fault in the Modern Mind," there seems to be a huge gap in our understanding of human experience. Terms such as "macrocognition" can be valuable, because they draw attention to this gap, but I am skeptical about whether they will actually help us to bridge it. The research motivated by terms such as macrocognition is raising essential questions for a full understanding of human experience, as it is lived.

However, segregating this research behind a label like macrocognition may hide the significance of this research, rather than amplify it. So, the goal for this chapter is to draw attention to some of the gaps in science with respect to both understanding human experience and generalizing that understanding to the design of socio-technical systems. In the end, I will argue that the rift stems from a metaphysics that breaks reality in a way that makes it impossible to integrate our understanding of mind and matter into a science of *what matters*.

Some Obvious Gaps

One thread that cuts across the chapters in this volume is a concern for applying an understanding of cognition to the design of sociotechnical systems. In the same way that the science of aeronautics has implications for the design of aircraft, it seems quite reasonable to assume that the science of cognition might be applied to the design of tools for information processing and problem solving. And in fact, many would argue (for example, Wickens et al., 2004) that this is the case—that cognitive scientists, cognitive systems engineers, and human factors psychologists have generated a wealth of data relevant to the design of effective human–machine systems. However, others (for example, Hutchins, 1995) find that cognition as it occurs in the "wild" (that is, in the context of natural work situations) has little resemblance to what is observed in the context of paradigmatic laboratory research.

Within cognitive science, laboratory tasks are typically designed to isolate one "stage" of processing from other stages (for example, a perceptual task, a memory task, or a decision task). These tasks are "micro" relative to most of the "macro" activities of everyday life that require integration across the stages. As Klein (1998) discovered, putting out fires is not a perceptual, decision, memory, or action task – it involves a dynamical integration across the stages to connect perception with action.

Laboratory researchers tend, at least implicitly, to assume that cognition happens "in the head." And, thus, they tend to generalize from the laboratory based on the assumption that the "head" is the same, whether in the psychological laboratory (for example, choosing among gambles, or tracking a cursor) or in the workplace (for example, performing surgery, or landing an aircraft). In other words, the assumption seems to be that "awareness" can be understood independently from "situations." Basic theories of awareness abound, but theories of situations (of the stimulus) are sorely lacking (for example, Gibson, 1960; Flach, Mulder, and Van Paassen, 2005). I doubt that many would question whether the implicit theory of situations represented by laboratory research is naïve with respect to the complexity of work experiences. The question is whether this is a positive or negative strategic choice for science. On the positive side, the "micro" tasks of the laboratory allow precise control and measurement. On the negative side, there is the possibility that we may be precisely measuring things that are largely irrelevant with respect to generalizing to the "macro" level properties of natural work environments. For example, when stimuli are a row of lights without any context other than their relative frequency, performance can be well predicted by information statistics (Hick, 1952; Hyman, 1953). In this case, the possibilities are fully specified by the probability distributions, without a need to consider semantics, expertise, agendas, context, organizational goals, and so on. As James Jenkins argued in 1978, this may reflect the constraints of the stimulus situation rather than a fundamental principle about how humans process information. Should we generalize based on characterizing the human as an information-processing channel? Or should we generalize only to tasks with similar characteristics where the alternatives are only differentiated by their relative frequencies?

Thus, one obvious gap is between the view of cognition that shapes experimental paradigms and laboratory tasks and the view of cognition derived from more ethological studies of human performance. This is, as I understand it, one of the features that is used to distinguish NDM and macrocognition from "traditional" psychological research.

The laboratory research program implicitly suggests a linear dynamic that can be understood as a concatenation of isolated elementary processes. In contrast, field observations suggest a nonlinear dynamic where processes are tightly and adaptively coupled both internally (for example, the "stages" are not independent but reflect richly connected networks) and externally (that is, to the specific constraints and opportunities provided by the work environment). The laboratory view tends to create the impression of a rigid, brittle limited-capacity information-processing system. On the other hand, observations of experts in their field of work create the impression of a flexible, adaptive system that is highly sensitive to the semantic context. In the laboratory, the content of stimuli can be described as information in a purely statistical sense. In the natural environment, researchers are faced with issues of meaning that in many cases will reflect the dynamics of physical and social systems (for example, the performance boundaries of a vehicle or the norms of an organization).

The idea of cognition as an activity in the head of an individual creates a second gap. This is the gap between the individual and society—between microcognition as a process associated with a single person and macrocognition as a process of an organization (for example, a management team). This would suggest that the dynamics by which an individual and an organization make sense of and behave in the world are qualitatively different. This is questionable. A strong case can be made that the dynamics of cognition are qualitatively similar, whether they happen in the context of an individual or in the context of a larger organization. In my view, Weick's (1995) description of "sensemaking in organizations" could be equally appropriate as a description of how individuals make sense of the world. Similarly, Hutchins (1995) frames the computations associated with navigation as happening in the sociotechnical space of the ship and crew, not in the individual heads of the crew-members.

Perhaps the contrast between cognition as a micro-activity of individuals and sensemaking as a macro-activity of social organizations is simply another facet of the first gap. That is, in the everyday world of work, the social organization is a very salient part of the situation (that is, the task environment; the stimulus). This is the point of the situated theorists (for example, Clancy, 2006; Suchman, 1987). This dimension of the situation tends to be under-represented in a "basic" laboratory approach to cognition, leaving a gap between observations in the field and in the laboratory, and between our theories of cognition and our experiences of life. If you assume that cognition happens inside a head, then the social context of work can be left to sociologists. But if you assume that cognition is embedded in the social and physical context, then it is impossible for a program of research on cognition to completely ignore this context.

A third gap between the classical view of cognition and the experiences of life and work has to do with the connection between rationality and emotion/motivation. Although there are notable exceptions, cognitive science has tended to adopt a micro view that the rational components of experience can be understood in isolation from the affective/emotional components of experience. While emotion is not completely dismissed, the micro-strategy seems to treat emotion as an exogenous disturbance with respect to the fundamental dynamic of making sense of the world.

In contrast to the classical view in cognitive science, others wonder whether emotional centers in the nervous system play a fundamental role in processes of sensemaking. For example, Zull (2002) observed that "sensory signals go directly to the amygdala, bypassing the sensory cortex *before we are even aware of them.* This so-called 'lower' route begins to make meaning of our experience before we have begun to understand it cognitively and consciously. Our amygdala is constantly monitoring our experience to see how things are" (p. 59, emphasis in original). If Zull and others are correct, then emotions may have a critical role in the dynamics of meaning processing.

Thus, the distinction between "micro-" and "macro-" cognition draws attention to important concerns with respect to generalizing from classical laboratory research on human performance to the problems of natural human experiences:

1. Micro-research and theory tends to be organized around isolated stages of processing. In contrast, success in everyday life tends to demand integration across stages.
2. Micro-research and theory tend to focus on humans in isolation. In contrast, the challenges of everyday life tend to involve rich social contexts.
3. Micro-research tends to treat emotion and affect as exogenous disturbances with respect to cognition. In contrast, the experiences of everyday life suggest an intimate connection between emotion and cognition with respect to the meaning of information.

While it is important to recognize such gaps, I believe that these are only symptoms of a much deeper problem that has roots in the metaphysics that guides Western science. The term "macrocognition" reflects an important concern with the phenomena of human experience as it is lived, particularly in relation to work. It reflects a concern that micro approaches miss significant aspects of this phenomenon. It reflects a belief that human experience involves intimate interactions across "stages" of processing, intimate interactions with the physical and social context, and intimate interactions with emotion and motivation. These are important concerns, too important to be isolated or marginalized with a label that suggests a distinct variety of human experience. I believe that the questions raised by the construct of macrocognition are fundamental to understanding cognition as a natural part of human experience. Can anyone give me an instance of cognition that is not integrative? That does not depend on context? (Even in the laboratory we must consider the demand characteristics of the experimental setting.) Can anyone give me an instance of cognition that is independent of emotional and motivational factors? I claim that all cognition is macro!

The San Andreas Fault

> It starts with Plato's separation of the intellect or rational soul from the body and its skills, emotions, and appetites. Aristotle continued this unlikely dichotomy when he separated the theoretical from the practical, and defined man as a rational animal—as if one could separate man's rationality from his animal needs and desires. If one thinks of the importance of the sensory-motor skills in the development of our ability to recognize and cope with objects or of the role of needs and desires in structuring all social situations, or finally of the whole cultural background of human self-interpretation involved in our simply knowing how to pick out and use chairs, the idea that we can simply ignore this know-how while formalizing our intellectual understanding as a complex system of facts and rules is highly implausible. However incredible, this dubious dichotomy now pervades our thinking about everything, including computers. [Dreyfus, 1992, p. 64]

As Dreyfus notes, the dubious dichotomy that we inherited from Plato between body and soul (mind and matter; rationality and emotion; objective and subjective) pervades our thinking about everything – including computers and brains. The gaps discussed above are perhaps merely symptoms of a flawed metaphysics that ensures that mind and matter can never meet. This dichotomy leads us to use one language to address issues of mind and another language to address issues of matter—and thus conversations about *what matters* become incoherent!

In another paper, Flach, Dekker, and Stappers (2008) have attempted to argue for and outline an alternative metaphysics. Here, I would like to take a different tack, and offer an example that I hope will illustrate the problem. The example, illustrated in Figure 2.1, was inspired by Kahneman's (2003) lecture upon receiving the 2002 Nobel Prize in economics. As Kahneman writes:

> An ambiguous stimulus that is perceived as a letter in a context of letters is seen as a number in a context of numbers. The figure also illustrates another point: the ambiguity is *suppressed* in perception. This aspect of the demonstration is spoiled for the reader who sees the two versions in close proximity, but when the two lines are shown separately, observers will not spontaneously become aware of the alternative interpretation. They "see" the *interpretation* that is the most likely in its context, but have *no subjective indication that it could be seen differently.* [2003, p. 455, emphasis added]

In the classical approach to examples such as the ambiguous figure, there are two conflicting realities, objective and subjective. In the objective reality, the two objects are identical. In the subjective reality, the two objects are different. In the objective reality, the "true" state of the object is independent of the context. In essence, the objective state of the object is the same whether banked by letters or by numbers. In the subjective reality, the "false" impression suggests that the object is not independent of the context. So, to explain this, it is necessary to posit that the "context" is somehow mentally added to the object to create the subjective reality.

In the physical reality, the ambiguous figures are identical black shapes on a field of white, with no connection to the other black shapes in the field. In this reality, there is no such thing as a "B" or a "13." The only measures that count are length, thickness, shape, and so on. This is considered to be the physical truth, while the interpretation of either "B" or "13" is treated as an illusionary construction. In the

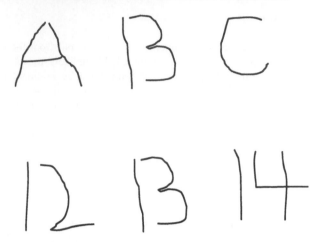

Figure 2.1   An example of an "ambiguous figure"

Note: This is an example of what is typically described as an "ambiguous figure." The middle items in the two lists are physically identical, but are perceived as the letter "B" in one case and as the number "13" in the other case.

Source: Adapted from Kahneman (2002).

mental reality of the literate Western world, the ambiguous figures are symbols defined in relation to a cultural history with written language and numbers. Not only does the context of the other black shapes in the field matter, but the unfolding context of the observer's experience also matters! Here is the rift! Important forces and constraints in one reality have no correspondent in the other reality. There really are two completely different realities. Another way of putting this is that there are two disconnected ontologies—systems of beliefs about what exists: the "physical" reality with one ontology and the "symbolic" reality with another ontology. This necessarily raises questions about the connection between the physics and the symbols. Here is the gap! Which science is responsible for connecting the physics with the symbols? How do we explain the connections between mind and matter needed to explain how fires are put out or aircraft are safely landed?

Since the dominant metaphysics starts with the assumption of two realities—one for the objective physical world and the other for the subjective mental world—no one is surprised that there is a gap between the way the world "is" and what people "see" and "think" about it. The surprise is that we ever get it right! This core ontology has had lasting repercussions in psychology, dating at least from Helmholtz's claim that perception is based on isolated, limited cues that must somehow be integrated in a process of "unconscious inference." This approach has generated an enormous collection of literature describing the illusions of perception, the biases of human judgment, the limitations of attention, and failures in problem solving. We have very detailed explanations of how and why people get things wrong. Yet, there is very little in this literature that addresses the successes reflected in everyday life. Human expertise and skill remain one of the deepest mysteries of science! How does the mind bridge the gap between the representations and the physical systems?

Certainly it is essential that the experiences of humans be connected with the physical context of that experience. The mental reality of the driving experience (for example, expectations) must be connected to the physical reality of the road (for example, consequences of actions). But are these two different realities or are the experiences and consequences aspects of a single reality? Can we address the experiences with one science or ontology (for example, psychology) and the consequences with another science or ontology (for example, physics) and expect to construct a coherent understanding of driving? If our goal is to understand "driving," we may need a single science or ontology that spans the experiences and the consequences.

Bridging the Gap

Metaphysics means only an unusually obstinate attempt to think clearly and consistently… And as soon as one's purpose is the attainment of the maximum of possible insight into the world as a whole, the metaphysical puzzles become the most urgent ones of all. [James, 1909, pp. 461–2]

Is there a viable alternative to this dichotomy that offers any hope of bridging this rift between physical and mental realities? I see two tacks that seem promising. One is the tack taken by Ecological Psychology (for example, Gibson, 1979) and the Distributed/Situated Cognition group (for example, Clancey, 2006; Hutchins, 1995; Suchman, 1987). The essence of this approach is to expand our view of the stimulus (that is, physical reality). In the classical view, the ambiguous stimulus is arbitrarily disconnected in time and space from all other objects. The ambiguous stimuli in Figure 2.1 are only identical if the larger spatial context that includes the surrounding figures and the larger temporal context (for example, the development of language and mathematics) is ignored as irrelevant features of the true "reality." In the classical view, the larger spatial and temporal patterns only exist in the mental representations. But somehow, their existence as patterns in the physical world is overlooked.

In the ecological view, the ambiguous figures are in fact "objectively" different, due to the patterns created in relation to the surrounding physical (other objects) and temporal (for example, observer's history with alphabets) objects and events. In this context, the research question is not "How do mental operations cause people to see something that is not there?" But rather, the question is "What properties of the physical reality specify 'B' and what properties specify '13'?" It should be clear that to find these properties we must expand our view of the stimulus to include more than the "ambiguous" figure in isolation from its context.

The ecological/situated cognition approach has important implications for the theory that cognition is a computational process. In the classical view, inferential/computational processes that exist only in the mental reality become the bridge between the two separate realities (the identical shapes on the page and the different symbols in the head). Somehow, these computational processes take a disconnected subjective/mental reality (with only an ambiguous connection to the "true" physical reality) and use them to coordinate actions with the true physical state of the world. In contrast, with the ecological/situated cognition view, the computational processes are intrinsic to the coupling with the ecology. Ambiguities are resolved through action, through exploration, through tool building and use. The computational processes themselves are grounded in the physical reality through action, *not* through inference! In fact, the inferences themselves are pragmatically grounded using abductive processes. The ultimate test of reality is not "rational" but empirical! Beliefs that lead to successful actions persist, while beliefs that lead to failure are extinguished. Note that it is not necessary to deny mental life—the key is that this mental life is intimately connected and adapted to the demands of the physical reality. The ambiguous figures are different due to the objective differences in the spatial-temporal patterns that they are embedded within.

In essence, the ecological/situated cognition perspective argues for one reality and claims that the reality is grounded in the physics of ecologies or situations. From this perspective, the differences in our experiences reflect real differences in the ecology or situation. The best strategy for bridging the gaps involves expanding our physics to include those things that make a difference. Practically, this means considering how well our laboratory paradigms represent the larger temporal spatial dynamics of everyday experience. This means considering the social dynamics of organizations as a reality of the world of experience that should be represented in our laboratories and our theories. This means exploring the connections between patterns in our ecologies and affect and emotion. We need to consider how the "quality" (for example, good or ill) of experience is specified by patterns in our ecologies (for example, Pirsig, 1974, 1991).

A second tack for escaping the dichotomous metaphysics of the classical approach is James's (1912) Radical Empiricism. The essence of this approach is to accept "experience" as the only reality. Many have dismissed this view as simple idealism or sophistry. However, this is a misunderstanding of James. In many ways, radical empiricism is the logical extension of an ecological approach. It doesn't deny a physical world in favor of a mental world. Rather, the point is that the realities of both physical and mental exist only in relation to their constraints on experience. In other words, this approach makes the dynamics of perception/action fundamental and the isolated separate images of objective physical and subjective mental worlds as two abstractions from this dynamic. From the view of radical empiricism, the idea of a physical reality apart from our observations and actions is a convenient fiction, as is the idea of a mental reality disconnected from the consequences of perception and action.

It seems to me to be quite ironic that while cognitive scientists have ignored and dismissed the philosophy of one of psychology's founding fathers, evidence for this position is emerging in physics—although it is true that many physicists are also trapped in the metaphysics of Plato and Descartes. In my view, one of the best arguments for an ontological stance consistent with radical empiricism was made by the physicist John Wheeler (1980) in his description of the "surprise" version of the Twenty Questions Game:

You recall how it goes—one of the after-dinner party sent out of the living room, the others agreeing on a word, the one fated to be questioner returning and starting his questions. "Is it a living object?" "No." "Is it here on earth?" "Yes." So the questions go from respondent to respondent around the room until at length the word emerges: victory if in twenty tries or less, otherwise, defeat.

Then comes the moment when we are fourth to be sent from the room. We are locked out unbelievably long. On finally being readmitted, we find a smile on everyone's face, sign of a joke or a plot. We innocently start our questions. At first the answers come quickly. Then each question begins to take longer in the answering—strange when the answer itself is only a simple "Yes" or "No." At length, feeling hot on the trail, we ask, "Is the word 'cloud'?" "Yes," comes the reply, and everyone bursts out laughing. When we were out of the room, they explain, they had agreed not to agree in advance on any word at all. Each one around the circle could respond "yes" or "no" as he pleased to whatever question we put to him. But however he replied he had to have a word in mind compatible with his own reply—and with all the replies that went before. No wonder some of those decisions between "yes" and "no" proved so hard!

In the real world of quantum physics, no elementary phenomenon is a phenomenon until it is an observed phenomenon. In the surprise version of the game no word is a word until that word is promoted to reality by the choice of questions asked and answers given. "Cloud" sitting there waiting to be found as we entered the room? Pure delusion! [Wheeler, 1980, pp. 392–8]

Thus, the idea that the ambiguous figures in Figure 2.1 exist as realities independent of the observer's experience of either "B" or "13" is pure delusion! The reality of the "B" or the "13" was created by the dynamics that includes the observer and the black marks on the page, but it includes so much more. The "B" and the "13" are real and they were created through a long process of interactions in which mind and matter participated. This dynamic is extended in time to include the evolution of visual systems and alphabets. This history is both physical and mental. Yet, it is exclusively neither. Thus, when we study the experiences of an expert pilot or fire ground commander, we are studying the reality of flying or firefighting. It is not sufficient to explicate the internal representations; the links between these representations and success must also be explicated for full understanding. We need to understand the experiences in relation to the dynamics of flying and firefighting.

Another metaphor that can help us to imagine the nature of a dynamic experience is that of a river (for example James, 1909; Rayner, 2006). The mental and physical dimensions of experience are linked like a river and its banks. The river is certainly not just the water; and it is not just the materials that shape the banks. Both the flowing water and the constraining materials of the bank contribute to the reality of the river. Even more, these two dimensions, water and bank, are intimately coupled with each constraining and shaping the other. The reality of the river is a creation of the interacting constraints.

Thus, the philosophy of radical empiricism does not deny either the waters of mental life or the banks of physical consequences. But it says that the dynamic of the river is the foundation for this reality. It is one river. It is one reality. It has facets that are mental (water) and facets that are physical (bank), neither of which has a reality independent of the other.

Finally, consider Gibson's (1979) construct—"affordance":

> The affordances of the environment are what if *offers* the animal, what it *provides* or *furnishes* either for good or ill. The verb *to afford* is found in the dictionary, but the noun *affordance* is not. I have made it up. I mean by it something that refers to both the environment and the animal in a way that no existing term does. It implies the complementarity of the animal and the environment. [1979, p. 127]

This affordance construct does not fit well with the old metaphysics, because the question arises: Are affordances objective or subjective? Are they physical or mental? The point of the construct is to get past the dichotomy to focus on the dynamic relations between actor and object. Affordances are neither objective nor subjective! Yet they are both. In the classical paradigm this is a conundrum. However, such non-Cartesian constructs are demanded by radical empiricism. Affordances reflect the possibilities for good or ill that an ecology offers an actor. These possibilities jointly reflect aspects of the world and of the actor.

In many respects, Pirsig's (1974, 1991) Metaphysics of Quality is consistent with Radical Empiricism and his construct of "Quality" is similar (perhaps identical) to the construct of affordance. Pirsig (1974) writes:

> "Man is the measure of all things." Yes that's what he is saying about Quality. Man is not the *source* of all things, as the subjective idealists would say. Nor is he the passive observer of all things, as the objective idealists and materialists would say. The Quality which creates the world emerges as a *relationship* between man and his experience. He is a *participant* in the creation of all things. The measure of all things—it fits. [1974, p. 384]

Research Implications

> It is not true that "the laboratory can never be like life." The laboratory *must* be like life! [Gibson, 1979, p. 3]

What are the implications of an ecological approach or radical empiricism for research on cognition? The first implication is to start with the phenomena of human experience as it is lived. Look to studies of firefighting, meteorologists, piloting, ship navigation, military command and control, and so on as representatives of the natural phenomena of cognition. These are not simply "applied" problems to be marginalized from the basic science of cognition. They are fundamental exemplars of the phenomenon of cognition.

The second implication is that in studying these exemplars we should not simply focus on the internal representations and thoughts of the people involved. Rather, we must explicitly include the contexts of the work as intrinsic to the phenomena. That is, in studying firefighting, we must include the dynamics of fires as a dimension to be studied. In studying meteorology, we must learn about weather systems. In studying piloting, we must not only consider how pilots think, but we must know how airplanes work. We need to talk to the aeronautical engineers as well as to the pilots. In studying ship navigation, we need to learn about the evolution and design of maps. In studying military command and control, we must learn about principles of war.

In sum, in studying naturalistic decision making or cognition in the wild, we must consider the constraints of the work as intrinsic or endogenous to the phenomenon of cognition, not as extrinsic or exogenous. The system of interest is not the decision maker or even the organization of decision makers, the system is the work ecology that includes psychological, social, organizational, and physical constraints.

A third implication concerns the relation between naturalistic observations, laboratory studies, and generalizations. The point is not to abandon laboratory research, but rather to consider the ecological context in the design of laboratory studies and in generalizing from those studies. In designing laboratory studies and generalizing from these studies, we need to give as much consideration to the stimulus context, as we do to the cognitive mechanisms. Hoffman and Deffenbacher (1993) decried this by carving the simple applied-basic distinction into multiple dimensions, including ecological validity (the task and materials used in experiments should correlate to real-world tasks), ecological relevance (the laboratory tasks should be pertinent to real-world cognition and behavior), ecological salience (the experimental tasks should relate to real-world tasks that are somehow important), and ecological utility (the research should help you do useful things). We should be skeptical of any generalization or principle framed in either purely psychological or purely physical terms. Cognition is adaptive in nature. Thus, principles of cognition should reflect the fit between organism and environment. Any principle that matters must reflect this fit or balance between mind and matter (or to avoid the dualistic language, between the objective and subjective poles of a single reality of experience).

Conventional wisdom seems to suggest that the further we abstract from real life, and the more we reduce phenomena into elementary components the more basic or general will be the principles that we learn. Researchers who are exploring macrocognition are providing insights that challenge that conventional wisdom. To find general principles, it is essential that researchers immerse themselves deeply in the phenomena of interest, as NDM researchers do, and as do sociologists and ethnographers who study the modern workplace. We need to start with good intuitions about the "eco-logic" of systems in order to make sensible abstractions or to learn how to parse the phenomena in meaningful ways. It is important that this basic research not be marginalized as "applied" or as a special case (that is, macrocognition). This work needs to be acknowledged as central to a science of cognition.

Conclusion

> Phaedrus remembered a line from Thoreau: "You never gain something but that you lose something." And now he began to see for the first time the unbelievable magnitude of what man, when he gained the power to understand and rule the world in terms of dialectic truths, had lost. He had built empires of scientific capability to manipulate the phenomena of nature into enormous manifestations of his own dreams of power and wealth—but for this he had exchanged an empire of understanding of equal magnitude: an understanding of what it is to be a part of the world, and not an enemy of it. [Pirsig, 1974, p. 387]

In my view, explorations of cognition in the wild, naturalistic decision making, ecological psychology, cognitive systems engineering, or macrocognition are essential to the success of cognitive science. However, the tactic of labeling this research, or the situation of researchers *having* to label it, seems to me to be unfortunate. My fear is that such labels cause the empirical research and theoretical insights generated in these explorations to be segregated from the general science of cognition. Time and energy that should be spent observing the bird to see how it behaves, end up being devoted to debates over what to name it (for example, is this an example of naturalistic decision making or macrocognition? Is this ecological psychology or macrocognition?).

The more niches we create for specialized kinds of cognition or specialized kinds of research, the less likely we are to converge onto a satisfactory science of cognition. Rather than developing separate theories associated with each of the different labels/ niches (for example, naturalistic decision making, macrocognition, ecological psychology), we should be looking for the convergent themes and common threads needed to stitch together a unified theory of cognition—a theory that does justice to the everyday experiences of humans. We need a theory that connects cognition to life as it is lived. We need a theory that addresses the psychical and physical aspects of that experience as intimate facets of a single system. We need a single ontology that spans both mind and matter.

I hope that the concerns raised here will not be dismissed as armchair philosophy. It is important to realize that we can't sit and wait for the answers to the ontological questions in order to begin the empirical research or in order to venture generalizations to improve the design of work systems. In fact, it is the empirical research and applied problems that are motivating the search for an alternative ontology. On the other hand, if we bury our heads in the empirical work and the applied questions and ignore the ontological questions we may find ourselves circling the same issues over and over again with little convergence and little hope for coherent theories to guide either our basic science or our applied generalizations. The metaphysics must be dynamically coupled to research, theory, and application. Again, the image of the surprise version of the Twenty Questions Game needs to be our guide. Our metaphysics can't stand apart from our search for the right word. Our metaphysics must evolve along with the search. Every observation and every design is simultaneously a test of and a challenge to our metaphysics.

Without an appropriate ontological foundation, there is little hope of building a science of *what matters*. Without such a foundation, we are doomed to continue stacking bricks on the pile without constructing any houses as suitable homes for life as it is experienced. We need a metaphysics that brings mind back into the world, before we can go forward to understand cognition. This is necessary for a basic science of cognition and it is especially true if we hope to generalize from this science to solve practical problems associated with the design of effective work organizations.

In sum, it is important to consider what we gain and what we risk losing by carving cognition into many different niches. I would be happier if the title of this volume were "cognition" instead of "macrocognition." Rather than differentiating the problems of macrocognition as a special form of cognition, we should be looking for the common threads among the different phenomena associated with human experience. This is not because I think the issues raised by the construct of macrocognition are unimportant. In fact, the opposite is true. The issues raised are far too important to cloak them in words that might move them to a sidetrack with respect to mainline cognitive science. All cognition is *macro*. It is only the paradigms of science that are *micro*.

Feynman, as one of the great teachers of science gets the first and last word. Again, Feynman is talking about his father, and how he trained his son to be a scientist:

> He had taught me to notice things and one day when I was playing with what we call an express wagon, which is a little wagon which has a railing around it for children to play with that they can pull around. It had a ball in it—I remember this—it had a ball in it, and I pulled the wagon and I noticed something about the way the ball moved, so I went to my father and I said, "Say, Pop, I noticed something: When I pull the wagon the ball rolls to the back of the wagon, and when I'm pulling it along and I suddenly stop, the ball rolls to the front of the wagon," and I says, "why is that?" And he said, "that nobody knows," he said. "The general principle is that things that are moving try to keep on moving and things that are standing still tend to stand still unless you push on them hard." And he says, "this tendency is called inertia but nobody knows why it is true." Now that's a deep understanding—he doesn't give me a name, he knew the difference between knowing the name of something and knowing something, which I learnt very early. [1999, pp. 4–5]

Acknowledgements

Sincere thanks to Robert Hoffman—his extensive feedback helped me to round some of the sharp corners off my rhetoric. However, I take full responsibility for any sharp corners or intemperate remarks that remain.

References

Clancey, W.J. (2006). "Observation of work practices in natural settings." In K.A. Ericsson, N. Charness, P.J. Feltovich, and R.R. Hoffman (eds), *The Cambridge Handbook of Expertise and Expert Performance*. New York: Cambridge University Press (pp. 127–45).

Dreyfus, H.L. (1992). *What Computers Still Can't Do: A Critique of Artificial Reason.* Cambridge, MA: MIT Press.

Feynman, R.P. (1999). *The Pleasure of Finding Things Out.* London: Penguin Books.

Flach, J.M. (1995). "Situation awareness: Proceed with caution." *Human Factors*, 37: 149–57.

Flach, J.M., Dekker, S., and Stappers, P.J. (2008). "Playing twenty questions with nature (The surprise version): Reflections on the dynamics of experience." *Theoretical Issues in Ergonomics Science*, 9:2: 125–45.

Flach, J.M., Mulder, M., and Van Paassen, M.M. (2004). "The concept of the 'situation' in psychology." In S. Banbury and S. Tremblay (eds), *A Cognitive Approach to Situation Awareness: Theory, Measurement, and Application.* Aldershot, England: Ashgate (pp. 42–60).

Gibson, J.J. (1960). "The concept of the stimulus in psychology." *American Psychologist*, 16: 694–703.

—— (1979). *The Ecological Approach to Visual Perception*. Boston, MA: Houghton-Mifflin.

Hick, W.E. (1952). "On the rate of gain of information." *Quarterly Journal of Experimental Psychology*, 4: 11–26.

Hoffman, R.R. and Deffenbacher, K.A. (1993). "An analysis of the relations of basic and applied science." *Ecological Psychology*, 5: 315–52.

Hutchins, E. (1995). *Cognition in the Wild*. Cambridge, MA: MIT Press.

Hyman, R. (1953). "Stimulus information as a determinant of reaction time." *Journal of Experimental Psychology*, 45: 188–96.

James, W. (1909). *Psychology*. New York: Holt.

—— (1912). *Essays in Radical Empiricism* (ed. R.B. Perry). New York: Longmans, Green, and Co.

Jenkins, J.J. (1978). "Four points to remember: A tetrahedral model of memory experiments." In L. Cermak and F. Craik (eds), *Levels of Processing and Human Memory*. Hillsdale, NJ: Lawrence Erlbaum Associates (pp. 429–46).

Kahneman, D. (2003). "Maps of bounded rationality: A perspective on intuitive judgment and choice." In Tore Frangsmyr (ed.), *Les Prix Nobel, The Nobel Prizes 2002*. Stockholm: Nobel Foundation.

Klein, G.A. (1998). *Sources of Power: How People Make Decisions*. Cambridge, MA: MIT Press.

Percy, W. (1991). *Signposts in a Strange Land*. New York: Farrar, Straus.

Pirsig, R.M. (1974). *Zen and the Art of Motorcycle Maintenance: An Inquiry into Values* (Perennial Classics Edition). New York: Harper Colllins.

—— (1991). *Lila: An Inquiry into Morals*. New York: Bantam Books.

Rayner, A. (2006). "Inclusional Nature: Bringing Life and Love to Science." Unpublished manuscript.

Suchman, L. (1987). *Plans and Situated Actions: The Problem of Human–Machine Communication*. Cambridge: Cambridge University Press.

Weick, K.E. (1995). *Sensemaking in Organizations*. Thousand Oaks, CA: Sage.

Wheeler, J.A. (1980). "Frontiers of Time." In G. Toraldi di Francia (ed.), *Problems in the Foundations of Physics*. Amsterdam: North–Holland.

Wickens, C.D., Lee, J.D., Liu, Y., and Gordon-Becker, S.E. (2004). *An Introduction to Human Factors Engineering*. Upper Saddle River, NJ: Prentice Hall.

Zull, J.E. (2002). *The Art of Changing the Brain*. Sterling, VA: Stylus.

Chapter 3

Technology and "Naturalistic" Decision Making: Myths and Realities

Kathleen L. Mosier

In the last decades, a body of research has documented the role of naturalistic decision making (NDM) within dynamic and challenging environments such as firefighting, combat operations, medicine, and aviation (see, for example, Klein, 1993; Zsambok and Klein, 1997). Much of this work has been concerned with the question of how people make successful decisions in highly complex scenarios with ill-structured problems, ambiguous cues, time pressure, and sometimes rapidly changing conditions.

Accurate situation assessment is critical in naturalistic environments, and one of the most important findings of NDM research is that experienced decision makers typically exhibit high intuitive competence within their domains (for example, Klein, 1993; Mosier, 1991; Orasanu and Fischer, 1997). They can quickly identify a subset of cues that is critical to accurate diagnosis of a situation and the subsequent decision and action. Expert pilots, for example, can look out the window and judge the speed and descent rate required to reach the runway, or use the shape, shade, and size of clouds to determine whether they should fly through them or around them. Expert fire-fighters use the sponginess of the floorboards or the color of smoke to assess the origin and extent of a fire. Naturalistic decision making research has advanced the field of judgment and decision making by identifying and describing expert decision processes in these and other naturalistic decision contexts. These processes include macrocognitive functions such as problem detection, sensemaking, planning and replanning, and coordination.

The Evolution of "Naturalistic" Decision-Making Environments into Hybrid Ecologies

Many naturalistic domains, such as aviation, medicine, military operations, or nuclear power, have undergone profound changes. A decade ago, I wrote about a set of myths surrounding expert decision making and automated decision aids (Mosier, 1997). Since that time, technology has become more sophisticated and more prevalent in dynamic decision environments. The increasing availability of sophisticated technological systems has profoundly altered the character of many "naturalistic" decision-making environments, and has changed macrocognitive processes such as assessing and making sense of the situation, coordinating data, and dealing with ambiguities. The use of technology "helps us past some barriers but introduces others" (Klein et al., 2003, p. 82), and inserts new cognitive requirements into the decision process. The technological environment creates a discrete rather than a continuous space (Degani, Shafto, and Kirlik, 2006), and as such demands more formal cognitive processes than does the continuous ecology of the naturalistic world.

Expert intuition, the foundation of naturalistic decision frameworks, may not be effective in meeting the requirements of a hybrid environment and in some cases may even be hazardous to the decision process. Moreover, technological aids typically are not designed to aid macrocognition in terms of eliciting situation-appropriate processes such as analytical data management or information search.

Unfortunately, research in NDM has not fully taken into account the impact of technology on the "naturalistic" decision environment, and several myths concerning the environment and associated cognitive demands persist. These erroneous beliefs have the potential to undermine essential macrocognitive functions and to lower the accuracy of resultant judgments and decisions. The purpose of this chapter is to discuss and "debunk" five existent NDM myths in the context of macrocognition in the technology-enhanced natural world, and to discuss the realities of macrocognition in hybrid domains. I make the case that high-tech decision environments are significantly different from purely naturalistic domains and that this difference impacts all macrocognitive functions. The first three myths have to do with the cognitive processes required in hybrid ecologies, and the last two myths concern the capability of current systems to support these processes.

Myth #1: Technology does not Significantly Alter the Character of "Naturalistic" Decision Environments

Toto, I've got a feeling we're not in Kansas anymore. [Dorothy, in *The Wizard of Oz*]

Reality:

Sophisticated technological aids have transformed many domains that were once solely "naturalistic" into complex ecological systems, combining probabilistic cues with electronic data and information. High-tech domains such as aviation create very different decision environments than those of lower technology domains. Automated cockpits, for example, contain a complex array of instruments and displays in which information and data are layered and vary according to display mode and flight configuration. The flight control task, once an active "stick and rudder" control process, is today essentially a complex automation problem involving supervising, programming, monitoring, thinking, and deciding (Mosier, 2002).

Technological aids in decision environments reduce the ambiguity inherent in naturalistic cues. They process data from the outside environment, and display them as highly reliable and accurate information rather than probabilistic cues. For example, exact location in space can be read from cockpit displays, even when no cues are visible outside of the cockpit. Some aircraft even provide auditory altitude callouts during landings. The location and severity of storm activity in and around clouds is displayed on color radar. Today's aircraft can fly from A to B in low (or no) visibility conditions—and once initial coordinates are programmed into the flight computer, navigation can be accomplished without any external reference cues. In contrast to earlier aviators, glass cockpit pilots may spend relatively little time looking out the window to update their situation awareness, and expend most of their effort integrating, updating, and utilizing information inside the cockpit (see Figure 3.1 for a look at the high-technology cockpit). In fact, in adverse weather conditions, cues and information outside of the aircraft may not be accessible.

Figure 3.1   High-technology cockpit (Boeing 747)

The environment, then, is a *hybrid* ecology—it is deterministic in that much of the uncertainty has been engineered out through technical reliability, but it is naturalistic in that conditions of the physical and social world—including ill-structured problems, ambiguous cues, time pressure, and rapid changes—interact with and complement conditions in the electronic world. In a hybrid ecology, cues and information may originate in either the naturalistic environment (external, physical), or the deterministic systems (internal, electronic). Input from both sides of the ecology must be integrated for situation assessment and decision making. Low-tech cues such as smoke or sounds, for example, can often provide critical input to the diagnosis of high-tech system anomalies.

Hybrid environments, then, *are* significantly different from naturalistic environments. What is constant across naturalistic and hybrid naturalistic/

deterministic environments is the emphasis on situation assessment in decision making. Unquestionably, decision making in high-tech/naturalistic environments is still heavily front loaded and accurate situation assessment is the most critical component. For experts in hybrid environments, accurate situation assessment will typically reveal a high-quality, workable action option. What does situation assessment entail in this ecology? What is the role of expert intuition in these hybrid naturalistic environments?

Myth #2: Expert Intuitive Processes are Effective and Sufficient in Hybrid Naturalistic Environments

> The development and introduction of modern automation technology has led to new cognitive demands…The result is new knowledge requirements (for example, understanding the functional structure of the system), new communication tasks (for example, knowing how to instruct the automation to carry out a particular task), new data management tasks (for example, knowing when to look for, and where to find, relevant information in the system's data architecture), and new attentional demands (tracking the status and behavior of the automation as well as the controlled process). [Amalberti and Sarter, 2000, p. 4]

Reality:

Expertise in hybrid ecologies does include the kind of intuitive know-how that is its hallmark in naturalistic environments. The intuitive facet of expertise, however, does not offer all of the same advantages to the decision maker in the hybrid naturalistic/electronic world as it does in the naturalistic world. Information in hybrid environments often cannot be managed intuitively, and relying solely on expert intuition may not only be insufficient but may also be counter-productive.

Making sense of data in an electronic world, searching for hidden information, maintaining coordination between system states, and planning and programming future states are macrocognitive activities that require something more than expert intuition. Processes such as pattern matching, the cornerstone of NDM frameworks of expert decision making such as Recognition-Primed Decision Making (Klein, 1993), do not work when data are in digital format, are hidden below surface displays, or mean different things in different modes. In a technological environment, a recognitional style of processing may even leave experts particularly vulnerable to errors if recognition is attempted before they have sought out and incorporated all relevant information into the pattern.

A critical component of expertise in hybrid environments, then, is knowing whether and when the situation is amenable to being intuitively recognized. Experience may offer hints as to where to look for anomalies in technological ecologies, but it does not insulate domain experts from the need to navigate their way through layers of data when anomalies occur. Numbers, modes, indicators, and computer readouts are some of the types of data that must be checked, processed, and integrated. Shortcuts may short-circuit the situation assessment process. Electronic data can help decision makers assess situations and make decisions much more accurately than ever before; however, the potential for heightened accuracy comes with a cost in terms of process—intuition simply is not sufficient. What kind of supplemental processing, then, is necessary in hybrid dynamic environments?

Myth #3: Analysis has No Place in Expert Decision Making in Dynamic Environments

> US aviation authorities are particularly curious about how the Gulf Air pilots handled the A320s' complex computer-operated control and navigation systems. The planes' software is designed to prevent even the most incompetent pilot from accidentally launching the plane into a fatal stall or dive. But records indicate that some Airbus crashes occurred when pilots misjudged the planes' limitations or made errors entering data into the A320s' computer system. And because the system is so complicated, US experts say, if something goes wrong only a mathematical genius could figure out the problem. [Mark Hosenball, *Newsweek*, 4 September 2000]

Reality:

Analysis is a necessary component of decision making in hybrid dynamic ecologies. Macrocognitive functions in these environments cannot be accomplished without it. At this point, however, clarification of the term "analysis" is required.

Naturalistic decision theorists have long recognized the inability of formal analytical theories such as Multiattribute Utility Analysis (MAUA) to describe judgment processes in dynamic, changing decision environments. The *analysis* prescribed by this and other normative theories includes assigning of weights and comparison of all options via mathematical computation. These processes are cumbersome and time consuming, and are typically related to the back side of decision making—the comparison of alternatives, assessment of probabilities, or evaluation of tradeoffs between utilities of outcomes. We know that experts in "real-world" dynamic decision contexts do not follow these normative analytical prescriptions, and in all probability would not survive in many environments if they tried to do so.

It is important, however, to distinguish the term "analysis" from the formal models that have been used to describe it. Analysis can also be defined in a broader context, as the use of a "conscious, logically defensible, step-by-step process" (Hammond, 1996, p. 60). This definition presents analysis as opposite to intuition, which produces a solution *without* the use of such a process. Analysis in this sense encompasses a broader range of cognitive behaviors than is specified in normative decision theories such as MAUA. It includes rationality, internal consistency, and appropriate use of cues and information. This characterization of analysis unquestionably fits into expert decision making in dynamic hybrid environments, In fact, as naturalistic domains become more technological, analysis in this sense becomes more critically important.

Analysis in situation assessment (the front side of the process) necessarily comes into play whenever decision makers have to deal with numbers, text, modes, or translations of cues into information (for example, via an instrument or computer).

Pilots of high-tech aircraft, for example, use analysis proactively and reactively when they discern and set correct flight modes, compare display data with expected data, investigate sources of discrepancies, program and operate flight computer systems, or evaluate what a given piece of data means when shown in this color, in this position on the screen, in this configuration, in this system mode.

In sum, while completely naturalistic ecologies are conducive to the intuitive use of probabilistic cues for situation assessment, hybrid naturalistic environments require analytical examination of data and information and the integration of cues and information from both sides of the hybrid ecology. In comparison with naturalistic cues, technological data and information are often not amenable to intuitive shortcuts or pattern recognition processes, but rather have to be assessed analytically to ensure that they form a cohesive, coherent representation of what is going on. Once interpreted, data must be compared with expected data to detect inconsistencies, and, if they exist, analysis is required to resolve them before they translate into unexpected or undesired behaviors.

Clearly, analysis has a place in expert decision making in hybrid dynamic environments. The management of data in naturalistic/deterministic ecologies such as the high-tech cockpit is a highly complex process. Designers in many hybrid environments are working to facilitate this process by creating "intuitive" displays—can these eliminate the need for analysis?

Myth #4: "Intuitive" Displays in Hybrid Environments Eliminate the Need for Analysis

Apparent simplicity, real complexity. [Woods, 1996, p. 15]

Reality:

System designers in hybrid naturalistic/deterministic environments have often concentrated on "enhancing" the judgment environments by providing decision aids and interventions designed to make judgment more accurate, and in doing so have made assumptions about what individuals need to know and how this information should be displayed. From the beginning of complex aircraft instrumentation, for example, the trend has been to present data in pictorial, intuitive formats whenever possible. The attitude indicator presents a picture of "wings" rather than a number indicating degrees of roll. Collision alert and avoidance warnings involve shapes that change color, and control inputs that will take the aircraft from the "red zone" to the "green zone" on the altimeter. Many high-tech cockpit system displays are designed to be holistic and pictorial, allowing for quick detection of some out-of-normal-parameter states. This design philosophy seems to be consistent with the goals of workload reduction and information consolidation for decision making—and, indeed, many features of cockpit displays do facilitate detection of some anomalies.

In their efforts to highlight and simplify automated information, however, designers of technological aids have inadvertently led the decision maker down a dangerous path by fostering the assumption that the systems they represent can be managed in an intuitive fashion. As discussed in previous sections, this is a false assumption. Within seemingly "intuitive" displays often reside layers of complex data. What the decision maker in a high-tech/naturalistic environment often sees is an apparently simple display that masks a highly complex combination of features, options, functions, and system couplings that may produce unanticipated, quickly propagating effects if not analyzed and taken into account (Woods, 1996). Technology serves as both a facilitator and an obstacle to accurate situation assessment. Because technological aids in hybrid environments have high internal reliability, the ecological validity of the information they present approaches 100 percent—but *only* if and when complete consistency and coherence among relevant indicators exists. This is not always easy to discern.

The cognitive processes required in hybrid systems, then, may be apparently intuitive, but they are really analytical. The glass cockpit, for example, is an electronic world in which accuracy of judgment is ensured *as long as* system parameters, flight modes, and navigational displays are consistent with each other and with what should be present in a given situation. When flight crews have ensured that *all* information inside the cockpit paints a consistent picture of the aircraft on the glide path—they can be confident that the aircraft *is* on the glide path. The pilots do not need to look out the window for airport cues to confirm it, and, in fact, visibility conditions often do not allow them to do so. They do, however, need to examine data with an analytical eye, decipher their meaning in the context of flight modes or phases, and interpret their implications in terms of outcomes. Failure to do so will result in inadequate or incorrect situation assessment and potential disaster.

Perhaps the most well-known example of the consequences of insufficient analysis in a hybrid environment (and possibly an attempt to manage a technological system intuitively via pattern-matching) occurred in Strasbourg, France (Ministre de l'Equipement, des Transports et du Tourisme, 1993), when an Airbus-320 confused approach modes and misinterpreted the apparently simple descent display:

> It is believed that the pilots intended to make an automatic approach using a flight path angle of -3.3° from the final approach fix...The pilots, however, appear to have executed the approach in heading/vertical speed mode instead of track/flight path angle mode. The Flight Control Unit setting of "-33" yields a vertical descent rate of -3300 ft/min in this mode, and this is almost precisely the rate of descent the airplane realized until it crashed into mountainous terrain several miles short of the airport. [Billings, 1996, p. 178]

Intuitive displays in hybrid domains, then, do not eliminate the need for analysis, but may make it more difficult to accomplish. Current technological systems are not set up to support either appropriate cognition or the development of expertise.

Myth #5: Technology in Hybrid Environments Supports Human Cognition

> ...errors have been considered to emerge from either the technology or the human, but seldom from the joint human-technology. The weakness in this approach becomes evident when considering the influence technology exerts upon human...performance. [Maurino, 2000, p. 955]

Reality:

Technology in hybrid environments *should* support human cognition, but often does not do so effectively. The way in which data and information are presented may have a powerful impact on the judgment strategy that will be induced, and thus whether or not the appropriate cognitive strategy will be elicited or facilitated. Because decision makers in hybrid naturalistic environments are required to use both analytical and intuitive processing, decision-aiding systems should recognize and support this requirement. In many hybrid ecologies, however, a discrepancy exists between the type of cognition required and what is fostered by current systems and displays. Many technological decision aids are designed on the basis of incomplete or inaccurate suppositions about the nature of human judgment and the requirements for metacognitive awareness and sound judgment processes.

Moreover, system data needed for analysis and judgment must be located before it can be processed. This is often not an easy process, because in many cases the data that would allow for analytic assessment of a situation may not only not be obvious, but not be presented at all, or may be buried below surface features. Technological decision-aiding systems, in particular, often present only what has been deemed "necessary." Data are preprocessed, and presented in a format that allows, for the most part, only a surface view of system functioning, and precludes analysis of the consistency or coherence of data. In their efforts to provide an easy-to-use intuitive display format, designers have often buried the data needed to retrace or follow system actions. Calculations and resultant actions often occur without the awareness of the human operator. System opaqueness, another facet of many technological systems, also interferes with the capability to track processes analytically, a phenomenon that has often been documented and discussed with respect to the automated cockpit (for example, Woods, 1996; Sarter, Woods, and Billings, 1997).

As an example of system complexity, Sherry and his colleagues (Sherry et al., 2001) decomposed the functions and displays of the vertical navigation system (VNAV) and of the flight mode annunciator (FMA) of the electronic cockpit. They found that the selection of the VNAV in the descent and approach phases of flight results in the engagement of one of *six* possible trajectories—and that these trajectories will change autonomously as the situation evolves. Moreover, the same FMA display is used to represent several different trajectories commanded by the VNAV function. It is not surprising, given these sources of confusion, that decision errors occur. It is impossible to utilize this system intuitively! However, the interface does not provide the necessary information to establish or reinforce correct mental models of system functioning (Sherry et al., 2001). The spontaneous generation of poor mental models of system functioning results in mistaken perceptions of a situation, and consequently incorrect decisions and actions.

In sum, the opaque electronic interface and sequential, layered presentation of data inherent in many hybrid naturalistic/electronic environments may short-circuit rather than support macrocognitive processing because it does not elicit appropriate cognitive responses. In fact, the design of many high technology systems encourages decision makers—including experts—to focus on the most salient, available cues and to make intuitive, heuristic judgments before they have taken into account a broader array of relevant cues and information.

Moving Past the Myths: Expertise and Macrocognition in Hybrid Environments

NDM and Expertise

The characteristics of expertise vary according to the cognitive demands that must be met. Expertise in hybrid ecologies entails the ability to recognize not only patterns of cues and information, but also potential pitfalls that are artifacts of technology, such as mode errors, hidden data, or non-coupled systems and indicators. Expertise includes knowledge of how a system works, and the ability to describe the functional relations among the variables in a system (Hammond, 2000). It entails being able to evaluate what a given piece of data means when shown in a particular color, position on the screen, configuration, and/or system mode. When technological data need to be processed, recognitional strategies characteristic of NDM have to be supplemented by careful scrutiny and attention to detail. Experts must examine data with an analytical eye, decipher their meaning in the context of modes or phases, and interpret their implications in terms of outcomes.

Because hybrid "naturalistic" environments blend natural and electronic components, they demand both intuition and analysis for situation assessment and decision making. Expertise entails knowing which strategy is appropriate at a given time in a given situation. Often, decision makers will be able to employ both strategies—while landing an aircraft, for example, a pilot may first examine cockpit instruments with an analytical eye, and set up the appropriate mode and flight parameters to capture the glide slope, and may then check cues outside of the window to see if everything "looks right." The physician may interpret the digital data on a set of monitors, and then examine the patient to see if the visual appearance corresponds to his or her interpretation.

In other cases, decision makers may opt to stay in one mode or the other. The pilot may be forced to rely on analytical processing of cockpit information if visibility is obscured, or may choose to disengage the automatic aircraft functions to intuitively "hand fly" an approach and landing if visibility is good. Jacobson and Mosier (2004) noted in pilots' incident reports that different decision strategies were mentioned as a function of the context within which a decision event occurred. During traffic problems in good visibility conditions, for example, pilots tended to rely on intuitive strategies, reacting to patterns of probabilistic cues; for incidents involving equipment problems, however, pilots were more likely to use more analytical decision strategies, checking and cross-checking indicators to formulate a diagnosis of the situation.

Problem Detection and Sensemaking

Strategies for problem detection and sensemaking in a hybrid environment include both analysis and intuition. To a great extent, problem detection and sensemaking is a matter of ensuring and maintaining consistency among indicators. Inconsistencies trigger the need for sensemaking to resolve the situation. Accomplishing this demands analytical cognitive processing, as experts are required to think logically and analytically about data and information. They need to know what data are relevant, integrate all relevant data to come up with a "story" of the situation, and ensure that the story that the data present is rational and appropriate. Sensemaking proceeds until cues and information from both sides of the naturalistic/deterministic system are integrated and the story is complete. Macrocognitive strategies for problem detection such as attention management and developing mental models are critically important for recognizing inconsistencies in data, as well as for understanding the limits of technological systems and recognizing their strengths and inadequacies.

Experts in hybrid ecologies must be aware of the dangers of relying too heavily on technology for problem detection. Parasuraman and Riley (1997) have discussed a tendency toward over-reliance on technological systems as "automation misuse." Trust in electronic systems, a byproduct of their high internal reliability, exacerbates the tendency to delegate analysis as well as task performance to the systems (for example Lewandowsky, Mundy, and Tan, 2000; Mosier and Skitka, 1996). Information from technological sources is often accepted with little scrutiny. In fact, "the more advanced the technology is thought to be, the more likely are people to discredit anything that does not come through it" (Weick, 1995, p. 3).

A byproduct of this is *automation bias*, a flawed decision process characterized by the use of automated information as a heuristic replacement for vigilant information seeking and processing, as a factor in professional pilots' decision errors. Two classes of technology-related errors commonly emerge in hybrid decision environments: (1) omission errors, defined as failures to respond to system irregularities or events when automated devices fail to detect or indicate them; and (2) commission errors, which occur when decision makers incorrectly follow a automation-based directive or recommendation without verifying it against other available information, or in spite of contradictions from other sources of information (for example Mosier et al., 2001; Mosier et al., 1998).

The automation bias phenomenon illustrates both the lure and the danger of non-analytical processing in hybrid environments. Even highly experienced pilots displayed high susceptibility to automation bias. In fact, experience was found to be *inversely* related to problem detection and decision making—the higher the amount of experience pilots had, the more automation-related errors they made (Mosier et al., 1998).

Planning and Replanning

Planning and replanning in a naturalistic environment typically entail evaluating options using mental simulation and story building. In hybrid environments, technological systems may be able to facilitate these processes by creating a visual projection of future states.

However, planning in hybrid environments is also likely to entail some form of programming as both the formulation and the initiation of a course of action. This introduces a new source of errors, as the analytical processes involved in data entry or human-computer interaction are much more fragile than intuitive processes—a single small error can be fatal to the process, and one small detail can destroy the coherence of the plan selected. Incident and accident reports in many hybrid domains, including the Airbus-320 accident cited earlier, document the hazards of one tiny programming mistake. The cockpit setup for a flight path angle of -3.3° in one mode looked very much like the setup for a -3300 ft/min approach in another mode.

The planning and replanning processes in the hybrid ecologies, then demand not only expertise in intuitive evaluation of options, but also analytical attention to detail and careful tracking of data that will be used by technological systems to implement decisions.

Coordination—Technology as Team Member

Decision makers in hybrid ecologies are part of a human-technology team. The need for coordination of activities and actions is always present, whether or not other human team members are involved. Coordination in hybrid environments requires the development of shared mental models of not only the situation on both sides of the hybrid system, but also of the technology and its functions and limitations, integration of cues and information from both the naturalistic and deterministic sides of the environment, and awareness of all interactions among human and electronic team members.

Facilitating Metacognition in Hybrid Environments: Designing Adaptive Decision Aiding Systems

In order to be effective, technological innovations to enhance human judgment in hybrid naturalistic environments should be based on correct assumptions about human judgment processes and should provide support for required cognitive processes. At least three potential avenues for facilitating appropriate metacognitive processes exist.

Facilitating the Development of Expertise

Few hybrid environments are designed to facilitate the development of expertise. In fact, high-technology system design often caters to inexperienced decision makers on the assumption that technological systems can make experts out of novices by augmenting inexperience and compensating for their lack of experiential knowledge and expertise. Data are often preprocessed, and presented in a format that allows, for the most part, only a shallow view of system states.

The problem with this approach to system design is that models and patterns of system functioning are not available to decision makers, with the result that, even after much experience, they are unable to develop the appropriate models or the informed experiential base that is a prerequisite to expertise. In order to facilitate the development of expertise in decision makers, technology in hybrid ecologies would have to be not only more transparent, as is typically noted in research literature, but also much more informative about the functional relationships that underlie the displays.

Presentation of Data and Information

Many researchers have pointed out the need for more transparency and less hidden complexity in decision aiding systems (for example, Woods, 1996; Sarter, Woods, and Billings, 1997). I would carry this a step further and suggest that systems provide a roadmap for tracking their recommendations, help the decision maker interpret them, and point out sources of confirming or disconfirming information. Through appropriate design in the presentation of information and cues, for example, technological aids could assist meta- and macrocognitive processes by highlighting relevant data and information, helping individuals determine whether all information is consistent with a particular diagnosis or judgment, and providing assistance in accounting for missing or contradictory information.

Importantly, data that cannot be used intuitively (for example, that requires interpretation or takes on different meanings in different modes) should not be presented in a format that fosters intuition rather than analysis. If what is being represented is really as simple as the pictorial display, then the design is consistent with the cognitive requirements. If the display is representing a complex system of layered data, however, a simple pictorial display may in fact be leading decision makers astray. This is an especially potent trap for domain experts, who are used to navigating the naturalistic domain via expert intuitive processes.

Aiding Metacognition

In addition to presenting data and information appropriately, decision-aiding technology should be designed to enhance individual metacognition—that is, it should help people to be aware of how they are thinking and making judgments, and whether their macrocognitive strategies match the characteristics of the environment. A metacognitive intervention, then, would *prompt process* (for example, *verify mode, check engine parameters*). Metacognitive interventions may also take the form of training for process vigilance. In our aviation research, for example, we have found that the perception that one is directly *accountable* for one's decision-making processes fosters more vigilant information seeking and processing, as well as more accurate judgments. Students who were made accountable for either overall performance or accuracy (which demanded checking several sources of information) displayed more vigilant information seeking compared with non-accountable students (Skitka, Mosier, and Burdick, 2000). Expert pilots who reported a higher internalized sense of accountability for their interactions with automation, regardless of assigned experimental condition, verified correct automation functioning more often and committed fewer errors than other pilots (Mosier et al., 1998). These results suggest that aiding the metacognitive monitoring of judgment processes may facilitate more thorough information search, more analytical cognition, and more coherent judgment strategies.

In summary, adaptive decision aids in hybrid naturalistic environments need to be adaptive not only in the *when* of aiding, but also in the *how*. Perhaps one of—if not *the*—most important functions of technology and technological aids may be to help domain experts to monitor the processes by which they make decisions by prompting process as well as providing information in an appropriate manner. In order to be truly adaptive, technology must help experts to navigate within and between naturalistic and electronic decision contexts, especially when they are initiating a new task or shifting contexts, and to adapt their own cognition effectively to the demands of hybrid naturalistic environment.

References

Amalberti, R. and Sarter, N.B. (2000). "Cognitive engineering in the aviation domain—Opportunities and challenges." In N.B. Sarter and R. Amalberti (eds), *Cognitive Engineering in the Aviation Domain*. Mahwah, NJ: Lawrence Erlbaum Associates (pp. 1–9).

Billings, C.E. (1996). *Human-centered Aviation Automation: Principles and Guidelines*. (NASA Technical Memorandum #110381) NASA Ames Research Center, Moffett Field, CA.

Degani, A., Shafto, M., and Kirlik, A. (2006). "What makes vicarious functioning work? Exploring the geometry of human–technology interaction." In A. Kirlik (ed.), *Adaptive Perspectives on Human–Technology Interaction: Methods and Models for Cognitive Engineering and Human–Computer Interaction*. NY: Oxford University Press (pp. 179–96).

Hammond, K.R. (1996). *Human Judgment and Social Policy*. New York: Oxford Press.

Jacobson, C. and Mosier, K.L. (2004). "Coherence and Correspondence Decision Making in Aviation: A Study of Pilot Incident Reports." *International Journal of Applied Aviation Studies*, 4: 123–4.

Klein, G.A. (1993). "A recognition primed decision (RPD) model of rapid decision making." In G.A. Klein et al. (eds), *Decision Making in Action: Models and Methods*. Norwood, NJ: Ablex (pp. 138–47).

Klein, G., Ross, K.G., Moon, B.M., Klein, D.E., Hoffman, R.R., and Hollnagel, E. (2003). "Macrocognition." *IEEE Intelligent Systems*, 18(3): 81–5.

Lewandowsky, S., Mundy, M., and Tan, G.P.A. (2000). "The dynamics of trust: Comparing humans to automation." *Journal of Experimental Psychology: Applied*, 6(2): 104–23.

Ministre de l'Equipement, des Transports et du Tourisme (1993). *Rapport de la Commission d'Enquete sur l'Accident survenu le 20 Janvier 1992 près du Mont Saite Odile à l'Airbus A320 Immatricule F-GGED Exploite par la Compagnie Air Inter*. Paris: Author.

Mauino, D.E. (2000). "Human factors and aviation safety: What the industry has, what the industry needs." *Ergonomics,* 43: 952–9.

Mosier, K.L. (1991). "Expert Decision-Making Strategies." In R. Jensen (ed.), *Proceedings of the Sixth International Symposium on Aviation Psychology*, Columbus, OH.

—— (1997). "Myths of expert decision making and automated decision aids." In C.E. Zsambok and G. Klein (eds), *Naturalistic Decision Making*. Mahwah, NJ: Lawrence Erlbaum Associates (pp. 319–30).

—— (2002). "Automation and cognition: Maintaining coherence in the electronic cockpit." In E. Salas (ed.), *Advances in Human Performance and Cognitive Engineering Research.* New York: Elsevier Sciences Ltd (pp. 93–121).

Mosier, K.L. and Skitka, L.J. (1996). "Human decision makers and automated aids: Made for each other?" In R. Parasuraman and M. Mouloua (eds), *Automation and Human Performance: Theory and Applications.* Hillsdale, NJ: Lawrence Erlbaum Associates (pp. 201–20).

Mosier, K.L., Skitka, L.J., Dunbar, M., and McDonnell, L. (2001). "Air crews and automation bias: The advantages of teamwork?" *International Journal of Aviation Psychology*, 11: 1–14.

Mosier, K.L., Skitka, L.J., Heers, S., and Burdick, M.D. (1998). "Automation bias: Decision making and performance in high-tech cockpits." *International Journal of Aviation Psychology*, 8: 47–63.

Orasanu, J. and Fischer, U. (1997). "Finding decisions in natural environments: The view from the cockpit." In C. Zsambok and G. Klein (eds), *Naturalistic Decision Making.* Mahwah, NJ: Lawrence Erlbaum Associates (pp. 343–58).

Parasuraman, R. and Riley, V. (1997). "Humans and automation: Use, misuse, disuse, abuse." *Human Factors*, 39: 230–53.

Sarter, N.B., Woods, D.D., and Billings, C. (1997). "Automation surprises." In G. Savendy (ed.), *Handbook of Human Factors/Ergonomics* (2nd edn). New York: Wiley (pp. 1926–43).

Sherry, L., Feary, M., Polson, P., Mumaw, R., and Palmer, E. (2001). "A cognitive engineering analysis of the vertical navigation (VNAV) function." *Journal of Human Factors and Aerospace Safety*, 1(3).

Skitka, L.J., Mosier, K., and Burdick, M.D. (2000). "Accountability and automation bias." *International Journal of Human–Computer Studies*, 52: 701–17.

Weick, K.E. (1995). *Sensemaking in Organizations.* CA: Sage Publications.

Woods, D.D. (1996). "Decomposing automation: Apparent simplicity, real complexity." In R. Parasuraman and M. Mouloua (eds), *Automation and Human Performance: Theory and Applications.* Hillsdale, NJ: Lawrence Erlbaum Associates (pp. 3–18).

Zsambok, C.E. and Klein, G. (1997). *Naturalistic Decision Making.* Mahwah, NJ: Lawrence Erlbaum Associates.

PART II
The Heartland of Macrocognition

Chapter 4

Macrocognition, Mental Models, and Cognitive Task Analysis Methodology

Gary Klein and Robert R. Hoffman

Introduction

In one view of macrocognition (Klein et al., 2003), the "primary functions" are things that domain practitioners say they need to accomplish: replanning, sensemaking, decision making, and so on. The "supporting processes" are cognitive capacities that make possible the achievement of the primaries. Mental modeling is such a supporting process, especially critical for sensemaking (Klein et al., 2006). This chapter is an exploration of some methods for gathering data, representing, and studying mental models. (In this chapter, we do not review methods of knowledge elicitation, broadly conceived. Reviews have been provided by Cooke (1992, 1994), Crandall, Klein, and Hoffman (2006), Hoffman and Lintern (2006), Hoffman et al. (1995), and Olson and Reuter (1987).)

Our purpose in this chapter is fairly immediate and practical: To discuss methods and methods for studying mental models. We discuss tried-and-true methods, but we also present some ideas about new methods, with an invitation for researchers to apply and evaluate them. This is a practitioner's account, that is, a description of a battery of techniques that researchers have developed for dealing with the problems associated with eliciting and refining meaningful and usable accounts of expert cognition. Our stance is "practitioner confessional" rather than a dogmatic "this is how it should be." This chapter is basically an invitation to naturalistic empirical inquiry: If one were conducting cognitive task analysis (CTA) on experienced domain practitioners, for purposes such as cognitive systems engineering, what sorts of methods might be used to reveal mental models?

First, however, we do need to deal with some philosophical issues. Any treatment of mental models could go into great length about issues of scientific meaning, validity, relation to phenomenology, and so on, and could rehash a considerable literature on the scientific status of mental models. With this in mind, we take pains to be succinct.

A History of Mental Models

The mental model notion, or something very similar, can be found in the works of Thomas Aquinas' *Summa Theologica* (1267), where he argued that, phenomenologically speaking, concepts are "in-formed" in the mind and "re-presented" to consciousness. In his application of the mental models notion to the field of business organization and management, Jay W. Forrester (1971) made Aquinas' point using modern language:

> A mental image or verbal description in English can form a model of corporate organization and its processes. The manager deals continuously with these mental and verbal models of the corporation. They are not the real corporation. They are not necessarily correct. They are models to substitute in our thinking for the real system that is represented. [Forrester, 1961, p. 40]

The mental image of the world that we carry in our heads is a model. We do not have a city or government, or a country in our heads. We have only selected concepts and relationships, which we use to represent the real system (Forrester, 1971, p. 213).

It has been suggested that Ludwig Wittgenstein was proposing a mental model notion in his discussion of the language, in which he contrasted a "picture" theory with rule-based theories (Mental Models website, 2007, p. 1). E.C. Tolman's (1948) notion of a "cognitive map" is often cited in the mental models literature. Kenneth Craik, a pioneer of the information processing viewpoint, talked in his book *The Nature of Explanation* (1943) about explanatory models as having a "relational structure" similar to the thing being modeled (p. 51):

> The idea that people rely on mental models can be traced back to Kenneth Craik's suggestion in 1943 that the mind constructs "small-scale models" of reality that it uses to anticipate events. Mental models can be constructed from perception, imagination, or the comprehension of discourse. They underlie visual images, but they can also be abstract, representing situations that cannot be visualised. Each mental model represents a possibility. Mental models are akin to architects' models or to physicists' diagrams in that their structure is analogous to the structure of the situation that they represent, unlike, say, the structure of logical forms used in formal rule theories. [Mental Models website, 2007, p. 1]

The mental model notion can be seen in the emergence of American cognitive psychology, in the debates about the "psychological reality" of theories of mental representation in the 1970s and 1980s (for example, Pylyshyn, 1981). The two seminal books on mental models (Johnson-Laird, 1983; Gentner and Stevens, 1980) used the notion to explain high-level cognitive capacities, especially expert–novice differences in knowledge content and organization. Since then, the mental model notion has seemed useful in many studies of cognition and learning, including research on sentence comprehension, research on deductive reasoning and argumentation, and research on causal reasoning (for example, Geminiani, Carassa, and Bara, 1996; Green, 1996; H.A. Klein and Lippa, in press, 2008; Legrenzi and Girotto, 1996). Mental models (of various stripes) have been regarded as fundamental to the field of cognitive engineering, in such topic areas as "user models" in software and interface design (for example, Ehrlich, 1996; Norman, 1986; van der Veer and Melguzio, 2003) and mental models as the basis for knowledge bases and rule systems in the development of expert systems (Scott, Clayton, and Gibson, 1991). The mental model notion has spilled over to other communities of practice in addition to the knowledge acquisition community, such as business systems dynamics research:

"Systems dynamics researchers have devoted a substantial portion of their research effort to developing a wide variety of techniques and procedures for eliciting, representing and mapping mental models to aid model building...[Mental models] are the 'products' that modelers take from students and clients" (Doyle and Ford, 1998, pp. 3–4).

For a full view on the nature of mental models one can also look to the large literature in philosophy of science on the relation of theories, metaphors, and models, and various kinds of models (mechanical, analog, conceptual, isomorphic, and so on) as mediators in scientific understanding (for example, Black, 1962; Hoffman, 1980).

In addition, within a number of disciplines one can find independent "discovery" of the mental models notion, as for instance in weather forecasting, where the notion of the forecaster's "conceptual model" has been used for some decades to distinguish understanding from the computational weather forecasting models (see Hoffman, Trafton, and Roebber, in press).

What Mental Models Are, Are Not, Might Be, and Might Not Be

There are several reasons why the mental model notion is controversial.

Mental Models are Mental

The mental model notion 'wears its mentalism on its sleeve,' making it a target for criticism and debate, especially in the United States where we still experience a lingering hangover from behaviorism. The argument is that mental models cannot be observed directly in "pure" behavior, and so like all phenomena of mind, they are of dubious scientific status (Rouse and Morris, 1986). For Europeans, less affected by the scourge of behaviorism, the mental model concept emerged gracefully. Our view is that the "mental model" designation/metaphor is invoked out of recognition of a phenomenon, not to be brushed aside for being subjective, or to be avoided because research on mental models fails to qualify as good science. To assert that the mental models notion has no explanatory value is, to us, to merely choose to ignore the obvious. Those who do that are walking a different road than the one we walk. Our challenge is to empirically explore, understand, and explain things that phenomenology dishes up—and that methodology makes difficult—not explain them away at some altar of methodolatry.

Mental Models Bring Mental Imagery into the Mix

A second reason why the mental model notion is controversial is because it points to the phenomenon of mental imagery. More than that, it points to the complication that for some people some of the time, mental phenomena can be abstract whereas for some people some of the time mental phenomena can be imagistic (Doyle and Ford, 1998; Hoffman and Senter, 1978). Jay W. Forrester (1961) has referred to this mix of mental images with concepts and relationships. Psychology has had great struggles over this (see Hoffman, 1979). The unfortunate fact is that any theorizing that invokes a notion of mental imagery in order to explain things will upset some people some of the time, and if it acknowledges that mental imagery is itself a thing in need of understanding and explanation, these people will get doubly upset.

Mental Models are in a Fuzzy Relation to a Host of Other Mentalistic Concepts

A third reason why mental models are controversial is that they relate, in fuzzy ways, to a variety of terms used to capture various aspects of versions of mental models: scripts, schemata, cognitive maps, prototypes, frames, and so on. Mental models relate, in fuzzy ways, to a variety of terms used to denote hypothetical "types" of knowledge (declarative, tacit, procedural, implicit, verbalizable, non-verbalizable, and so on). Although history and philosophy of science might benefit from extended treatments, research might come to a quick halt were one to dwell on such matters and not get on with the work. Our goal in this chapter is immediate and practical. Above all, we seek to avoid ontological paralysis that, as we will show, can stem all too easily from theorizing about mental models.

Mental Models Fit the Macrocognition Paradigm, not the Microcognition Paradigm

A fourth reason why the mental model notion is controversial is that it does not fit comfortably with the agenda of mainstream cognitive psychology. In fact, the mental model notion can be used to highlight the distinction between microcognition and macrocognition. Many of the seminal theories in cognitive science are essentially linear causal chains. Examples are the Shannon-Weaver theory of communication (encoder→transmitter→receiver→decoder) which helped stimulate information-processing psychology (see Shanon, 1948), and Colin Cherry's (1955) filter model of attention. Though some early, and most later models had to have at least one loop (top-down as well as bottom-up processing), the basic search has been for causal sequences of mental operations that are believed to be somehow fundamental (for example, memory access, shifts of attention, and so on). The beginning assumption in macrocognition is that the primary functions and the supporting functions are parallel and highly interacting, with subtly different mixtures of primary and supporting functions involved for certain aspects or types of cognitive work. It is difficult to think of mental models in any "causal chain" sense, given their fluxing conjunctions of concepts, images, beliefs, and so on. And computational cognitive modeling seems to lie well beyond some far-off horizon. In fact, this aspect of mental models may be one reason why some in cognitive psychology might have trouble with the notion— they are trying to fit a macrocognitive phenomenon into the microcognitive mold.

Mental Models are Limited

A fifth point for contention has to do with the fact that mental models are said to be selective and incomplete (Forrester, 1971; Rouse and Morris, 1986). Scientists have asserted that mental models are limited to seven chunks (see Doyle and Ford, 1998, pp. 18–19).

We know that with sufficient practice at immediate recall for various kinds of materials (for example, strings of numbers, restaurant orders, and so on) the 7 ± 2 can be surpassed. How many state variables can a nuclear control-room operator consider at one time? After the human-factors practitioner has confidently offered 7 ± 2 as the answer, the engineers or designers ask, "What's a chunk?" The response is typically, "It depends on the domain and the person's experience." Studies of expertise have shown clearly that the amount of information people can integrate into chunks is both flexible and domain-dependent. This challenges the researcher to dig deeper to help the designer or engineer discover what might be the basis for "chunking" information in that particular domain. So although there might be constraints on the number of chunks people can deal with effectively in working memory, and although 7 ± 2 might be a good ballpark estimate of that constraint, 7 ± 2 is hardly a limitation in the sense of a practical bound on the span of immediate apprehension. Rather, it reflects how material comes to be meaningfully organized. The challenge for research is *precisely* to study cognition across the proficiency continuum to see how experts develop larger, organized, and functional chunks (Flach and Hoffman, 2003).

It would set an impossible benchmark indeed to say that only perfect mental models merit scientific scrutiny. At any one point in time, anyone's mental model, even that of an expert, is bound to have some simplifying and incorrect aspects. This has been called the "reductive tendency" (Feltovich, Hoffman, and Woods, 2004), a consequence of what it means to learn, rather than a bias or limitation.

Mental Models are Wrong

A sixth reason why the mental model notion is controversial has to do with the fact that mental models are often wrong. Specifically, in research on laypersons' and students' mental models, it is generally shown that mental models can be fuzzy, implicit, mostly wrong, vastly simplified, dynamically deficient, broad, amorphous, and so on. The stance of cognitive psychology, the psychology of reasoning, and related fields is that "the most noteworthy characteristics of mental models are the deficiencies that arise from bounded rationality" (Doyle and Ford, 1998, p. 11).

The "flawed" nature of mental models is of course the reason they came under scrutiny in cognitive psychology in the first place, especially for instructional design (see McCloskey, 1983). Johnson-Laird's mental model theory was in fact proposed in order to explain the errors people typically make when trying to answer even fairly simple logical syllogisms. In the human–machine interaction field, Norman (1983) has described mental models as "incomplete," "unstable," and "unscientific," and that people's ability to simulate is "severely limited." Doyle and Ford (1998) provide a litany of instances where people's mental models are incomplete and incorrect, on topics ranging from the causes of global warning, to the workings of computers, to people's understanding of risk. Indeed, mental models have even been defined simply as sets of misconceptions (Atman et al., 1994).

When transported into applications, this leads some to suggest that mental models should not be studied at all. In cognitive psychology, Kieras (1988; Kieras and Bovair, 1990) argues that since mental models are incomplete, they would not permit correct inference of the steps required to solve problems or operate devices, and so should not be used to form instructional methods and materials. In the field of human factors, Vicente (1999, Ch. 2) argues that ecological interface design should not be "driven by" an analysis of cognitive constraints, that is, interface design should not be based on compatibility with the user's mental model. The reason is that mental models, perhaps even those of experts, are limited, incorrect, and so on.

Is this characteristic of mental models a reason to not examine mental models at all? Or is it throwing out the baby with the bathwater? From the standpoint of Expertise Studies and Cognitive Systems Engineering, 'flawed' mental models are definitely worth studying to enhance our understanding of novice–expert differences, or knowledge across the proficiency scale. Furthermore, Keil (2003) has shown that people have a less complete understanding of the workings of the world around them than they realize, but that these compact accounts permit them to track causal structures and get the gist of key relations without getting swamped by details. Thus, mental models can be thought of as lean cognitive representations rather than cracks in the veneer of rationality.

Paralysis by Analysis

What all this comes down to is that mental models are hard to define; "available definitions are overly brief, general, and vague" (Doyle and Ford, 1998, p. 3). Indeed, mental models have been defined so as to embrace all of knowledge: "deeply ingrained assumptions, generalizations…that influence how we understand the world" (Senge, 1990, p. 8).

We do *not* define mental models so broadly as to be equivalent to world knowledge or "collections of beliefs" (Maharik and Fischhoff, 1993), but we believe that anything that a person knows may be foundational to a mental model, including some beliefs. (Philosophically speaking, "beliefs" are a mere fuzzy boundary away from "knowledge" anyway.) We do *not* ignore the root phenomenon by retreating to a behavioristic-ecological view that mental models are merely things that "mediate between perception and action" (Wild, 1996, p. 10). That does not explain, it explains away. We do *not* ignore the phenomenon by asserting simply that mental models are "mechanisms." That does not explain, it merely substitutes one reductionistic metaphor (machine) for another (model) which obscures the phenomenon.

Could our entire discussion of CTA methods proceed without any reference to the notion of "mental model?" Perhaps. We could substitute "organized knowledge structures" or some such term, or otherwise and try and reduce all the conceptual baggage (images, scripts, and so on) to a small set of terms people might agree upon and nail down. One might conclude therefore that the mental model notion offers no value added. But it would be a vain hope that everyone would agree, and that the concepts would help us out by gladly remaining immutable. Furthermore, the approach of semantic de-boning would not allow one to actually escape any of the philosophical conundrums. It would just shift the issues from one turf to another. What do you mean by "organized knowledge"? What is a "structure"? Our view is that the notion of mental model (the aspects or features that various scientists and scholars have noted) gets at the core of a salient and important mental phenomenon. Further, we believe that the "mental model" designation serves at least as well as other (equally problematic) designations cognitive scientists might prefer. In this light, this chapter can be understood simply as an articulation of some of the methods for going beyond observation of "pure behavior."

Moving from philosophical paralysis to methodological paralysis, mental models can be hard to study. Psychologists generally view the detailed study of mental models as a difficult and complex, if not impossible, task. According to this view, mental models are continually changing, and furthermore, efforts to elicit, measure, or describe them can themselves induce changes in mental models. When people are asked to report their mental models, they may fail to report them accurately for any of several reasons, for example, they simply may not be aware of all of the "contents" of their mental models; they may feel compelled to invent explanations and answers on the spot that did not exist until the question was asked; or they may deliberately or unconsciously change the answers to correspond to the answer they think the researcher wants to hear (Norman, 1983). (We can set aside for the moment the fact that *all* knowledge elicitation is a co-constructive process; see Ford and Bradshaw, 1993.) The methods that cognitive psychologists believe are necessary to address these problems and to minimize measurement error (replication, verification, validation, and so on) are labor-intensive, time-consuming, and expensive, and are therefore only rarely applied (Doyle and Ford, 1998, p. 10).

This can be quite paralyzing, with some scientists arguing that mental models should not be studied at all. For instance, Kieras (1988) argues that as people become more proficient, they rely less and less on mental models (the knowledge becomes tacit); hence, the construct of mental model is useless in explaining expert performance. All of our experience and research on domains of expertise cuts against this bleak claim. Our view is that expert performance cannot *possibly* be understood fully without studying practitioner knowledge and reasoning. The phenomenon of mental models is manifest in every domain we have studied, and in some domains experts can be quite articulate in talking about how they imagine events and project them into the future (Hoffman, Trafton, and Roebber, in press).

Furthermore, we have no qualms about presenting *new* methods that researchers might try out, methods that await further study and validation. If a psychological method had to undergo rigorous validation, verification, replication, and successful use *before* it was even discussed, science would never have gotten off the ground. Indeed, many of the classic paradigms of cognitive psychology only became classic *after* replication, cross-validation, and so on, when people tested and refined the new methods that had been presented in the seminal papers (for example, the Brown, Peterson, and Peterson paradigm, the Sternberg paradigm, and the priming paradigm).

Defining Mental Models

In the reach for clarity about mental models, a number of scientists have distinguished types of models (for example, Black, 1962; Norman, 1983; Hoffman and Nead, 1983) in order to clarify the relations of the model to the thing being modeled. There is the target system to be understood, the human's conceptual model of that target system, and the scientist's conceptualization of the human's mental model (a representation of a representation). We agree largely with Doyle and Ford (1988) who define mental models as mental representations, but this concept can be unpacked:

- Mental models are a phenomenon or "presentation" to consciousness, in other words, they are "accessible" or "declarative."
- Mental models emerge in the interplay of perception, comprehension and organized knowledge.
- Mental models are relatively enduring (that is, they are not strictly static, not strictly "structures," not strictly "stored" things).
- Mental models are not snapshots but are themselves dynamic.
- Mental models are representations of ("mappings") something in the world (often, some sort of dynamical system).
- The ways in which a mental model emerges is shaped by the regularities, laws, principles, dynamics that are known or believed to govern the "something in the world" that is being represented.
- We would add that mental models often have a strong imagery component, but this additional assertion links easily enough into the ones stated above.

This takes us to methodology by asserting further that:

- The mental representations can be inferred from empirical data. People may not be able to tell you "everything" about their mental models, and they may not be able to tell it well. But with adequate scaffolding in a cognitive task analysis procedure, people can tell you about their knowledge, or, the knowledge can be manifested in the cognitive work in which people engage.
- The mental representations can be analyzed in terms of concepts and their relations.
- The concepts and relations can express states of affairs and dynamics.
- The concepts and their relations can be depicted in some form (words, propositions, diagrams, and so on) that constitutes the scientist's "model" of the mental model.
- The depictions can be regarded as re-representations of organized knowledge.

This definition is not substantively incommensurate with other definitions, such as that of Johnson-Laird (1983), although we do not mix our phenomenology with information-processing metaphors (that is, we do not see much value in saying that mental models are representations of tokens of variables). Our definition is not substantively different from that offered by Greeno (1977), of mental models as relatively enduring representations that interact with new information in a constructive process of dynamic problem representation resulting in the phenomenon (comprehension or image) that is presented to consciousness.

What we focus on in this chapter is the *functionality* of mental models.

The Functions of Mental Models

Mental models have the function of description and comprehension, but also projection to the future (Rouse and Morris, 1986, p. 351). The word "model" can be taken as a metaphor that is used to designate the core phenomenon precisely because the act of mental modeling is the apperception of a dynamic "runnable" event that can be mentally inspected, thought about, and projected into the future. Many scientists (for example, Rasmussen, 1979) have asserted that mental models are used to predict future events, find causes of observed events, determine appropriate actions to cause changes, and engage in "what-if" reasoning. For Rouse and Morris (1986), mental models are used to predict as well as describe and explain. Mental models help people to generate expectancies and to plan courses of action.

This is the meaning of mental model as a macrocognitive supporting function (Klein et al., 2003). Without accurate mental models, we would not be able to make sense of events. That is why it is so important to develop cognitive field research methods to study mental models, particularly the mental models of domain experts. Thus, to our definition above we add the following assertion: The depictions (re-representations of organized knowledge) can be studied through the applications of CTA.

Practical Methodological Challenges

Granted, the assumptions one makes about the features of a mental model can affect the methods used to study it, but if the history of experimental psychology (that is, concepts such as script, schema, and so on) shows anything, it shows that methods can be easily adapted to fit different theoretical predilections. No method used in experimental and cognitive psychology is adequate for eliciting, analyzing, and representing the full range of knowledge structures, schemata, scripts, and so forth that a person may bring to bear in describing, explaining, and predicting events. If we consider a mental model to include everything a person knows that could be relevant to describing, explaining, and predicting an event, the range of beliefs becomes unmanageable. Paralysis again.

But perhaps the vagueness is a sign that we are thinking about mental models at too abstract a level. Often, when a concept seems too fuzzy to be useful, that means it needs to be examined in specific cases. In this chapter, we will explore ways researchers might reveal and understand the mental models of the domain practitioners they are studying, and use depictions of the mental models to good result in applied research. Thus, when we refer to mental models, we are referring to both the human's conceptual model of that target system and the scientist's conceptualization of the human's mental model. These are coupled because the ways in which mental models are depicted (diagrams, propositions, and so on) are loosely coupled to the methods that are used to reveal them. For instance, the Concept Mapping knowledge elicitation method results in Concept Maps, sometimes referred to as "knowledge models" (Hoffman and Lintern, 2006). A card-sorting task in which people place related concepts into piles results in semantic similarity judgments that can be analyzed as semantic networks (for example, Schvaneveldt, 1990). A task in which people represent their understanding of some dynamic process results in process or event diagrams. The coupling is neither necessary nor strict, in that any of a variety of CTA methods can yield knowledge about domain concepts, events, principles, reasoning strategies, and so on (see Hoffman et al., 1995). The primary objective of this chapter is to offer some suggestions for how CTA methods can be used to describe mental models.

CTA for the Study of Mental Models

Arguably, modern CTA began in the 1970s with the introduction of Hierarchical Task Analysis (see Shepherd, 2001). The phrase "cognitive task analysis" emerged in a number of communities of practice in the United States in the late 1970s. Task Analysis was never void of cognition, with even the earliest forms of task analysis of the early 1900s having such descriptive categories as "decide" and "perceive." (A full history of task analysis is Hoffman and Militello, 2008.) CTA methods as we know them today (reviewed in Crandall, Klein, and Hoffman, 2006) include a number of types of structured interview, methods of work observation, and experiment-like methods, to elicit data that are analyzed and represented as a description of how people think about different types of activities. CTA methods have been used to describe decision strategies (for example, Klein, 1998) and knowledge (for example, Hoffman, Trafton, and Roebber, in press). We should be able to apply CTA to elicit and represent mental models for the design of interfaces, software tools, training programs, and so on. Not only do the technologies need to accord with how people think but they must also help people accelerate the achievement of proficiency: We need to be able to help people develop better, richer mental models.

In the following set of claims we suggest just one way to direct CTA research to more effectively capture mental models. Our presentation of methods is an invitation to explore. Many methods we discuss are tried-and-true, but some are new. In this chapter we do not consider issues of method reliability and validity.

Claims About CTA for Mental Modeling

Claim #1: CTA Studies can be Focused on Relationships

The search for mental models may become more tractable if we move down a level, and distinguish between different *relationships* that a person has a mental model about. Instead of thinking about global mental models, we can treat mental models as descriptions of relationships. This approach follows from the basic notion of propositional representations (attributable to Thomas Aquinas) as expressed on modern cognitive psychology by Kintsch (1974), Greeno and Simon (1988), and others, who posited that knowledge can be represented as concepts (objects, attributes, and so on) linked by relations to form propositions rather than associations. Moray (1987) proposed that learning a mental model of a system depends on learning the relationships among subsystems (also see Samurçay and Hoc, 1996; Staggers and Norcio, 1993).

What sorts of relationships might we consider? Johnson-Laird (1983) contrasted the features of physical and conceptual mental models, and identified several types of relationships that needed to be captured in a representation: conceptual, temporal, spatial, and causal. Building on Johnson-Laird's framework, we can identify different types of relationships that could be captured by CTA: mental models about spatial relationships, conceptual relationships, causal relationships, organizational relationships, and so forth.

Claim #2: Each of These Kinds of Relationships is Captured Fairly Well by Different CTA Methods

Table 4.1   Matching CTA methods to conceptual relationships

- Conceptual domain—Concept Map
- Activity—cause-effect diagram or process diagram
- Spatial relationship—map
- Device activity—cause-effect sequences, diagrams, or stories
- Temporal relationship—script
- Organizational relationship—wiring diagram
- Dependency relationship—plan or storyboard

Table 4.1 illustrates how we might match CTA methods with types of relationships. Thus, a mental model of a domain might be captured by a Concept Map (Crandall et al., 2006; Novak, 1998). A mental model of a spatial relationship is best depicted by a topological diagram (for example, Gould and White, 1974). A mental model of a device—an image and related beliefs about the device's structure and functionality (the activity of the device)—might be described by diagrams accompanied by expressions of cause-effect sequences or even stories illustrating different device modes (for example, Norman, 1983). A temporal relationship might be illustrated by a script, an organizational relationship might be represented by a "wiring diagram" showing roles and functions, and a dependency relationship might be described by something like a plan.

Claim #3: We Can Select CTA Strategies by Clarifying the Purpose of the Research

Investigators rarely initiate a CTA study without some external motivation. The reason for the study can constrain and direct the research (Hoffman and Deffenbacher, 1993). One reason to conduct a CTA study is to identify critical cues that play a role in the formation and use of mental models. For instance, Crandall and Calderwood (1989) selected a CTA method to catalog critical cues in a study of neonatal intensive care unit nurses in spotting early signs of sepsis. Hoffman, Coffey, and Ford (2000) observed the weather briefings of forecasters to reveal the patterns in radar images that are indicative of tornado formation.

CTA methods can help the researcher to explain strategies. Thus, Klein, Calderwood, and Clinton-Cirocco (1986) explained how fireground commanders were able to make critical decisions under extreme time pressure and uncertainty.

CTA methods can account for errors by revealing flaws in mental models. As described earlier, McCloskey (1983) found evidence that people held an impetus theory that led them to expectations that violated principles of physics. Norman (1983) noticed that people were making unnecessary key presses on hand calculators (for example, clearing the registers by pressing the CLEAR key, not once, but several times). Norman found that people did not have accurate mental models of the devices; they acknowledged their inadequate mental models. Their inefficient behaviors were actually reasonable strategies that reduced memory burdens.

CTA methods can discribe the knowledge base of experts, which can be relevant for training and for system design. R. Hoffman, Trafton, and Roebber (in press) and Pliske et al. (1997) conducted CTA studies with weather forecasters for these reasons.

Claim #4: Researchers Draw on a Small Set of Paradigms

The challenge of designing CTA studies to capture mental models becomes more tractable when we identify the common paradigms used in these kinds of studies.

Mental models are foundational to a range of different types of cognitive performance, such as solving problems, generating predictions, anticipating events, and forming inferences. Investigators have developed many different types of experimental paradigms to measure success on these kinds of tasks. These are not the paradigms of interest here.

Claim 4 is about paradigms that uncover the content of a mental model, not just its impact on performance. Rouse and Morris (1986) listed several CTA approaches that could be useful in eliciting a person's mental models. These include the use of the Think-Aloud Problem Solving (TAPS) task, interviews, and questionnaires. Gentner and Stevens (1983) listed the methods they had encountered: TAPS, card-sorting tasks, cognitive psychology experiments, field observations, and so on. The taxonomy described below, which overlaps that of Rouse and Morris, as well as Gentner and Stevens, is based on a selective review of articles and chapters on mental model research:

- TAPS. Have participants think aloud as they try to solve problems.
- Directed inquiry. Ask participants to describe their reasons for adopting a regimen.
- Nearest Neighbor. Present several alternative mechanisms and have participants identify the one closest to their beliefs.
- Cognitive Intervention. Direct the participants to use a particular mental model (or metaphor) and see how that turns out.
- Glitch Detector. Uses bogus models of cognitive strategy to see if participants notice the error, and then compile improvements.
- Oddity. Look for unexpected activities during observations of cognitive work.

CTA Methods

By far, the most common CTA paradigm for revealing mental models is TAPS, usually combined with a data analysis method called protocol analysis. TAPS is not one single procedure but is a suite of methods, variations on a theme. The most widely cited method is the one preferred by Ericsson and Simon (1993). It involves TAPS, but has come to be equated with "protocol analysis" since that is its given name. In their procedure, the experimenter presents the participant with a problem and asks the participant to think aloud. In other variations on TAPS procedures, participants can be asked questions while they work. They can be interviewed afterwards, and so on. (For examples of such variations on TAPS methods, see Beach, 1992.)

Ericssonian protocol analysis is aimed at getting a running record of the "contents of working memory" from which the researcher derives a description of problem states and operators. But it can also be used to describe knowledge and reasoning, that is, mental models. The mental models serve to explain why participants did what they did in the course of problem solving (see Beach, 1992).

Another form of Think-Aloud Problem Solving can provide insights about the processes people use in building and revising mental models, as well as about the content of the mental model. Thus, for example, Collins and Gentner (1987) asked people questions about the physical process of evaporation, and found a frequent use of mini-analogies as the participants used mental simulations to construct story accounts. They described their participants as using visual images and having an "introspective feel of manipulating images" (p. 246).

Researchers commonly set up comparisons. They may recruit different types of participants, such as experts and novices, in order to contrast the sophistication of their mental models. For instance, a task in which people sort domain concepts according to their perceived semantic relatedness can reveal the deep versus superficial understanding of problems (for example, Chi, Feltovich, and Glaser, 1981). Researchers may contrast mental models formed for easy versus difficult problems. They may contrast participants from different cultures. They may contrast performance before and after instruction that is designed to alter mental models. Another type of comparison observes the nature of the mental models from the beginning of a training program to the end.

A great deal of preparation often goes into the design of the problems that are presented to participants. Problems must be carefully selected and designed in order to address the research purposes (for example, the mental models of pilots faced with a simulated malfunction; Smith, McCoy, and Layton, 1993). Furthermore, the process of transcribing the verbalization and encoding the transcript statements is effortful and time-consuming (Hoffman et al., 1995).

In addition, the researchers usually need to be familiar with the domain and the problem set, and should try to anticipate the mental models that participants might be using. This is particularly true in designs that permit the data collectors to question the participants as they go about solving the problems. The kinds of tip-offs they look for, the kinds of hypothetical questions they pose, the ways of phrasing an inquiry without leading the participant, all require skill, knowledge, and also sophistication. Even designs that postpone the questioning until after the participant completes the problem place demands on the data collectors to avoid leading questions or hints. This is a reason why some researchers shy away from the use of probe questions and advise others to shun such methods (for example, Ericsson, 2006). As we know from studies of expertise, expertise is achieved only after practice, and more practice. The same holds for those who would try out alternative methods of CTA, including interviews and TAPS tasks that rely on probe questioning. As we have pointed out, our choice is to explore rather than let ourselves be paralyzed because a method cannot be formed as a controlled experiment.

Directed Inquiry methods (see Table 4.1) use interviews to see if participants can explain their reasons for adopting a strategy. This method can rely on specific procedures such as the Critical Decision method (Klein et al., 1989). For example, Lippa, Klein, and Shalin (2008) have conducted interviews with patients diagnosed with Type II diabetes to learn how they control the disease. The interviews try to uncover the patients' belief systems about the causal relationships between the disease, the various interventions for controlling it, the dynamic life events that complicate any control regimen, and the effect these strategies and adaptations have on the patients' blood glucose levels. The interviews may elicit general beliefs, or they may probe beliefs in the context of specific incidents, such as recent self-management activities, adaptations to stress, illness, and other complications that the patients report.

The Nearest Neighbor technique is particularly helpful in cases where people cannot readily articulate their beliefs. For example, Klein and Militello (2001) describe using this method with housewives to examine their beliefs about how a common item works. The housewives gave blank stares when asked about how the item did its job. Subsequently, the researchers prepared a set of alternative diagrams, each depicting a different possible mechanism. The next group of housewives was asked which diagram came closest to matching their understanding for the product. This time, the housewives had little trouble selecting the best match, and then explaining where it didn't quite fit. Hardiman, Dufresne, and Mestre (1989) used a similar method with physics experts and novices. They presented the subjects with a target problem and asked them to judge which of two comparisons would be solved most similarly.

The Cognitive Intervention technique is often applied as a form of instruction. For example, Gentner and Gentner (1983) suggested analogs to people who were trying to make sense of electricity. The nature of the analog that people used (flowing water or teeming crowds) influenced the type of inferences they were able to accurately make.

The Glitch Detector method presents participants with a diagrammatic account of a process, but with a twist—the diagram has a subtle but important flaw. The participant is not told that the diagram is flawed. The task is simply to review the diagram. The question is whether the participant notices the flaw—to indicate the way the participant understands the process.

A variant of the Glitch Detector method is the Macrocognitive Modeling Procedure (perhaps more appropriately thought of as a cognitive mis-modeling procedure) (Hoffman, Coffey, and Carnot, 2000). The Macrocognitive Modeling Procedure was designed in an attempt to create a "fast track into the black box," in the sense of supporting the development and behavioral validation of macrocognitive models of practitioner reasoning in an efficient way, avoiding labor-intensive protocol analysis. The MMP evolved after Hoffman, Coffey, and Ford (2000) had created the "Base Model of Expertise," presented in Figure 4.1. This model combined the notion of the hypothesis testing refinement cycle, from Karl Duncker (Duncker, 1945; see also Newell, 1985), combined with other key macrocognitive functions that can be seen in Figure 4.1. The model seems to capture (as variations on a theme) a number of proposed hypothetical reasoning models that have been offered in studies of diverse domains of expertise. (See Hoffman and Militello, 2008, Ch. 8.)

The MMP was first conceived when a weather forecaster was shown the Base Model of Expertise, and was asked whether it seemed appropriate to the domain. It was felt that this would be sensible since, as we pointed out earlier, the weather forecasting community is comfortable with the distinction between

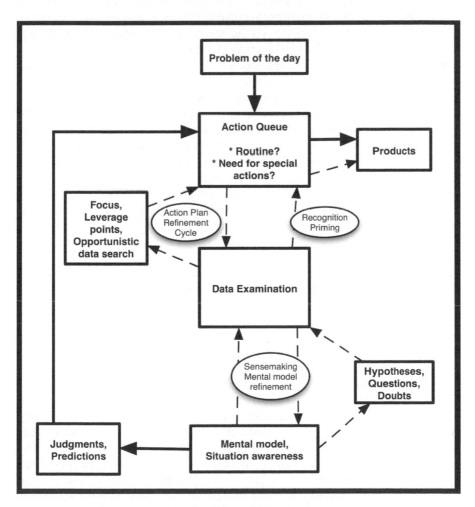

Figure 4.1   The "Base Macrocognitive Model" of expertise

Source: Adapted from Trafton and Hoffman, 2006.

conceptual models versus computational models. The informant spontaneously took the diagram as an opportunity to add domain-specific details to the process description, and modify some of the relations among the diagram elements. With this experience as the starting point, Hoffman et al. created a formal procedure, consisting of four steps:

- *Step 1*: The researcher takes the Base Model and adjusts it to make it directly pertinent to the domain. For example, the "Problem of the Day" would be specified as "The Forecasting Problem of the Day" and "Data Examination" would be specified as "Examination of images, data, radar." Next, two alternative "bogus models" are created. At least one of these includes some sort of loop, and both include a number of the elements of the Base Model. Taken together, the bogus models include all the core macrocognitive functions (for example, recognition priming, hypothesis testing, and so on). Ideally, bogus models are not too unrealistic, yet the researcher would expect the practitioner to not be entirely satisfied with either of them.
- *Step 2*: Domain practitioners, spanning a range of proficiency, are shown the bogus models and are invited to pick the one that best represents their strategy. Then, using the bogus models and their elements as a scaffold, the practitioners are invited to concoct their own reasoning diagram.
- *Step 3*: After the individual diagrams have been created, the researcher deliberately waits some weeks, or even months, and then each practitioner within the organization is shown all of the models and is asked to play a "guess who" game. This step in the MMP is a form of sociogram, revealing the extent to which practitioners share their knowledge and discuss their reasoning. This step also helps to identify individuals who possess special sub-domain expertise or skills. Thus, this step contributes to proficiency scaling.
- *Step 4*: After another delay, the researcher locates him or herself in the workplace and observes each practitioner as they arrive and begin their day's work. Some elements of the model can be validated by observing worker activities (for example, "examine satellite images"), whereas other elements that cannot be validated that way can be the focus of probe questions. Figure 4.2 shows results for just one of the practitioner models, indicating the results in Step 4. The call-out balloons indicate the results from the observation/probe questioning procedure.

Results from the procedure included models of the reasoning of journeymen and expert forecasters, affirmed based on observations of forecasting activities. It took an average of 52 minutes task time to develop a model. This is without doubt considerably less than the time taken in other methods. (For example, preparing and functionally coding a transcript of problem-solving protocol can itself take many hours.) The results conformed to the Base Model of forecaster reasoning that had been developed in an initial documentation analysis. The results also clearly revealed differences in proficiency, with less experienced forecasters relying in an uncritical way on computer forecasts, and less likely to think hypothetically and counterfactually.

This study was a first attempt, and limited in sample size, but promising enough for us to invite others to try it out. (Details on procedure, instructions, and so on can be provided by the second author upon request.) We feel that the MMP holds promise for the development of reasoning models and the testing of hypotheses concerning reasoning models in less time than taken by traditional experimentation. It also showed that it makes little sense to think that a single macrocognitive model will effectively capture practitioner reasoning. One forecaster, in Step 3, rejected his own

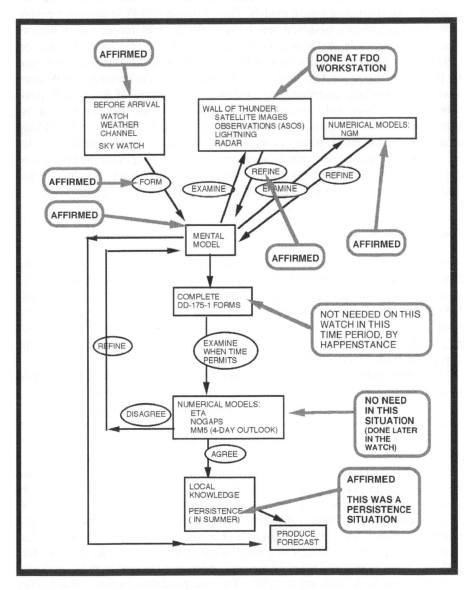

**Figure 4.2   One of the forecaster models, with annotations from
Step 4 of the MMP**

model, later explaining that the model that had been created weeks earlier was no longer his preferred strategy since the weather regime had changed. Hoffman et al. speculated that for a domain such as forecasting, many dozens of "strategy models" would be needed to present a rich and fair picture of practitioner reasoning.

Additional observational methods of work analysis are described in Crandall, Klein, and Hoffman (2006), and in Hoffman and Militello (2008). For example, the Oddity strategy is employed during field observation. In this method, the researchers remain on the lookout for activities that don't seem to make sense. These are windows into alternative mental models. Thus, Norman (1983), in observing people using hand calculators, observed seemingly inefficient and unnecessary behaviors that revealed gaps in the mental models of how the devices worked. Hutchins (1983) described the way that he was able to understand Micronesian navigation by being sensitive to practices that made little sense from the standpoint of Western navigation. Hutchins assembled the evidence, in the form of many individual, related observations, and synthesized these to formulate his hypothesis about the mental models of the Micronesians. Darnton (1984) described the value of an Oddity strategy for historians, who can never really understand the way people from earlier eras reasoned. Darnton deliberately used unusual and baffling events as his point of departure. He was able to break out from his own mental models to gain a deeper understanding of the mental models used during a given historical period.

Conclusions

The concept of mental models becomes somewhat manageable if we focus our inquiry on knowledge people have about specific types of relationships (for example, conceptual, spatial, temporal, organizational). CTA methods already exist, and have been applied in multiple contexts or domains for eliciting and representing knowledge of a variety of relationships. Moreover, researchers have developed a set of paradigms for studying mental models. We can identify a number of next steps that might improve our research into mental models.

It could be valuable to compare different CTA methods that are relevant to a type of conceptual relationship, and see their relative strengths and weaknesses regarding yield of information, validity, reliability, effort, efficiency, and so on (see Hoffman, 1987). Such work could improve the effectiveness of CTA methods. It might also suggest new forms of CTA that might describe important relationships that are not handled well by our existing methods.

Another direction is to survey the techniques used by researchers to code data. Ford and Kraiger (1995) identified several ways of scoring data. One method was to score levels of complexity, a measure that might reflect the level of sophistication of a mental model. Additional measures were the degree of differentiation among elements within a level, and the distance or number of links between key nodes. Recent developments in Concept Mapping (for example, Cañas, Novak, and Gonzalez, 2004) expand the range of potential measures reflecting the quality and complexity of mental models. As with any mapping of a conceptual definition to a measurable, caution is in order because the quantitative analysis is only a snapshot of a qualitative understanding (in this case, of the content of a mental model). Dekker (2003) has warned about categorization that appears to provide insight but is insensitive to the dynamics of each individual case. For that reason, we need to be careful to use the qualitative findings of CTA studies along with quantitative measures.

Finally, researchers can focus their work on specific criteria for effective mental models. For example, Lippa and H.A. Klein (in preparation) are finding variations in comprehensiveness (whether the mental model includes all the relevant processes involved in the causal relationships), depth (the level of detail of how each major factor operates), coherence (whether the mental model forms a compelling metaphor or story, as opposed to a set of fragmentary beliefs that are not well-integrated), accuracy (whether the mental model contains flawed beliefs), and utility (whether the mental model is helpful in maintaining a regimen for controlling the diabetes). Other researchers might use different criteria. We expect that the using of criteria can guide the examination of mental models, and that the nature of the findings will shape the criteria as the study progresses.

In a few places in this chapter we have referred to issues of validation and verification. Our stance is to avoid paralysis resulting from the refusal to try a method unless it has been proven to reliable and at least up to the methodolatrist standard of 'observation of pure behavior' (whatever that might be). All methods, even experimental procedures, have strengths and weaknesses, and the potential weaknesses should not prevent people from exploring the methods as opportunities. This being said, of course validation and verification are important. There are other issues as well, such as ethical ones. For instance, in some contexts, Step 3 of the MMP (the "guess who" game) may raise issues of disclosure.

We look forward to an expansion of research on mental models that can elaborate and extend the methods we have described, provide guidelines for using CTA methods to capture mental models, and also increase our understanding of macrocognitive functions.

Acknowledgements

We would like to thank Beng-Chong Lim for posing the initial questions about using CTA to study mental models—questions that started us down this path of exploration. The second author's contribution was through participation in the Advanced Decision Architectures Collaborative Technology Alliance, sponsored by the US Army Research Laboratory under Cooperative Agreement DAAD19-01-2-0009. Thanks to Jan Maarten Schraagen and Tom Ormerod for offering insightful and challenging comments on a draft of this chapter.

References

Aquinas, T. (c. 1267/1945). *Summa Theologica*. A. Pegis (trans). New York: Random House.
Atman, C.J.A., Bostronm, B., Fischhoff, B., and Morgan, M.G. (1994). "Designing risk communications: Completing and correcting mental models of hazardous processes, Part 1." *Risk Analysis*, 14: 779–88.

Beach, L.P. (1992). "Epistemic strategies: Causal thinking in expert and non-expert judgment." In G. Wright and F. Bolger (eds), *Expertise and Decision Support*. New York: Plenum Press (pp. 107–27).

Black, M. (1962). *Models and Metaphors*. Ithaca, NY: Cornell University Press.

Cañas, A.J., Novak, J.D., and Gonzalez, F.M. (2004). *Concept Maps: Theory, Methodology, Technology: Proceedings of CMC 2004, the First International Conference on Concept Mapping*. Pamplona, Spain: Dirección de Publicaciones del la Universidad Pública de Navarra.

Cherry, E.C. (ed.) (1955). *Information Theory*. London: Butterworths.

Chi, M.T.H., Feltovich, P.J., and Glaser, R. (1981). "Categorization and representation of physics problems by experts and novices." *Cognitive Science*, 5: 121–52.

Collins, A. and Gentner, D. (1987). "How people construct mental models." In D. Holland and N. Quinn (eds), *Cultural Models in Language and Thought*. Cambridge: Cambridge University Press (pp. 243–65).

Cooke, N.J. (1994). "Varieties of knowledge elicitation techniques." *International Journal of Human-Computer Studies*, 41: 801–49.

Cooke, N.M. (1992). "Modeling human expertise in expert systems." In R.R. Hoffman (ed.), *The Psychology of Expertise: Cognitive Research and Empirical AI*. New York: Springer Verlag (pp. 29–60).

Craik, K.M. (1943). *The Nature of Explanation*. Cambridge: Cambridge University Press.

Crandall, B. and Calderwood, R. (1989). *Clinical Assessment Skills of Experienced Neonatal Intensive Care Nurses* (No. Contract 1 R43 NR0191101 for The National Center for Nursing, NIH). Fairborn, OH: Klein Associates Inc.

Crandall, B., Klein, G., and Hoffman, R.R. (2006). *Working Minds: A Practitioner's Guide to Cognitive Task Analysis*. Cambridge, MA: The MIT Press.

Darnton, R. (1984). *The Great Cat Massacre and other Episodes in French History*. New York: Vintage Books.

Dekker, S.W.A. (2003). "Illusions of explanation: A critical essay on error classification." *The International Journal of Aviation Psychology*, 13: 95–106.

Duncker, K. (1945). "On problem solving." *Psychological Monographs*, 58: 1–113 (Whole No. 270) (L.S. Lees, Trans.).

Ehrlich, K. (1996). "Applied mental models in human-computer interaction." In J. Oakhill and A. Garnham (eds) (1996), *Mental Models in Cognitive Science: Essays in Honor of Phil Johnson-Laird*. London: Taylor and Francis (pp. 223–45).

Ericsson, K.A. (2006). "Protocol Analysis and expert thought." In K.A. Ericsson et al., (eds), *Cambridge Handbook of Expertise and Expert Performance*. New York: Cambridge University Press (pp. 223–42).

Ericsson, K.A. and Simon, H.A. (1993). *Protocol Analysis: Verbal Reports as Data* (2nd edn). Cambridge, MA: MIT Press.

Feltovich, P.J., Hoffman, R.R., and Woods, D. (2004). "Keeping it too simple: How the reductive tendency affects cognitive engineering." *IEEE Intelligent Systems*, May–June: 90–95.

Flach, J. and Hoffman, R.R. (2003). "The limitations of limitations." *IEEE Intelligent Systems*, January–February: 94–7.

Ford, J.K., and Kraiger, K. (1995). "The application of cognitive constructs to the instructional systems model of training: Implication for needs assessment, design and transfer." In C.L. Cooper and I.T. Tobertson (eds), *The International Review of Industrial and Organizational Psychology, Vol. 10*. Chichester, England: John Wiley and Sons, Inc. (pp. 1–48).

Ford, K.M. and Bradshaw, J.M. (eds) (1993). *Knowledge Acquisition as Modeling*. New York: Wiley.

Forrester, J.W. (1961). *Industrial Dynamics*. Portland, OR: Productivity Press.

—— (1971). "Counterintuitive behavior of social systems." In *Collected Papers of J.W. Forrester*. Cambridge, MA: Wright-Allen Press (pp. 2211–44).

Geminiani, G.C., Carassa, A., and Bara, B.G. (1996). "Causality by contact." In J. Oakhill and A. Garnham (eds) (1996), *Mental Models in Cognitive Science: Essays in Honor of Phil Johnson-Laird*. London: Taylor and Francis (pp. 275–303).

Gentner, D. and Gentner, D.R. (1983). "Flowing waters or teeming crowds: Mental models of electricity." In D. Gentner and A.L. Stevens (eds), *Mental Models*. Hillsdale, NJ: Lawrence Erlbaum Associates Inc. (pp. 99–129).

Gentner, D. and Stevens, A.L. (eds) (1983). *Mental Models*. Mahwah, NJ: Lawrence Erlbaum Associates, Inc.

Gould, P. and White, R. (1974). *Mental Maps*. Middlesex, England: Penguin Books.

Green, D.W. (1996). "Models, arguments, and decisions." In J. Oakhill, and A. Garnham (eds) (1996), *Mental Models in Cognitive Science: Essays in Honor of Phil Johnson-Laird*. London: Taylor and Francis (pp. 119–37).

Greeno, J.G. (1977). "Processes of understanding in problem solving." In N.J. Castellan, D.B. Pisoni, and G.R. Potts (eds), *Cognitive Theory*. Hillsdale, NJ: Lawrence Erlbaum Associates (Vol. 2, pp. 43–84).

Greeno, J.G. and Simon, H.A. (1988). "Problem solving and reasoning." In R.C. Atkinson et al. (eds), *Stevens' Handbook of Experimental Psychology, Vol. 2 Learning and cognition*. New York, NY: John Wiley and Sons.

Hardiman, P.T., Dufresne, R., and Mestre, J.P. (1989). "The relation between problem categorization and problem solving among experts and novices". *Memory and Cognition*, 17: 627–38.

Hoffman, R.R. (1979). "On metaphors, myths, and mind." *The Psychological Record*, 29: 175–8.

—— (1980). "Metaphor in science." In R.P. Honeck and R.R. Hoffman (eds), *Cognition and Figurative Language*. Mahwah, NJ: Lawrence Erlbaum Associates (pp. 393–423).

—— (1987). "The problem of extracting the knowledge of experts from the perspective of experimental psychology." *The AI Magazine*, 8(Summer): 53–67.

Hoffman, R.R., Coffey, J.W., and Carnot, M.J. (2000). "Is there a 'fast track' into the black box? The Cognitive Modeling Procedure." Poster presented at the 41st Annual Meeting of the Psychonomics Society, New Orleans, LA.

Hoffman, R.R., Coffey, J.W., and Ford, K.M. (2000). "A Case Study in the Research Paradigm of Human-Centered Computing: Local Expertise in Weather Forecasting." Report on the Contract, "Human-Centered System Prototype," National Technology Alliance.

Hoffman, R.R., Crandall, B., and Shadbolt, N. (1998). "A case study in cognitive task analysis methodology: The Critical Decision Method for the elicitation of expert knowledge." *Human Factors*, 40: 254–76.

Hoffman, R.R. and Deffenbacher, K.A. (1993). "An analysis of the relations of basic and applied science." *Ecological Psychology*, 2: 309–15.

Hoffman, R.R. and Lintern, G. (2006). "Eliciting and representing the knowledge of experts." In K.A. Ericsson et al., (eds), *Cambridge Handbook of Expertise and Expert Performance*. New York: Cambridge University Press (pp. 203–22).

Hoffman, R.R. and Militello, L. (2008). *Perspectives on Cognitive Task Analysis: Historical Origins and Recent Developments*. Boca Raton, FL: CRC Press.

Hoffman, R.R. and Nead, J.M. (1983). "General Contextualism, Ecological Science and Cognitive Research." *The Journal of Mind and Behavior*, 4: 507–60.

Hoffman, R.R. and Senter, R.J. (1978). "Recent history of psychology: Mnemonic techniques and the psycholinguistic revolution." *The Psychological Record*, 28: 3–15.

Hoffman, R.R., Shadbolt, N.R., Burton, A.M., and Klein, G. (1995). "Eliciting knowledge from experts: A methodological analysis." *Organizational Behavior and Human Decision Processes*, 62: 129–58.

Hoffman, R.R., Trafton, G., and Roebber, P. (in press). *Minding the Weather: How Expert Forecasters Think*. Cambridge, MA: MIT Press.

Hutchins, E. (1983). "Understanding Micronesian navigation." In D. Gentner and A.L. Stevens (eds), *Mental Models*. Mahwah, NJ: Lawrence Erlbaum Associates (pp. 191–227).

Johnson-Laird, P.N. (1983). *Mental Models: Towards a Cognitive Science of Language, Inference, and Consciousness*. Cambridge, MA: Harvard University Press.

Kieras, D. (1988). "What mental model should be taught?: Choosing instructional content for complex engineered systems." In J. Psotka, L.D. Massey, and S.A. Mutter (eds), *Intelligent Tutoring Systems*. Hillsdale, NJ: Lawrence Erlbaum Associates (pp. 85–112).

Kieras, D., and Bovair, S. (1990). "The role of a mental model in learning to operate a device." *Cognitive Science*, 8: 255–73.

Kintsch, W. (1974). *The Representation of Meaning in Memory*. Hillsdale, NJ: Lawrence Erlbaum Associates.

Klein, G. (1998). *Sources of Power: How People Make Decisions*. Cambridge, MA: MIT Press.

Klein, G., Calderwood, R., and Clinton-Cirocco, A. (1986). "Rapid decision making on the fireground." *Proceedings of the Human Factors and Ergonomics Society 30th Annual Meeting*, 1: 576–80.

Klein, G., and Militello, L.G. (2001). "Some guidelines for conducting a cognitive task analysis." In E. Sales (ed.), *Human/Technology Interaction in Complex Systems, Vol. 1*. Oxford: Elsevier Science Ltd (pp. 161–97).

Klein, G., Moon, B., and Hoffman, R.R. (2006). "Making sense of sensemaking 1: Alternative perspectives." *IEEE Intelligent Systems*, July–August: 22–6.

Klein, G., Ross, K.G., Moon, B.M., Klein, D.E., Hoffman, R.R., and Hollnagel, E. (2003). "Macrocognition." *IEEE Intelligent Systems*, May–June: 81–5.

Klein, H.A. and Lippa, K. (in press, 2008). "Type 2 diabetes self-management: Controlling a dynamic system." *Journal of Cognitive Engineering and Decision Making*, 1.

Legrenzi, P. and Girotto, V. (1996). "Mental models in reasoning and decision-making processes." In J. Oakhill and A. Garnham (eds) (1996), *Mental Models in Cognitive Science: Essays in Honor of Phil Johnson-Laird*. London: Taylor and Francis (pp. 95–118).

Lippa, K., Klein, H.A., and Shalin, V.L. (2008). "Everyday expertise: Cognitive demands in diabetes self-management." *Human Factors*, 50 (1).

Maharik, M. and Fischhoff, B. (1992). "The risks of using nuclear energy sources in space: Some lay activists' perceptions." *Risk Analysis*, 12: 383–92.

McCloskey, M. (1983). "Naive theories of motion." In D. Gentner and A.L. Stevens (eds), *Mental Models*. Hillsdale, NJ: Lawrence Erlbaum Associates (pp. 299–324).

Mental Models website (2007). Downloaded 24 March 2007 at <http://www.tcd.ie/Psychology/Ruth_Byrne/mental_models/>.

Moray, N. (1987). "Intelligent aids, mental models, and the theory of machines." *International Journal of Man-Machine Studies*, 27: 619–29.

Newell, A. (1985). "Duncker on thinking: An inquiry into progress in cognition." In S. Koch and D.E. Leary (eds), *A Century of Psychology as a Science*. New York: McGraw-Hill (pp. 392–419).

Norman, D.A. (1983). "Some observations on mental models." In D. Gentner and A.L. Stevens (eds), *Mental Models*. Hillsdale, NJ: Lawrence Erlbaum Associates (pp. 7–14).

—— (1986). "Cognitive engineering." In D.A. Norman and S.W. Draper (eds), *User-centered Design*. Hillsdale, NJ: Lawrence Erlbaum Associates.

Novak, J.D. (1998). *Learning, Creating and Using Knowledge: Concept Maps as Facilitative Tools in Schools and Corporations*. Mahwah, NJ: Lawrence Erlbaum Associates.

Oakhill, J. and Garnham, A. (eds) (1996). *Mental Models in Cognitive Science: Essays in Honor of Phil Johnson-Laird*. London: Taylor and Francis.

Olson, J. and Reuter, H. (1987). "Extracting expertise from experts: Methods for knowledge acquisition." *Expert Systems*, 4: 152–68.

Pliske, R.M., Klinger, D., Hutton, R., Crandall, B., Knight, B., and Klein, G. (1997). *Understanding Skilled Weather Forecasting: Implications for Training and the Design of Forecasting Tools. Technical Report AL/HR-CR-1997-0003*. Brooks AFB, TX: US Air Force Armstrong Laboratory.

Pylyshyn, Z. (1981). "The imagery debate: Analog media versus tacit knowledge." *Psychological Review*, 88: 16–45.

Rasmussen, J. (1979). *On the Structure of Knowledge—A Morphology of Mental Models in a Man-Machine System Context* (Tech. Rep. No. Riso-M-2192). Roskilde, Denmark: Riso National Laboratory.

Rouse, W.B. and Morris, N.M. (1986). "On looking into the black box: Prospects and limits on the search for mental models." *Psychological Bulletin*, 100: 349–63.

Samurçay, R. and Hoc, J.M. (1996). "Causal versus topographical support for diagnosis in a dynamic situation." *Le Travail Humain*, 59: 45–68.

Scott, A.C., Clayton, J.E., and Gibson, E.L. (1991). *A Practical Guide to Knowledge Acquisition*. Reading, MA: Addison-Wesley.

Senge, P. (1990). *The Fifth Discipline: The Art and Practice of the Learning Organization*. New York: Doubleday.

Shannon, C.E. (1948). "A mathematical theory of communication." *Bell System Technical Journal*, 27: 379–423, 623–56.

Shaw, A.G. (1952). *The Purpose and Practice of Motion Study*. New York: Harlequin Press.

Shepherd, A. (2001). *Hierarchical Task Analysis*. London: Taylor and Francis.

Smith, P.J., McCoy, E., and Layton, C. (1993). "Design-induced error in flight planning." *Proceedings of the Human Factors and Ergonomics Society 37th Annual Meeting*, pp. 1091–5.

Staggers, N. and Norcio, A.F. (1993). "Mental models: Concepts for human–computer interaction research." *International Journal of Man–Machine Studies*, 38: 587–605.

Tolman, E.C. (1948). "Cognitive maps in rats and men." *Psychological Review*, 55: 189–208.

Trafton, J.G. and Hoffman, R.R. (2007). "Computer-aided visualization in meteorology." In R.R. Hoffman (ed.), *Expertise Out of Control*. Mahwah, NJ: Lawrence Erlbaum Associates (pp. 337–57).

van der Veer, G.C. and Melguzio, M.C.P. (2003). "Mental models." In J.A. Jacko and A. Sears (eds), *Handbook of Human–Computer Interaction*. Mahwah, NJ: Lawrence Erlbaum Associates (pp. 52–80).

Wild, M. (1996). "Mental models and computer modeling." *Journal of Computer Assisted Learning*, 12: 10–21.

Chapter 5

Investigative Sensemaking
in Criminal Contexts

Thomas C. Ormerod, Emma C. Barrett, and Paul J. Taylor

Introduction

A critical, though as yet poorly understood, aspect of expertise in criminal investigation is the set of knowledge and skills associated with "sensemaking," whereby an investigator uses available information to construct an understanding of a "to-be-investigated" or ongoing incident. Criminal investigators are often faced with the task of making sense of a large amount of ambiguous and complex data, with a view to establishing which of the various plausible alternative explanations is likely to be the truth. Sensemaking is of central importance to identifying appropriate and promising lines of enquiry. Inappropriate lines of enquiry can hamper investigations, at best wasting time and resources and at worst proving fatal to the chances of resolving the case successfully.

Our interests in sensemaking are discussed here with reference to three different but related domains of criminal enquiry: understanding crime reports and scenes, monitoring and decision making during ongoing hostage taking and barricade incidents, and following up suspicious insurance claims to evaluate whether fraud has been committed. Despite differences in the scope and complexity of their respective investigative problems, police officers, negotiators, and insurance investigators face similar challenges in sensemaking. In each of these contexts, the professional investigator must deal with a large set of domain attributes that make sensemaking a difficult enterprise. These include:

- complexity (for example, there may be large amounts of potential evidence and many witnesses and/or perpetrators);
- incomplete data (for example, in a hostage situation, attending officers have limited channels of communication that are controlled by the hostage takers);
- ambiguity (for example, the same piece of information, such as bruising on a child, can be interpreted in more than one way, accidental or inflicted deliberately by an assailant);
- risk (for example, the failure to identify a claim as fraudulent may expose an insurance company to further claims once a route to fraud is established).

These features are by no means exhaustive, and nor do they, in themselves, distinguish investigative contexts from other domains of expertise. For example, process control domains such as the nuclear industry present similar challenges (Roth et al., 2001; Ormerod and Shepherd, 2004): a fault in a heat exchanger will cause widespread perturbations across a plant (complexity), potentially knocking out alarms and indicators or conversely bombarding the operator with too much information (both leading to incomplete data); the nature of heat exchange may reveal symptoms and warnings some distance from the site of the fault (ambiguity), and faults must be diagnosed and fixed under pressure of serious consequences (risk).

However, there are three challenges faced by criminal investigators that we suggest are in some respects different to those faced by operators in most domains studied hitherto in NDM research. First, at the heart of criminal investigation is the need to make sense of human action, reaction, and interaction (as opposed to, say, the abnormal operation of aircraft, likely paths of missiles or spread of fires). An investigator must deploy a sophisticated understanding of the behaviors of many types of individuals and groups (for example, suspects, offenders, victims, and witnesses). As Alison and Barrett (2004) point out, this "requires both a deep and a broad understanding of the properties of social systems that are inherently complex and unpredictable" (p. 68). We suggest that expert investigators are able to accomplish this feat by calling upon internalized cognitive frames relating to human behavior that allow them to generate expectations about the actions and responses of others in real time. Arguably, similar cognitive frames underlie all expertise: firefighters can call upon knowledge structures built up through experience that help them make real-time decisions. Where human-behavior frames differ is in being reflexive: an investigator holds a cognitive frame that models not just the situation to be understood but also the cognitive frame of a perpetrator.

The centrality to a criminal inquiry of human motives, intentions, actions, and reactions raises a second challenge for sensemaking: although there are archetypical crime scenes, modus operandi and clues, the range of scenarios that a criminal investigator might be faced with is, in principle, infinite. Moreover, what evidence there is may point to more than one possible explanation. Thus, a fundamental component of investigative expertise is explanation-building. Pennington and Hastie (1988) have shown how inexperienced jurors are guided by the presentation of evidence to create narrative explanations that subsequently determine their judgments. In the case of expert-led investigation, explanation-building through narrative construction plays a similar role, but tends to be guided as much by internal knowledge structures as by external evidence. As we illustrate below, explanation-building relies on stereotypical scripts that emerge both from personal experience and also from structures, particularly legal scripts, which pertain to the specific domain of investigation. However, explanation-building also appears to involve going beyond the available evidence to construct speculative narratives that can provide missing components such as motives, potential evidence trails, and so on. Moreover, in contrast to Pennington and Hastie's novices, the expert investigators we studied used explanation-building to search for alternative hypotheses rather than as a device for reifying a single account.

A third challenge is that in criminal investigations, unlike most other diagnostic contexts, one is likely to be working in an environment where there is intent to deceive. The deceptive characteristic of criminal contexts necessarily requires expertise that goes beyond straightforward situation assessment: at the very least there is a need for both evidence-based (the "given") and inferential (the "hidden") variants of situation assessment. We propose that expert investigators are adept at reasoning in the face of deception, and are able to turn it to their advantage. For example, in the context of insurance fraud investigation, we describe empirical evidence that suggests investigators deliberately adopt a framework of innocence, a cognitive frame that enables them to construct powerful tests of suspicious contexts.

Sensemaking is most closely related to the NDM concepts of situation awareness and assessment (for example, Endsley, 1995, 1997; Klein, 1989, 1993). For example, Endsley (1988) defines situation awareness as "the perception of the elements in the environment within a volume of time and space, the comprehension of their meaning and the projection of their status in the near future" (p. 97). Features of seeing, understanding, and prediction are common facets to any domain in which an expert encounters an incident or situation. Klein's (1989) recognition-primed decision theory of expert judgment presents an account of expertise that is based on situation awareness: recognizing a situation as appropriate for a particular course of action triggers an appropriate rule-based response decision.

Although sensemaking and situation awareness are related, we suggest that they are not precisely the same thing. In fact, situation awareness may be thought of as a subset of sensemaking, in which presented situations are such that experts are able to reach rapid perceptually based assessments. Sensemaking also includes complex and novel scenes that may never have been encountered but where experts can nonetheless bring to bear their knowledge and skills to make sense of a situation or incident in ways that a novice cannot. For example, Feltovich, Spiro, and Coulson (1997) describe how highly skilled medical diagnosticians are able to go beyond obvious but incorrect diagnoses reached by less skilled physicians to develop sophisticated accounts of rare symptom sets.

The three factors described above that make criminal investigations different from other expertise contexts also, we propose, make sensemaking in these contexts different from situation assessment. Given an infinite range of possible scenarios and a high likelihood that there will be few if any immediately recognizable perceptual cues that can trigger an appropriate rule-based response requires the investigator to create a potentially novel response in real time. Also, the fact that investigators are focusing upon understanding human action and reaction rather than the state or responses of physical processes or plant, introduces additional layers of complexity and unpredictability. Moreover, the deceptive nature of investigative domains means that what is seen first is unlikely to be a true picture of the underlying structure of a scenario. As a consequence, a rule-based response to an immediate and perceptually based judgment is unlikely to prove the most beneficial course of action in every case. Indeed, anecdotally at least, a feature that distinguishes successful from less successful investigators is an ability to hold back from making immediate judgments about likely causes, motives, and perpetrators. Early commitment to a specific hypothesis can shut off potentially useful lines of enquiry and lead investigators down blind alleys.

In the remainder of this chapter, we discuss three studies in which police and insurance claims investigators undertake real decision-making activities. The chapter provides an exemplar of applying an NDM approach to understanding expertise in criminal investigations, and highlights commonalities and differences in expert investigators' approaches to sensemaking.

Sensemaking in a "To-Be-Investigated Situation"

In criminal investigations, the initial assessment of a to-be-investigated (TBI) situation can sometimes have a profound influence on the course of an enquiry, as the classification of a situation as a particular type of crime (or non-crime) may lead officers to follow worthless lines of enquiry or to close off potentially fruitful lines of investigation. The classification of an event as a particular type of crime may also limit the resources that a force devotes to the investigation, which will also have an important impact on the likelihood of a successful resolution. However, the assessment of a TBI event is far from straightforward: the information available at the start of a criminal investigation is often complex, ambiguous and contradictory, and thus subject to multiple interpretations and multiple potential classifications. As such, a simple perceptually based model of situation awareness is unlikely to capture fully the expertise required for sensemaking in these contexts.

Elsewhere (Barrett, 2002), it has been suggested that detectives seek to build hypothetical investigative situation models (see Graesser, Mills, and Zwaan, 1997; Endsley, 1995, 2000). These are mental models of particular situations, as opposed to mental representations of general states of affairs based on stored domain and general knowledge. So, for instance, a detective may have generic mental representations of what occurs in the crime of rape. When faced with a TBI incident that may or may not be a rape, he or she must develop one or more plausible mental representations of that specific situation. Generic mental models provide a framework for the construction of situation models, but can also be used to detect anomalies in constructed situation models, a process which may trigger a search for additional data to resolve the anomalies, or prompt the rejection of a situation model.

The suggestion that detectives draw on generic mental models to make sense of TBI incidents raises two issues: the general form of mental models in investigative settings, and the factors that influence the selection of a particular model or model(s) as appropriate in a particular setting. Klein and his colleagues, in their Data/Frame theory of sensemaking (Klein et al., 2003) argue that sensemaking is an active, bidirectional (top-down/bottom-up) process. Individuals respond to ambiguous or anomalous information by seeking an explanation for it, but the repertoire of generic mental models (or, as Klein et al. term them, frames) influences both the data attended to, and the interpretation put on that data. In the investigative context, Klein et al.'s theory is supported by Innes (2003), who, in his qualitative study of UK homicide investigations, describes how officers identify and decode particular "signs of crime" in investigative information (p. 178). These "investigative signifiers" are the crucial elements of a TBI situation which enable the detective to label it as being of a particular type. This categorization of a TBI situation supports the detective by providing a framework that serves to explain the elements within the situation, to highlight deviations from the expected framework, and to guide investigative action.

The operator's goal in a particular situation will also, crucially, influence the set of frames which is likely to be activated. In criminal investigations, the detective's task is to make sense of that situation in relation to a specific goal: to determine whether or not a crime has been committed and if so, what crime, and by whom. More precisely, enough evidence must be gathered to convince a jury that the defendant is guilty according to the terms of a precisely worded legal charge. It is likely, therefore, that investigative sensemaking by police officers is guided by legally determined scripts, and that the most salient cues will be those relating to the nature of the crime committed (if any) and the identity of the offender. The goal of criminal investigation limits the set of scripts which it is necessary for a detective to apply in a TBI situation to those that are legally determined, and this in turn influences what data will be attended to in that TBI situation.

In an exploratory study (Barrett and Alison, submitted), the aim of which was to begin to understand the process by which detectives construct an understanding of potential crime in the early stages of an investigation, 44 detectives from a British police force gave written interpretations of crime-related vignettes, similar to reports received by detectives at the beginning of an investigation and of varying ambiguity. Answers were content-analyzed to determine participants' hypotheses about what had occurred in the vignettes and to examine what cues they attended to.

In one scenario, a concerned grandmother called the police to report that while bathing her four-year-old grandson she had noticed marks on his skin, which she likened to those made by a stick. She also volunteered the information that her daughter-in-law (her grandson's mother) had recently started a relationship with a man who had convictions for assault. In all, 88 percent of participants agreed that the most likely explanation of what had occurred in this case was that the child had been abused, and 65 percent suggested that the boyfriend had been responsible for this abuse. The cues mentioned most frequently, and which, therefore, might be considered to be particularly salient, were those relating to the nature of the injuries to the child, and whether or not they had been deliberately inflicted. If this scenario were to be investigated as a child abuse case, the prosecution in any subsequent court case would have to prove that the child's injuries were non-accidental.

Information concerning a plausible suspect—the mother's new boyfriend—was also salient for many participants. In particular, this man's violent tendencies, as evidenced by his convictions for assault, were mentioned by 35 percent of participants (including all of those who suggested that the boyfriend was responsible for assaulting the child). However, clearly not every participant believed that the boyfriend's criminal history was sufficient in itself to make him the sole or, in some cases, even the prime suspect. Indeed, there were several other potential suspects in this scenario, including the child's grandmother, mother and birth father. That 65 percent of participants did not mention the boyfriend's previous violence does not necessarily mean that they found it unimportant. But when investigating a crime, what must be proven is that a suspect committed a *particular* crime, not that they had carried out a related crime in the past. In this context, it is interesting that although in general participants made few elaborated inferences about the motives and characteristics of the individuals involved in this scenario, where they did, narrative elaborations concerned the factors that may have prompted the boyfriend to strike the child, indicating that for some participants, the boyfriend's violent tendencies were not in themselves strong enough to suspect that he had hit the child.

In another scenario, a man was found slumped unconscious in a chair in his flat, with knife wounds to his neck and chest, a bloody knife at his feet, with all the windows and doors in his home apparently secure. The police had been alerted by the man's business partner, who had gone to the flat when the man had failed to turn up to work that day. The business partner volunteered the information that he and the man had argued about company profits the previous evening. In their explanations, participants were in less agreement than in the previous scenario: 64 percent suggested that the man had sustained injuries during an assault, the majority of whom (and 48 percent of all participants) named the man's business partner as responsible, whereas 31 percent suggested that it was more likely or at least as likely that he had attempted suicide. In justifying their hypotheses, participants interpreted the same cues differently, making inferences about, and elaborating on, these cues to enhance the plausibility of the chosen hypothesis. For instance, the fact that the doors and windows to the flat appeared secure was used by almost all of those who argued for the suicide hypothesis to support their view that no one else had been involved. The same information was interpreted by those who argued that the man had been assaulted to indicate that his assailant was known to him. Another example was the argument about company profits, which was mentioned by 75 percent of those who suggested the business partner was responsible, and was interpreted as providing a motive for his actions. The same cue was mentioned by 71 percent of those who suggested the man had attempted suicide, but in this case was interpreted to mean that he was concerned about a serious problem with his company. Interestingly, none of those who suggested that the man had been assaulted by a person or persons unknown mentioned the argument about company profits.

In the third scenario, a taxi driver reported that he had been robbed twice by one of his male customers. He alleged that on the first occasion the passenger had touched his leg and suggested driving somewhere quiet. The driver said he had complied, but when they reached a nearby car park he was threatened by the passenger, to whom he handed a bag of money. On the second occasion, the same passenger had apparently accused the taxi driver of being frightened of him, and made a homophobic remark. The driver said that he believed that he was going to be assaulted by the man, and again gave his takings to the passenger.

Participants were even more divided in their theories of what had gone on in this scenario: 41 percent suggested that the driver had been the victim of a robbery, whereas 29 percent suggested he was being blackmailed, 10 percent suggested that he had stolen money for himself, and 20 percent refused to speculate on what had occurred. Unlike the previous answers, in which there were relatively few narrative elaborations and inferences, participants' answers for this scenario were dominated by conjecture and story-like reasoning. For those who suggested that the driver had been a victim of a robbery, the most common cues cited were the handing over of money by the driver, and the report by the driver that he was in fear. The most common inferences made were that the passenger had propositioned the driver sexually on the first occasion and had made a threat on the second. In their selection of cues and in their inferences, the participants were clearly influenced by the legal definition of robbery, in which (in English law) there is a need to prove that the person committing the offence used, or threatened, immediate force to steal property from the victim. Participants therefore made sense of the incident by generating a story in which the offender deliberately intimidated the driver, who, fearing for his safety, handed over cash on both occasions.

The cues most salient for those who suggested that the driver was being blackmailed, on the other hand, were those relating to the driver's sexual orientation, with 75 percent of participants in this group inferring that the driver was gay or intended to have sex with the passenger, compared to only 35 percent in the group of participants who suggested that robbery was most likely. This is suggestive of a readily available prototype script for illicit homosexual interactions, the recall of which is triggered by a few salient cues relating to the passenger's apparent propositioning of the driver and the driver's apparent willingness to drive to a quiet location. Despite the fact that the driver was reporting a robbery, for some participants, cues in the scenario seemed to trigger a blackmail script that overrode a robbery script.

While the striking degree of variation in participants' answers in this final scenario may be attributed in part to conscious or unconscious homophobic bias, an alternative explanation is that the scenario described a situation which was difficult to fit into a legal script. Although the report claimed to be of a robbery, many elements in the scenario were inconsistent: the lack of force used, the apparent willingness of the driver to hand over his money, and the failure of the driver to report the crime at the earliest opportunity. Since the incident did not conform to a prototypical robbery script, officers were forced to generate alternative explanations to account for the facts, and, in doing so, drew on lay theories of threatening interactions, sexual behavior and interpersonal manipulation.

This exploratory study suggests that officers look for particular cues to help them make sense of the incidents they investigate, and that these cues relate to elements in legally determined scripts: the nature of the crime, and the identity of an offender. The variation in investigative hypotheses appears to depend on the degree to which incidents match legally determined scripts. Where more than one legal script fits, cues are interpreted so as to be consistent with the chosen script. If cues are ambiguous or violate expectations, incidents do not easily conform to a script, officers struggle to make sense of the data, leading to more inferences and speculation, and more variation in the interpretation of particular scenarios. Further research is required to examine whether the kinds of frames that are being brought to bear are the result of a sophisticated expertise, or whether they call upon general knowledge about crime scenes. It seems likely, however, that the match between participants' accounts and legally determined scripts indicates that they reflect frameworks of expertise.

Sensemaking in Insurance Fraud Investigation

The second context, the investigation of fraudulent insurance claims, illustrates the deceptive nature of the scenarios in which investigators undertake sensemaking activities. The Association of British Insurers estimated the costs of fraud to the UK insurance industry to be around £1 billion in 2001, suggesting that the cost internationally may be tens of billions *per annum*. However, the problem affects all policyholders, through increased premiums and in some cases increased exclusions and difficulties in obtaining insurance cover. The majority of insurance claims are genuine, but a worryingly large number are fraudulent. The claimant committing fraud will deliberately construct a claim to appear genuine, and the investigator's job is to look beyond first appearances to see what might lie behind them. Detecting fraud is not easy. Fraud can vary in scale, from inflated claims for genuine incidents through to systematic multi-person "scams" that involve staged accidents, thefts, and so on. Fraud is also dynamic: as one scam is uncovered, a new one takes its place. Moreover, the need to maintain customer loyalty and efficient sales practices means that the kinds of information checks that can be carried out at policy inception and during the handling of a claim are limited and time-pressured.

The data we discuss in this section come from an empirical study conducted as part of a three-year project developing technologies to enable the early detection and subsequent investigation of potentially fraudulent insurance claims (Ormerod et al., 2003). The study adopted a mixed-methods approach (Ormerod et al., 2004). In particular, we conducted ethnographic studies (for example, Hammersley and Atkinson, 1983; Ball and Ormerod, 2000) of work practices in different parts of the claims management process. We spent up to eight weeks recording practices at each stage, from telephone-based claims handlers receiving initial claims from customers, through to investigators in specialist units and loss-adjustors from external companies following up claims that have either high value or some initial suspicion associated with them. The study of claims handlers, typically inexperienced staff employed for less than one year, is reported elsewhere (Morley, Ball, and Ormerod, 2006). Here we focus upon data from the more experienced specialist investigators. We also conducted experimental studies of reasoning and decision making using research paradigms more commonly found in the judgment and decision-making literature. However, participants in the studies were expert investigators and claims handlers (contrasted against an out-of-domain student control group) and the materials for the studies were collected during the ethnographic studies.

Ethnographic Studies of Insurance Fraud Investigation

In common with the data describing judgments of police officers given in the previous section, explanation-building played a key role in the sensemaking activities of skilled fraud investigators. The ethnography field notes contain many examples of investigators constructing explanations around suspicious claims. For example, an investigator followed up an unexpected match between identity details on new and existing motor policies for two separate addresses. The investigator inspected the existing policy and noted that the insured was a flight attendant. She went on to suggest that the insured probably parked her car at her mother's house while she was away at work. Thus, the investigator builds an elaborate but coherent explanation of the anomaly.

Interestingly, and a feature common to many of the explanations (or at least common to parts of them) that we saw generated by expert investigators, the explanation offers a hypothesis of innocence, that is, a way in which an anomaly might be explained as being consistent with a genuine claim or policy application. One of the key differences between the investigative activities of police officers and fraud investigators is the end goal: police officers are seeking a prosecution whereas fraud investigators are seeking repudiation of a claim (that is, a demonstration that the company is not liable to pay the claim). In essence, the core role of police officers is to endeavor to identify true positives as quickly and effectively as possible (that is, to prosecute the perpetrators). In contrast, the core role of fraud investigators is to endeavor to identify false positives as quickly and effectively as possible (that is, to establish that a claim is genuine so that the claimant can be paid and the company's good name upheld, since rejection of a genuine claim would be a false positive).

In practice, insurance frauds rarely lead to criminal prosecutions, since there is little for a company to gain financially from legal action. Repudiating a claim or refusing to insure a fraudster in the future have direct financial benefits, while incorrectly refusing to pay a genuine claim can be very damaging to a company's reputation. As a consequence, we hypothesize that explanation-building differs according to whether it is conducted in order to establish a framework of guilt or innocence, and this difference is likely to vary across investigative contexts. The police officers described above seemed almost exclusively to generate explanations imposing a framework of guilt, in contrast to the insurance investigators described here. However, a proper empirical test of this hypothesis has yet to be undertaken.

As a second example of explanation-building, an investigator examined a case that had thrown up an anomaly: checks made by a less experienced claims handler using a government vehicle registration database indicated that a car reported as stolen abroad was, according to records, actually a truck. This kind of mismatch might be taken as evidence that the vehicle is a "ringer" (that is, a vehicle that has been stolen and re-registered with the identity of a previously written-off vehicle). The claim was complicated by the fact that the owner could supply no information about the vehicle, except that he bought it from a friend while in Spain. The investigator built the following explanation in response to a query from a less experienced claims handler:

> So his friend takes the vehicle out to Spain, that may be subject to hire purchase [rent to own]. Gets it out there, doesn't want to take it back, throws it in as a debt, and then our man may well have found out its true pedigree and thinks how am I going to get out of this one—it's subject to hire purchase.

Much of this scenario (for example, the hire-purchase) was not information given in the case notes—it came from the investigator's efforts to build a plausible explanation of the evidence. The hire-purchase explanation is a speculation that does not follow in any way from the information given, but it provides a coherent narrative about why a UK-registered vehicle may have been sold abroad (illegally, since it was subject to a finance contract, hence disguising its true identity), and why the purchaser might have wanted to have the car stolen (since the ownership would revert to the finance company if the initial sale and identity switch were discovered). The investigator then went on to pursue an alternative account, in which the vehicle would have been a legitimate purchase.

The generation of alternative explanations is a feature that we found with expert investigators but not with less experienced claims handlers. For example, in one case a search of databases revealed an unexpected match between the details of a third party's vehicle reported in a claim and the vehicle reported in five claims with other companies (that is, the same "victim" of an accident turns up in six separate claims). The investigator generated a variety of explanations. First, he proposed that the matches all referred to separate individuals from the same accident (that is, a hypothesis of innocence, in which a single vehicle is the third party to a number of other vehicles all involved in the same crash). Second, he proposed that it might be duplication (either accidental or deliberate, perhaps by a company insider) of a single claim. Third, he proposed that there might indeed be five separate claims all involving the same third party (who is either very unlucky or involved in some kind of fraud network). Only once he had generated three explanations did he undertake a test (checking whether the claims were made at the same or different times).

Generating explanations under a framework of innocence may reflect the end goals of the fraud investigator, but it has another valuable property: it provides a way of testing the feasibility (and consequently of demonstrating the infeasibility) of suspicious claims. For example, two experienced investigators were dealing with a claim made by a couple who had allegedly been on a day-return shopping trip across the English Channel to northern France, where the insured claimed they had spent nearly $50,000 (an American couple, no doubt) on jewelry and electrical goods. They claimed that on the return journey, their vehicle had been stolen from a UK service station, along with all the documents concerning their purchases. Under UK law, the burden of disproof rests with the insurance company, even in the absence of any documentation. Inexperienced investigators had previously examined the claim, and had undertaken a trawl of major criminal and insurance databases, which had in turn thrown up large numbers of anomalies and matches against the claimants. However, none of these new pieces of data did the job of repudiating this particular claim, but simply cast more general suspicion on an already suspicious claim. The experienced investigators decided to treat the claim as a genuine one, in order to establish how the day's shopping might have taken place. To do this, they conducted a "virtual" shopping trip, using the Internet to find out the shortest distance that could have been traveled by the claimants in order to conduct the shopping trip. They were able to demonstrate that the purchases listed by the claimants could not physically have been purchased and collected within a single day in time to get the return ferry. Thus, by adopting a framework of assumed innocence, the investigators were able to repudiate the claim.

Experimental Study of Investigators' Reasoning About Insurance Fraud

We examined the role played by the end-goal of the fraud investigator in experimental studies of deductive reasoning. Previous studies (for example, Ellio and Pelletier, 1998) have found that individuals who are given conditional reasoning tasks often draw extra-logical inferences in preference to an invited logical inference. For example, given the conditional statement "If the airplane crashes then the pilot will die," and the minor premise "In one case, the pilot did not die," participants often fail to draw the logically valid *Modus tollens* inference "The airplane did not crash." Instead they draw an extra-logical inference such as "He bailed out before the plane hit the ground."

We hypothesized that, in contexts where a conditional statement invites confirmation of a hypothesis of suspicion, similar extra-logical inferences might be drawn by investigators in attempting to set up a framework of innocence. For example, one of the many indicators for a fraudulent claim (in this case, an indicator of a staged theft) that we encountered in the ethnographic studies can be summarized as follows:

> If a reported car theft is genuine then the insured will possess both sets of keys.

The minor premise "In one case, the insured did not possess both sets of keys" can be explained either in terms of a logically valid inference "The reported theft is not genuine" or in terms of an extra-logical explanation that creates an account of potential innocence, such as "Maybe the claimant's spouse had lost the spare set." The conditional statement above holds the hypothesis in the antecedent and the evidence that might corroborate this suspicion in the consequent. Combined with the minor premise given above, it invites a logical inference that would confirm a hypothesis of suspicion. The rule components can be reversed, as follows:

> If the insured possesses both sets of keys then a reported car theft is genuine.

Given this rule and the minor premise "In this case, the reported car theft was not genuine," the logical inference invites an explanation of an already confirmed suspicion. Thus, we hypothesized that, with rules in which the antecedent contains the evidence and the consequent contains the hypothesis of suspicions, investigators would be more likely to draw a logically valid *Modus tollens* inference than with conditional statements where the order of hypothesis and evidence components was reversed.

We compared three groups (fourteen participants in each group) comprising experienced investigators (more than five years' experience), claims handlers (with one to three years' experience, handling but not investigating claims) and a non-domain control (undergraduate students). We constructed conditional statements that embodied widely held beliefs concerning fraud indicators, and also a set of materials that embodied general knowledge (for example, "If the toilet is vacant then the door will be unlocked"). In each of sixteen trials, participants were shown a conditional statement and a minor premise designed to elicit a *Modus tollens* inference. They were asked "to reconcile the fact about this specific case with the given rule—in other words, outline how this set of circumstances might have arisen."

Figure 5.1 shows the frequency of *Modus tollens* inferences drawn by each of the three groups for insurance and general knowledge domains with conditionals of each order (hypothesis -> evidence or evidence -> hypothesis). The key finding of this study was that experienced investigators drew significantly fewer logically valid inferences with insurance materials when the premise and rule invited an inference to confirm a hypothesis of suspicion (p< .01). In the same study we also found that expert investigators made fewer logically invalid Denial of Antecedent inferences with conditionals where the evidence came in the antecedent and the hypothesis came in the consequent. Thus, the effect is not simply one of generalized logical competence. Moreover, it seems restricted to insurance materials only.

We interpret this finding as support for our view that expert investigators are cued into adopting a framework of innocence as a strategic refinement to testing suspicions. This framework makes sense in a domain where the key requirement is to repudiate false claims and discriminate them from genuine ones as quickly and reliably as possible. We are testing this hypothesis further in ongoing work in the domain of murder investigations. We predict that we are unlikely to get the same suppression of *Modus tollens* inferences according to the order in which conditionals are stated. The end-goal of the investigator in a murder investigation is to secure prosecution. Thus, hypotheses of suspicion are most salient and are unlikely to be ignored by investigators in this context whatever the form in which they are encountered.

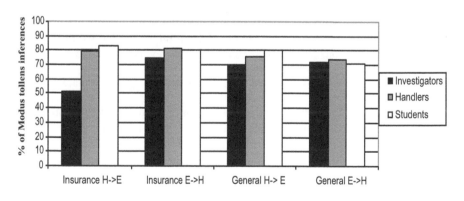

Materials and conditional rule order

Figure 5.1   Percentage of logical inferences drawn to insurance-related and general knowledge statements by different participant groups

Note: N–14 per group, four inferences per participant per condition.

Sensemaking in Hostage Negotiations

The third context, hostage/barricade scenarios, illustrates the dynamic and interactive nature of investigative sensemaking. It also illustrates the fact that investigative sensemaking involves understanding, predicting, and responding to the actions, beliefs, and justifications of other human agents. Hostage and barricade incidents are "crimes-in-action," in which the police engage in fast-paced dialogue with a hostage taker to resolve a high-stakes situation. In this scenario, police negotiators rely on what the hostage taker says to guide their decisions and actions. They use salient features of the dialogue to build an understanding of the hostage taker's perspective and concerns, and then attempt to craft their own responses in a way that addresses these concerns and reduces the tensions of the incident (Taylor, 2002). For example, when a hostage taker protests angrily about the actions of the firearms team, a police negotiator is unlikely to respond by highlighting the possibility of providing food for the hostages. Instead, the negotiator will actively listen to the hostage taker's complaints, empathize with their concern for personal safety, and reassure them that the police intend to harm nobody. Police negotiators do this because they recognize that the hostage taker's current concern is personal safety and not the instrumental issue of how to resolve the parties' various substantive interests.

To achieve such sensemaking, it is likely that negotiators draw on processes which are similar to those that found in the contexts described above. Using information derived from dialogue, negotiators seek to build a hypothetical situation model about the events that have led up to the current incident. From this model, they formulate a plausible explanation for what has happened and attempt to draw inferences about the hostage taker's current concerns and goals. As the dialogue unfolds, so they gain information that allows the situation model to be developed and their inferences refined to reflect the crime in action at that particular moment.

Existing negotiation research provides some insights into how negotiators make sense of dialogue. Many of the processes that are highlighted parallel those discussed in the previous two contexts. For example, negotiators often report that they use criteria to "classify" interactions in a way that enables them to quickly draw on experience-based, "off-the-shelf" negotiation strategies (see Amalberti and Deblon, 1992; DeFilippo, Louden, and McGowan, 2002). Similarly, the quantity and ambiguity of the information exchanged in dialogue makes it necessary for negotiators to restrict their focus to salient features of recent utterances (see Klein, 1993; Taylor and Donald, 2003). The features or cues that become salient are likely to be those that resonate with the police negotiator's explanatory model of the hostage taker's initial actions and current motivations. Any anomalies between what the police negotiator believes and what the hostage taker says are likely to prompt further investigative dialogue on that issue.

One process that seems to be particularly prominent in the hostage/barricade context is cognitive framing. At any one time, a hostage taker will communicate about a single issue and it is important for the police to identify and address this issue. This assessment is driven by negotiators' predominant interpersonal perceptions and beliefs about the current dialogue—their motivational framing (Drake and Donohue, 1996) or interpretative schemata (Green, Smith, and Lindsey, 1990). The extent to which negotiators align their framing determines the extent to which they can make sense of each other's dialogue. Inappropriate framing may lead a police negotiator to interact in ways that make the hostage taker feel misunderstood or unvalued. Appropriate framing, which connects with the hostage taker's perceptions, may enable a police negotiator to demonstrate understanding and present alternative solutions in an effective, persuasive way.

The extent to which police negotiators and hostage takers align their framing of dialogue is therefore expected to play a central role in sensemaking and the way in which a negotiation unfolds (Drake and Donohue, 1996). Successful sensemaking will be characterized by greater synchrony in the way police negotiators and hostage takers frame their messages, particularly when this synchrony is maintained over a significant number of utterances. When negotiators correctly understand how the other is approaching a particular issue, they are able to exchange information and problem solve. This exchange of information allows them to better understand and adjust to the other's perspective, which in turn facilitates sensemaking in the future. Sensemaking in interactive contexts is therefore self-reinforcing, and we should expect longer periods of frame synchrony to occur during later stages of negotiation.

While negotiators' overall sensemaking is arguably driven by cognitive frames, a second, more immediate process may be responsible for negotiators' choice of frame in the first place. This immediate sensemaking is likely to involve a rapid assessment of salient information in a form similar to that identified in the contexts described above. Negotiators must respond to information as it presents itself in the dialogue, and the speed and accuracy with which they evaluate this information determines how well the negotiator is able to appropriately frame the interaction. Research from other contexts suggests that experienced individuals may achieve this sensemaking by recognizing salient features of the other party's actions (Klein, 1993).

This is consistent with the fact that negotiators are trained extensively in "phase" models of negotiation, which map out the typical changes in framing that occur as a negotiation unfolds (Donohue et al., 1991). Phase models are essentially external (partly legally defined) scripts about how to move through the negotiation process. For example, the Michigan State Police model proposes four stages of interaction: i) a contact stage, in which the police negotiator begins structuring the relationship and minimizing the hostage taker's anxiety; ii) a problem-clarification stage, in which interaction begins to concentrate on defining the actual problem; iii) a problem-solving stage, where the focus is on developing a mutually acceptable solution; and iv) a resolution stage, where the goal is to implement the solution and allow the hostage taker to surrender agreeably. Combined with experience, such models provide an explicit framework that negotiators can draw on to anticipate and interpret changes in dialogue. They provide a way of managing complexity, and we should expect to find evidence of negotiators' framing following the pattern of development prescribed by such models.

To test our expectations about the role of cognitive framing and scripts in negotiator sensemaking, we coded and analyzed dialogue from recordings of nine actual hostage negotiations. Specifically, we coded each utterance of the dialogue as one of three motivational frames that have repeatedly been found to reflect the major ways in which negotiators use dialogue over time (see Taylor, 2002; Taylor and

Donald, 2004 for more information). By examining the ways in which negotiators adjusted their framing across utterances, it was possible to derive an indirect picture of how negotiators were "making sense" of the other's perspective. For example, evidence of rapid shifts in the police negotiator's frame to match the hostage taker's frame would indicate good short-term situational awareness, but if such adjustment occurred consistently over the interaction it would also suggest that the police negotiator was unable to generate any longer-term synchrony in framing.

Figure 5.2 provides an example of actual interaction from one of the examined negotiations. The example comes from a hijacking incident in which a single male held two pilots hostage in order to speak with his girlfriend and get adequate help for his drug addiction. At this point in the incident, the police negotiators have successfully persuaded the hostage taker to accept food, but they continue to find him agitated and suspicious of their actions. As can be seen in Figure 5.2, the hostage taker frames his dialogue around his identity, asserting that he is an individual to be respected and that the police are "not dealing with your average hijacker." His concerns centre on the possibility of the police tricking him to storm the aircraft, and with the fact that they may not be taking him, and his demands, seriously. Given this focus, it is perhaps no surprise that the hostage taker reacts badly when the police negotiator makes only a passing reference to his concerns and moves on to dictate how the police are going to get food to the plane. The hostage taker, still framing the interaction in identity terms, takes this suggestion as a challenge to his position in the negotiation and reacts aggressively. Thus, in this interaction, the framing of the police negotiator and hostage taker diverges and becomes "misaligned," which leads to conflict between the parties.

One of the most striking findings to emerge when examining negotiations in this manner is how often police negotiators and hostage takers match one another's frame. Framing was found to be consistent for an average of 8.89 consecutive utterances, with sequences of three or more equivalently framed utterances occurring

---

Hostage Taker: If someone comes through the pilot's window and sticks a gun up, it all goes up. I'm being calm. You all are being a little bit irrational in a type situation like this. You're not dealing with your average highjacker.

Police Negotiator: Well, I understand the fact that you're very intelligent, and I appreciate that. We're just about ready to bring the food out to you. Here's the way it's gonna be done, now listen up.

Hostage Taker: I told you how it was gonna be done! You tell me how you want to do it and then I'm gonna tell you how it's gonna be done.

Figure 5.2   An example of frame divergence in dialogue from a hijacking incident

in 78 percent of the nine negotiations. However, as expected, the extent to which negotiators matched one another's framing varied between negotiations that ended peacefully and those that did not. Negotiations that ended unsuccessfully showed a small gradual decrease in the length of synchronous framing over time, while successful negotiations were associated with an increase in frame alignment over time. This increase in frame alignment was four times the magnitude of the decrease associated with unsuccessful incidents. In the interactive context, how negotiators made sense of issues in the initial stages seems to significantly effect their ability to make sense of the other's utterances during later stages of interaction.

The evident importance of early framing on later sensemaking highlights the need to uncover the process that underlies negotiators' initial assessment of dialogue. To derive an understanding of this process, we examined the 22 percent of dialogue in which negotiators did not have congruent frames. These dialogues reflect transitional periods during which negotiators seek a common frame, such that understanding these periods may inform our understanding of how frames are chosen by negotiators. To examine the make-up of these transitions, we counted the contingencies among different utterance frames and noted which of the two negotiators initiated the subsequent sequence of synchronous frames. Results showed that police negotiators typically adopted a significantly less dominant role during the transitional periods. They switched their personal framing to match the hostage taker's frame more frequently than was the case in non-transitional periods, and they reduced the average length of their utterances. In effect, they reduced their role in the interaction, which potentially allowed for increased prominence and greater recognition of the hostage taker's perspective.

However, while police negotiators took a passive role in moving through transitional periods, they often took a dominant role in determining the next period of synchronous framing. How they acted depended on the stage of the negotiation. Specifically, during early stages of negotiation, the police negotiators remained passive when hostage takers focused on instrumental (substantive) issues, but took an active role in promoting dialogue framed around relational or identity issues. In contrast, during later stages of interaction, police negotiators typically took the lead when responding to instrumentally framed utterance, but remained passive when dialogue focused on relational or identity issues. This was particularly true of successful negotiations, where police negotiators were significantly more likely to "pick up" on a hostage taker's temporary instrumental framing of dialogue in a way that generated a prolonged period of frame synchrony. This change in the focus of frame initiation is consistent with phase models of interaction, and it is consistent with the possibility that negotiators draw on cognitive scripts to focus their sensemaking efforts.

The analyses in this section sought to provide a preliminary investigation of sensemaking in hostage negotiation. In such a dynamic scenario, sensemaking involves more than a heuristic-led reaction to possible negotiation payoffs (for example, Bazerman et al., 2000). It involves a fast-paced, evolving assessment of hostage taker's actions and reactions, which draws on cognitive frames that simplify the necessary interpretation of the hostage-taker's utterances, and learned negotiation scripts that help negotiators anticipate the motivations underlying hostage takers' dialogue.

Conclusions

Investigators working in criminal contexts build complex explanations to flesh out possible crime scenarios, motives and *modus operandi*. They do so by structuring available evidence around internalized cognitive frames and also externally imposed legal scripts that determine likely courses of events. In addition, they appear able to adopt different inferential stances that allow them to evaluate multiple suspicion hypotheses against potential frameworks of guilt or innocence. A central feature of explanation-building in experts' sensemaking seems to be creating narratives that include more evidence than is strictly necessary to build up an account, or even that go beyond the immediate evidence. Extending explanations to develop a rich narrative allows the expert to test the plausibility of potential hypotheses in a rapid depth-first exploration (see Ball and Ormerod, 2000). It also presents a method for seeking alternative hypotheses or motives. The empirical studies described above point to a set of skills underlying investigative sensemaking that seem more complex than perceptually based situation awareness that triggers single decision rules. In adopting this view, we are rising to the challenge set by Yates (2001) in his review of NDM research, where he identifies explanation construction as potentially a key component of expert judgment.

Interestingly, some of the empirical examples of explanation-building that we present above share characteristics of the so-called "conjunction fallacy" identified in the JDM literature (Kahneman, Slovic, and Tversky 1982). So, for example, in the Linda problem, participants are given a stereotypical description of an individual ("Linda") and asked to rate the likelihood of two or more conclusions (for example, "Linda is a bank teller" versus "Linda is a feminist bank teller"). Participants rate the conjunctive conclusion ("Linda is a feminist bank teller") as most plausible when it is consistent with the stereotype given in the description. This judgment is a logical "fallacy" if the task is simply to derive a probabilistic judgment of separate conclusions, since the singular conclusion is a subset of the conjunctive conclusion. However, reconceived as an explanation-building task, the selection of the conjunctive conclusion makes perfect sense: it offers a coherent explanation of the largest subset of evidence available in the description. So, for example, in the vignette where a grandmother reports bruising on a child, logically the most probable conclusion is simply that the child has been abused, but officers tended to draw a conjunctive conclusion that the child had been abused by the mother's violent boyfriend. In this context, it behoves investigators to consider the conjunctive possibility rather than the singular conclusion.

That said, part of the sensemaking skills of an investigator is derived from an ability to step back from conjunctive conclusions and to consider alternative and weaker hypotheses (for example, the child may have been abused but maybe not by the violent boyfriend). One strategy for generating alternatives is to treat each of the propositions in a conjunctive conclusion as a separate building block for a line of enquiry. So, for example, we described above an insurance investigator developing three separate explanations for the conjunction of evidence in which a third-party vehicle appears in multiple claims. A standard tenet of problem-solving is the notion of satisficing (Simon, 1981), the idea that individuals will expand on one solution possibility at the expense of alternatives as a way of controlling search through a potentially infinite problem space. Satisficing is implicit in recognition-primed decision making (Klein, 1989), since the expert makes a single rapid and seemingly intuitive judgment from which decisive action follows. This kind of sensemaking works particularly effectively in contexts such as emergency response, where to dwell on consideration of multiple possible views might delay action, with potentially negative consequences. In contexts of criminal investigation, however, we suggest that satisficing is an inappropriate approach (though quite possibly one found with less effective or experienced investigators). Expert investigators should generate alternative explanations: investigation is rewarded by divergent thinking not convergent recognition.

A key factor that differentiates investigation contexts from contexts such as emergency response is time: typically, investigators (even at times those involved in real-time hostage negotiations) are able to spend time amassing evidence that might test alternative hypotheses. In emergency response contexts, on the other hand, time pressures lead to needs for immediacy of action. Thus, not only is divergent thinking desirable in investigation, the context may also make it possible. In contrast, the context of emergency response may restrict expert decision-making performance to recognition-based methods.

In the course of investigating crimes being planned (for example, disruption of drug traffickers), crimes under commission (for example, during a hostage/barricade incident), or crimes that have already been committed (for example, a murder investigation), police officers and insurance investigators take actions that impact on the behavior of their target. Sensemaking is, therefore, a dynamic, ongoing process of understanding and anticipating the target's likely reaction to investigative action. This dynamic characteristic is shared by some other expertise domains. For example, the actions a pilot takes to remedy a fault can change the situation that is faced sometimes quite dramatically, a point neatly exemplified by Orasanu, Martin, and Davison (2001) in their description of the Kegworth Boeing 737–400 crash of 1989 in which the pilot isolated the wrong (that is, functioning) engine in trying to deal with a suspected engine fire (p. 212).

What makes the dynamism of investigative domains different is the deliberate intent on behalf of the perpetrators of crimes to mask the true causal structure of an incident. It turns out that expert investigators can use an expectation of deliberate deception to good effect. If you expect to be deceived, then you can construct tests that force the perpetrator to expand upon their deception until a point where the deceptive story becomes unsustainable. In the domain of insurance fraud, this process is called setting "elephant traps," where seemingly innocuous lines of questioning can be pursued that inevitably demonstrate a claimant is lying. In some respects, accidental (that is, unexpected) deception is much more problematic because the expert cannot prepare for it, and indeed, their expertise is undermined by it. For example, in the Kegworth air crash, the action of turning off a functioning engine had the effect of removing a vibration that served as a cue to malfunction, thereby falsely confirming the pilot's diagnosis. Knowing that you will be deceived in a criminal context provides a powerful investigative lever, and it is a lever that is used to good effect in sensemaking by skilled investigators.

Acknowledgements

The first author's research was funded by grant no. GR/R02900/01 from the UK EPSRC/DTI Management of Information Programme. He gratefully acknowledges the contributions of his colleagues Linden Ball and Nicki Morley to the research, and thanks representatives of insurance companies who took part in the studies. Support to the second author's research was provided by the Economic and Social Research Council, grant PTA-030-2002-00482.

References

Alison, L.J. and Barrett, E.C. (2004). "The interpretation and utilization of offender profiles." In J.R. Adler (ed.), *Forensic Psychology: Concepts, debates and practice*. Cullompton, Devon: Willan Publishing (pp. 58–77).

Amalberti, R. and Deblon, F. (1992). "Cognitive modelling control: A step towards an intelligent on-board assistance system." *International Journal of Man–Machine Systems*, 36: 639–71.

Ball, L.J. and Ormerod, T.C. (2000). "Putting ethnography to work: the case for the cognitive ethnography of design." *International Journal of Human–Computer Studies*, 53: 147–68.

Barrett, E.C. (2002). "Towards a theory of investigative situation assessment: An examination of the psychological mechanisms underlying the construction of situation models in criminal investigations." Unpublished Master's thesis. University of Liverpool, Liverpool.

Barrett, E.C. and Alison, L.J. (submitted for publication). "Detective's hypotheses in criminal investigations: An exploratory study."

Bazerman, M., Curhan, J., Moore, D., and Valley, K. (2000). "Negotiation." *Annual Review of Psychology*, 51: 279–314.

DeFilippo, C., Louden, R., and McGowan, H. (2002). "Negotiation under extreme pressure: The 'mouth marines' and the hostage takers." *Negotiation Journal*, 18: 331–43.

Donohue, W., Kaufmann, G., Smith, R., and Ramesh, C. (1991). "Crisis bargaining: A framework for understanding intense conflict." *International Journal of Group Tensions*, 21: 133–54.

Drake, L. and Donohue, W. (1996). "Communicative framing theory in conflict resolution." *Communication Research*, 23: 297–322.

Elio, R. and Pelletier, J. (1997). "Belief change as propositional update." *Cognitive Science*, 21: 419–60.

Endsley, M.R. (1988). "Design and evaluation for situation awareness enhancement." In *Proceedings of the Human Factors Society 32nd Annual Meeting*. Santa Monica CA: Human Factors Society (pp. 97–101).

—— (1995). "Toward a theory of situation awareness in dynamic systems." *Human Factors*, 37: 32–64.

—— (2000). "Theoretical underpinnings of situation awareness: A critical review." In M.R. Endsley and D.J. Garland (eds), *Situation Awareness: Analysis and Measurement*. Mahwah, NJ: Lawrence Erlbaum Associates Inc. (pp. 3–32).

Feltovitch, P.J., Spiro, R.J., and Coulson, R.L. (1997). "Issues of expert flexibility in contexts characterized by complexity and change." In P.J. Feltovitch, K.M. Ford, and R.R. Hoffman (eds), *Expertise in Context*. Cambridge, MA: MIT Press (pp. 125–46).

Graesser, A.C., Mills, K.K., and Zwaan, R.A. (1997). "Discourse comprehension." *Annual Review of Psychology*, 48: 163–89.

Green, J., Smith, S., and Lindsey, A. (1990). "Memory representations of compliance-gaining strategies and tactics." *Human Communication Research*, 17: 195–231.

Hammersley, M. and Atkinson, P. (1983). *Ethnography: Principles in Practice*. London: Routledge.

Innes, M. (2003). *Investigating Homicide: Detective Work and the Police Response to Criminal Homicide*. Oxford: Oxford University Press.

Kahneman, D., Slovic, P., and Tversky, A. (1982). *Judgment Under Uncertainty: Heuristics and Biases*. Cambridge: Cambridge University Press.

Klein, G. (1989). "Recognition-primed decisions." In W.B. Rouse (ed.), *Advances in Man–Machine System Research* (Vol. 5). Greenwich, CT: JAI Press (pp. 47–92).

—— (1993). "A recognition-primed decision (RPD) model of rapid decision making." In G. Klein et al. (eds), *Decision Making in Action: Models and Methods*. Norwood, NJ: Ablex (pp. 138–47).

Klein, G., Phillips, J., Rall, E.L., and Paluso, D.A. (2007). "A Data/Frame Theory of Sensemaking." In R.R. Hoffman (ed.), *Expertise Out of Context*, Mahwah NJ: Lawrence Erlbaum Associates (pp. 78–94).

Morley, N.J., Ball, L.J., and Ormerod, T.C. (2006). "How the detection of insurance fraud succeeds and fails." *Psychology, Crime, and Law*, 12: 163–80.

Orasanu, J., Martin, L., and Davison, J. (2001). "Cognitive and contextual factors in aviation accidents: Decision errors." In E. Salas and G. Klein (eds), *Linking Expertise and Naturalistic Decision Making*. Mahwah NJ: Lawrence Erlbaum Associates (pp. 209–25).

Ormerod, T.C., Morley, N.J., Ball, L.J., Langley, C., and Spenser, C. (2003). "Using ethnography to design a Mass Detection Tool (MDT) for the early discovery of insurance fraud." In G. Cockton and P. Korhonen (eds), *CHI 2003: New Horizons-Extended Abstracts*. New York: ACM Press (pp. 650–51).

Ormerod, T.C., Morley, N.J., Mariani, J.M., Lewis, K., Hitch, G., Mathrick, J., and Rodden, T. (2004). "Doing ethnography and experiments together to explore collaborative photograph handling." In A. Dearden and L.Watts (eds), *Proc. HCI2004: Design for Life* (Vol. 2). 6–10 September 2004, Leeds (pp. 81–4).

Ormerod, T.C. and Shepherd, A. (2004). "Using task analysis for information requirements specification: the Sub-Goal Template (SGT) method." In D. Diaper and N. Stanton (eds), *The Handbook of Task Analysis for Human–Computer Interaction*. Mahwah, NJ: Lawrence Erlbaum Associates (pp. 275–98).

Pennington, N. and Hastie, R. (1988). "Explanation-based decision-making: Effects of memory structure on judgment." *Journal of Experimental Psychology: Learning, Memory and Cognition*, 14: 521–33.

Roth, E.M., Lin, L., Kerch, S., Kenney, S.J., and Sugibayashi, N. (2001). "Designing a first-of-a-kind group view display for team decision-making: A case study." In E. Salas and G. Klein (eds), *Linking Expertise and Naturalistic Decision Making*. Mahwah, NJ: Lawrence Erlbaum Associates (pp. 113–35).

Simon, H.A. (1981). *The Sciences of the Artificial* (2nd edn). Cambridge, MA: MIT Press.

Taylor, P. (2002). "A cylindrical model of communication behavior in crisis negotiations." *Human Communication Research*, 28: 7–48.

Taylor, P. and Donald, I. (2003). "Foundations and evidence for an interaction based approach to conflict." *International Journal of Conflict Management*, 14: 213–32.

—— (2004). "The structure of communication behavior in simulated and actual conflict negotiations." *Human Communication Research*, 30: 443–78.

Yates, J.F. (2001). "'Outsider': Impressions of Naturalistic Decision-Making." In E. Salas and G. Klein (eds), *Linking Expertise and Naturalistic Decision Making*. Mahwah, NJ: Lawrence Erlbaum Associates (pp. 9–33).

Chapter 6

Learning from Experience: Incident Management Team Leader Training

Margaret T. Crichton, Kristina Lauche, and Rhona Flin

Introduction

Management of critical incidents in complex sociotechnical systems, such as industrial accidents, large fires, traffic disasters, or terrorist attacks, places high requirements on the decision-making skills of those in charge. The situation is complex, hazardous, constantly changing, and characterized by incomplete information on the current state and the underlying problem. Incident managers need to diagnose, consult, decide, delegate, monitor—all under extreme time pressure and with immense responsibility for the livelihood of others (Flin, 1996). They may also have scarce resources, and the criteria for a good resolution of the incident are multidimensional and shifting. In this sense, incident management is an example of macrocognition (Cacciabue and Hollnagel, 1995; Klein et al., 2003) *par excellence*: Incident management is a case of human cognition in interaction with the environment in real life. Its success depends on the expertise of skilled practitioners, and the conditions under which it occurs are literally impossible to control and manipulate for the purpose of research. Suitable methods of inquiry are therefore field observations, interviews with practitioners (Crandall, Klein, and Hoffman, 2006) and, once some of the influencing factors have been understood, simulation studies such as Omodei, Taranto and Wearing's (2003) simulation of bush fires.

The special challenge of an industrial incident is that those in charge need to switch from their ordinary technical and managerial activities into the role of on-scene commanders, as Flin (1996) has pointed out in her analysis of the Piper Alpha disaster and transport accidents. A manager's day-to-day-role tends to be governed by the paradigm of rationalist decision making, combined with the need for consultation and political arguments. This does little to prepare them for unexpected events and decision making under high time pressure, and unlike professional emergency service personnel, they rarely have the opportunity to practice for incident command.

The Naturalistic Decision Making approach has made major contributions to current understanding of decision making in real-life environments (Klein et al., 1993; Zsambok, 1997). We have learned that incident commanders assess the situation to identify what type of event they are dealing with, and draw on their repertoire of appropriate responses, and then continue to assess the situation to modify their course of action accordingly, rather than making deliberate choices between options. Yet the focus has traditionally been on individuals and less on teams, and more on short-term decisions, for which the recognition-primed decision-making approach has been particularly apt. Yet in many high-hazard industries, the recovery from an incident can pose equally complex challenges in containing long-term damage and financial loss, reestablishing technical equipment and managing public perceptions, multiple rescue organizations and governmental bodies.

There are two main questions from the existing body of research for the type of incident management we are interested in here. First, what are the different requirements posed by the initial short-term response and longer, protracted incidents? Training typically prepares for the response phase, yet the recovery phase can throw up equally complex challenges—what is appropriate in the initial response might not be the best skills for long-term recovery. This occurs in many high-hazard industries. Training therefore should be targeted at the potentially longer-term situation assessment and consequence management. NDM has normally been directed at the immediate, hands-on decision making in dynamic situations. Recovery from an incident is still dynamic, but over a longer timescale, and the decision making, while not quite as immediate, needs to occur as other aspects of the management and organization may be waiting.

Second, how can practitioners for whom incident management is only part of their role be prepared for unexpected events, so that they can develop the required knowledge, skills, and attitudes? We have learned from macrocognition that people behave differently in real-life, complex, dynamic situations than they do in laboratories, and that we must opt for the more challenging research environment in order to obtain valid findings. Yet it is these very same challenges of the uncontrolled conditions and high stakes that make the realistic situation problematic as a training environment. So in which way can we emulate realism and the stress of making decisions, under time pressure and with high risks involved, in a manner that does not put people at risk or consume massive resources?

This case study, which involved interviews with members of an Incident Management Team (IMT) who had dealt with a significant incident in the oil and gas industry, provides the opportunity to examine the level of transfer from training to actual experiences. The following section explains the context of the research in terms of the industry in general and the history of disasters, and the incident. We then report the study, in which we interviewed participants on their experience of the training situation compared to the real incident. The chapter concludes with requirements for further research and training.

Context of Research: Incident Management in the Oil and Gas Industry

The oil and gas industry, and in particular its exploration and production sector, is characterized by risk mitigation and lagging feedback. Drilling for potential sources of hydrocarbons requires a lot of planning and detailed geological data, as well as very expensive equipment. Only if the planned targets can actually be reached and the predicted hydrocarbons have been found, can the investment be recuperated from selling oil to downstream markets. The drilling process has therefore been subjected to more and more sophisticated methods of risk assessment in the planning stage, but still drillers have to deal with unexpected events, for instance, if equipment fails or there is a sudden influx of gas. A common saying among drillers is "you are dealing with Mother Nature," indicating that there is a limit to control and prediction. Offshore installations, which combine a chemical plant, an industrial shop floor, a heliport and living accommodation in a very confined space, continue to be a work environment with hazards that have to be managed.

Over the past decades, a number of crises have occurred during operations, resulting in loss of life and injury (see Table 6.1).

Table 6.1   A chronology of some of the major oil industry accidents in the last thirty years

Date	Rig	Location	Fatalities	Incident
1976 Apr.	Ocean Express	Gulf of Mexico	13	Sinking
1976 Mar.	Deep Sea Driller	Norway	6	Grounding
1979 May	Ranger 1	Gulf of Mexico	8	Sinking
1979 Nov	Bohai 2	Bay of Bohai, China	72	Sinking
1980 June	Bohai 3	Bay of Bohai, China	70	Blowout/fire
1980 Mar.	Alexander Kielland	North Sea, Norway	123	Capsize
1980 Oct.	Ron Tappmeyer	Saudi Arabia	19	Blowout
1982 Feb.	Ocean Ranger	Hibernia, N. Atlantic	84	Capsize
1983 Oct.	Glomar Java Sea	South China Sea	81	Sinking
1983 Sept.	60 Yrs of Azerbaijan	Caspian Sea	5	Sinking
1984 Aug.	Petrobras Enchova	Enchova Field, Brazil	36	Fire
1984 May	Getty Platform A	Gulf of Mexico	1	Explosion
1985 Jan.	Glomar Arctic II	North Sea, UK	2	Explosion
1985 Oct.	West Vanguard	Haltenbanken, Nor.	1	Blowout
1985 Oct.	Penrod 61	Gulf of Mexico	1	Sinking
1988 July	Piper Alpha	North Sea, UK	167	Explosion/fire
1988 Sept.	Ocean Odyssey	North Sea, UK	1	Blowout/fire
1988 Sept.	Viking Explorer	SE Borneo	4	Blowout/sinking
1989 Apr.	Santa Fe Al Baz	Nigeria	5	Blowout/fire
1992 Mar.	Cormorant A	North Sea	11	Heli crash
1995 Jan.	Mobil rig	Nigeria	13	Explosion
1996 Jan.	Morgan Oil Field	Gulf of Suez	3	Explosion

Table 6.1 continued   A chronology of some of the major oil industry accidents in the last thirty years

Date	Rig	Location	Fatalities	Incident
1998 July	Glomar Arctic IV	North Sea	2	Explosion
1999 Nov.	Mighty Servant 2	Indonesia	5	Sinking
2000 Apr.	Al Mariyah	Persian Golf	4	Collapse
2001 Jan.	Petrobas P37	Campos Field, Brazil	2	Platform fire
2001 Mar.	Petrobras P36	Campos Basin, Brazil	11	Explosion/ sinking
2005 July	Bombay High North	Bombay High, Indian Ocean	12	Fire

Source: Based on <http://home.versatel.nl/the_sims/rig/losses.htm>, accessed 1 October 2006.

One of the worst disasters was that of the Piper Alpha disaster (Cullen, 1990) in the UK, where the command and control abilities of Offshore Installation Managers (OIMs) were criticized in the official inquiry following the incident.

Incident command skills are, according to Cullen (1990), a level of management skills that are not a feature of normal management posts, especially at the tactical and strategic levels. Although training in crisis management has since been introduced to the oil industry, this typically involves knowledge of incident management systems and procedures, and practice using full-scale emergency exercises (Borodzicz and van Haperen, 2002), but incident command skills for managers are seldom specifically addressed, especially at the tactical command level. The following section describes the case study incident illustrating the command skills required by the Incident Management Team members to manage an incident.

Incident Description

Command skills, particularly decision making, were required by members of an IMT that was established in an oil company's head office, to manage an incident on one of their drilling rigs in the Gulf of Mexico. The incident occurred as this rig was drilling for oil and gas in extremely deep water.

At 3:59 a.m. on 21 May 2003, the 6,000-ft drilling riser suddenly parted at a depth of 3,200 feet sub-sea (see Figure 6.1). The drilling riser is a large-diameter pipe that connects the wellhead equipment on the seabed to the rig at the sea surface and which returns drilling chemical "mud" to the surface (see Figure 6.2a). Without the drilling riser, the chemical mud used for drilling purposes could spill and contaminate the seabed. The riser was under 2.1 million pounds of top tension and the parting produced not only a loud bang but everyone on board the drill ship felt the subsequent sudden jarring (Kirton, Wulf, and Henderson, 2004). Inspection by a Remote Operated Vehicle (ROV) showed that the drill pipe (which is internal to

Figure 6.1   Example of a drill ship

**Figure 6.2a   Illustration showing a riser connecting the vessel (in this case a
semi-submersible drilling rig) to the sub-sea wellhead equipment**

the drilling riser) was intact, and was actually supporting the broken riser. Wellhead
equipment is the part of the oilwell where it reaches the surface (whether on land
or sub-sea) and comprises valves and spools that contain the pressure and the oil
within the well. The wellhead equipment on the seabed (see Figure 6.2b) functioned
properly, preventing substantial leakage from the sub-sea reservoir into the Gulf.

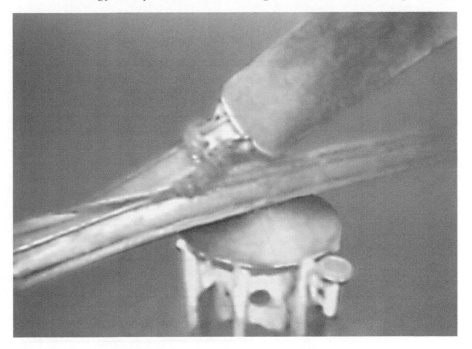

Figure 6.2b   Riser joints resting on sub-sea wellhead equipment

The Incident Management Team

The first call reporting the incident was received, in Houston, at 4.15 a.m., and the Incident Command System was established. Within the next hour, the Incident Management Team (IMT) was assembled in the onshore office. The Incident Command System provided a starting point for the management of the incident, and the command structure was modified to specifically meet the unique needs of this incident. The IMT (shown in Figure 6.3) was designed to meet the requirements for decision making during the incident.

Team leaders, at the strategic and tactical levels, immediately began to assess the situation, to give instructions to the rig to control damage, to respond to any potential threat to people, wildlife, or the environment, and to plan a longer-term response. After a period of 68 days, with a financial cost of $100,000 per day, the rig was reattached to the wellhead and the well was stabilized (Kirton, Wulf, and Henderson, 2004). Fortunately, unlike those incidents described in Table 6.1, no fatalities or injuries occurred on this occasion. However the IMT was faced with managing a prolonged incident, which had the potential to cause a release of hydrocarbons creating an incident impacting the environment.

Command-and-control teams function in a stressful environment characterized by decisions being made in situations characterized by time pressure, uncertainty and ambiguity, and dynamic and evolving conditions (Cannon-Bowers and Salas, 1998). In addition, incident management involves multiple players to gather,

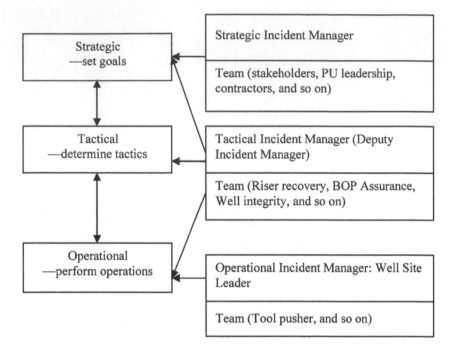

Figure 6.3   Incident command team structure

process, integrate, and perform based on accumulated data (Salas, Cannon-Bowers, and Weaver, 2002). Effective performance in a command-and-control environment depends on an interaction of the training of team members, leaders, and team interactions.

Shortly before this incident, key team leaders of the IMT had participated in a full-scale simulated emergency exercise, based on a scenario where a completions riser (similar but not identical to a drilling riser) had broken. This then allowed the comparison of participating in an emergency exercise to an actual incident, where the scenarios were very similar, and involved many of the same participants. This case study provides the opportunity to examine the level of transfer from training to actual experiences for members of an IMT, and whether such training achieves the training principles proposed by Salas, Cannon-Bowers, and Weaver (2002).

Method

Semi-structured interviews, based on the Critical Decision Method (Hoffman, Crandall, and Shadbolt, 1998), were conducted with seven key members of the IMT (six tactical-level IMT leaders and the strategic-level Incident Manager). Each interview was conducted individually, and lasted for approximately one hour. Interviewees initially described the actual riser breakage incident, focusing on their role and decisions made. In particular, interviewees were asked to describe what "went through their minds" when they received the initial call, and the information they received. Probes were used to target specific aspects such as the goals during the incident, communication (what helped and hindered the information flow), and situation awareness (how updating information was used to maintain an overview of the situation).

Next, their role and decisions made in the actual event were compared to the simulated exercise. Differences were explored between their experiences between participation in an emergency exercise and dealing with a comparable actual incident. The format of the interview included:

- Description of the event in his[1] own words.
- Identification of the most challenging decisions and what made them challenging.
- The extent to which previous experiences of training in incident command were found to be useful.
- Determination of what they felt could be learned from this incident, that is, lessons learned.

The interviews were recorded on a digital voice recorder, then de-identified and transcribed into a summary protocol (Breakwell, Hammond, and Fife-Schaw, 2000). Each interview transcript was examined to identify any of the differences between the training event and the actual experience. Particular attention was paid to the skills used by these roles during the management of the incident. The skills of interest here are the non-technical skills, namely the cognitive and social skills of team members, not directly related to equipment, system management, and procedures, for example, Standard Operating Procedures (SOPs) (Flin et al., 2003; Flin, O'Connor and Crichton, 2008). The skills used by the members of the IMT were analyzed and a taxonomy of incident command skills for drilling and completions teams was developed (see Crichton, Lauche, and Flin, 2005, for further details).

It is accepted that many aspects of command and control, required when managing an unplanned event, can be reduced to the functional, involving systems and methods (Flin and Arbuthnot, 2002). However, the skills as well as the operational techniques, of the IMT leaders help to make more effective use of systems and methods that can prevent an undesirable event deteriorating further, and to return the situation to normality. The descriptions of skills and organizational issues were collated, and common themes were then generated. A second rater then cross-checked the interview transcripts, resulting in a 96 percent agreement on those identified.

In addition, team members also rated themselves on their effectiveness and performance during the first seven days of the incident on a scale of -5 (highly ineffective) to 5 (highly effective).

---

1   Note that the masculine term is used throughout as all participants in the interviews were male.

Table 6.2   Benefits and limitations of incident management exercises

Benefits	Limitations
• Understand the Incident Management System (IMS)	• Focus on what to do rather than how to do it
• Set up an organization to deal with specific incident	• Emphasis on the initial stages of response, with less emphasis on the recovery phase
• Practice using checklist of what to do during initial stages	• Lays out prescribed or scripted responses rather than practicing coping with a unique situation
• Awareness of communication/ information flow necessary for managing incident	• Less opportunity to make command decisions under extreme pressure, for example, incessant demands for information, intense media attention, multi-agency demands
• Provide experience of challenging events and more difficult decisions	
• Develop documentation that would be required in an actual incident	• Less ability to replicate or enact consequences of decisions and actions
• Opportunity to practice considering and implementing options but in low-risk situation	• Focus on completing response to incident within half a day—but no objective detailing how to deal with next seven days or so for start of recovery phase
• Appreciation of the resources (personnel and equipment) that might be required in an actual incident	
• Increased understanding of major issues, that is, what to look for (critical cues), what questions to ask, where to look for information	• Does not rehearse how certain procedures must be completed by certain times (in complex, dynamic event with uncertain information) or how to structure recovery period on a daily basis
• Identification of possible differences in specific incident management procedures in various parts of the globe	• Lack of emphasis on communicating information to wide variety of people within certain timescales
	• Focus on specific details of the exercise rather than generic learning during debrief
	• Less intense than a real incident and does not mirror fatigue and its potential effects on performance
	• The duration of an actual incident is usually much longer
	• Debrief does not include review of decisions and how they were made

Results

Participants contrasted their experiences from the emergency exercise compared to the actual event as listed in Table 6.2. They highlight some of the exercise artificialities that are recognized by emergency exercise designers.

Challenges that differentiated between the exercise and the actual incident according to the IMT members are shown in Table 6.3. In both instances, as new problems and dilemmas arose, IMT leaders had to make decisions about what to do and which course of action to follow. However, in the actual incident, the more challenging decisions followed the successful recovery phase of the incident rather than the response phase.

Table 6.3   Challenges that differed between actual incident and emergency exercise

Challenges

• Decision making:
 	
    —Uncertainty
    —Novelty
    —Time pressure
    —Communication

• Anticipation of future states
• Consequences of decisions and actions
• Fatigue

Response and Recovery Phases

Decision making during the response phase of the actual incident was considered to be less demanding, as most decisions related to establishing the IMT and procedures and guidance were provided by the Incident Management System (IMS). These aspects are typically rehearsed during an emergency exercise, but there tends to be less emphasis on managing the recovery. In the actual incident, however, challenging decisions arose during the recovery phase of the incident as the return to normality had to be planned and implemented. The IMT members noted that they went through a period of transition between the response and recovery phases.

The two phases of the incident, response and recovery (illustrated in Figure 6.4), resulted in the use of different decision strategies—more rule-based during the response phase, and more analytical during the recovery phase, that is, identifying longer-term critical path activities, scheduling of activities, and writing new procedures as necessary.

Over the initial few days, teams were tasked with specific responsibilities in order to resolve the incident, systems were in place, and a proactive period commenced involving consequence management. Any inefficiencies in the IMT organization were identified and dealt with. This recovery period also involved planning of what needed to be done, identifying and assessing risks, and determining the way forward.

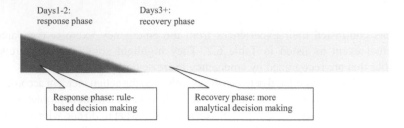

Days1-2:
response phase

Days3+:
recovery phase

Response phase: rule-
based decision making

Recovery phase: more
analytical decision making

Figure 6.4   Response and recovery phases

Team members also indicated their concerns over their effectiveness and performance during the first seven days of the incident. As Figure 6.5 illustrates, team members rated their performance as diminishing during the period of Days 2 to 6, until reaching their perceived optimal level from Day 7 onwards. This time period coincided with the transition from the response to the recovery phase.

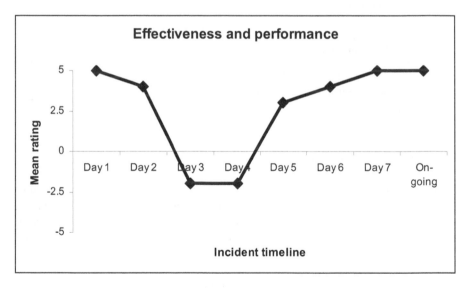

**Figure 6.5   Mean rating of effectiveness during first seven days
of the incident management**

As the incident progressed, more facts and details were sought to confirm or alter an initial assessment, and project forward into the future. A series of "what if" questions were considered, for example: "What if the BOP (blow-out preventer) hasn't closed? What if it has?" A common reaction was also to speculate about why the incident occurred, that is, feedback, rather than a feed-forward response of considering how the situation could evolve. Dealing with unknowns could destabilize any progress, therefore efforts had to be made to anticipate what could happen and to plan on the basis of the worst-case scenario, following the philosophy of "over-react and pull back." In other words, deploy resources on the basis that it is better to ramp up in terms of resources and then stand them down if they are not needed.

Stressors Affecting Command Skills

Incident command can be affected by increased acute stressors/pressure, which can arise from (Flin, 1996):

Table 6.4   Acute stressors

Unfamiliar, dynamic, chaotic events	Responsibility/fear of failure
Communication problems	Information (missing, ambiguous, overload)
Time pressure	Team management and coordination
Consequences and accountability	Dealing with media demands for information, updates, and background

Stressors that emerged during the management of the actual incident tended to be related to problems of communication. For example, participants described a number of situations where information was not exchanged as effectively between different teams. Initial difficulties with shift handovers were smoothed with a formal handover routine being established to ensure that sufficient details of what had happened during each shift period were passed on. Shift handovers are particularly crucial in assisting incoming team members to gain an adequate and appropriate assessment of the situation (Lardner, 1996). However, these difficulties were possibly due to the novelty of the situation, and the IMT quickly learned from the situation and introduced necessary processes as the incident management progressed. Novelty, uncertainty, and time were reported as being the key pressures that impacted mostly on decision making. Finally, the effects of fatigue on team members were noted. The IMT initially consisted of two shifts of personnel on a 12-hour rota, but after the initial response phase, such a work schedule was found to be creating fatigue for the immediate team members. A further lesson for the actual incident was the need to introduce a three-shift system to manage a longer-duration incident. In the meantime, a separate investigation team was set up to establish the cause of the incident, to identify potential lessons learned, and to implement them.

Organizational Processes

Incident management processes, such as checklists, are valuable for providing guidance in the initial response phase of an incident. How and where to access these, as well as knowledge and understanding of these processes, also assists individual and team functioning during incident management. During this incident, the formal company Incident Management System (IMS) acted as a foundation for the initial response, and described the roles and responsibilities of individuals in the IMT. This included a web-based checklist that was used by the IMT during the critical early stage. However, the IMS also allows a degree of flexibility in that the Incident Manager could determine the location of the Incident Control Room and how specific tasks within the IMT would be established. Using the IMS, the organizational structure of the IMT was established, a documentation system was in place, sub-teams were delegated with specific tasks, and regular meetings were held whereby the strategic goals were checked.

Under the IMS, roles and accountabilities and reporting relationships are defined; however, the Incident Manager had to ensure that these were understood and followed, as one of the key tasks for the Incident Manager was the delegation of activities to appropriate teams or personnel. A critical issue identified during this incident was that of resources, including numbers of IMT personnel, such that IMT member availability had to be checked, especially to ensure that adequate manpower was available if and when required. An incident log was established, which was used to record communication, key decisions, and actions taken. Furthermore, formal communication channels with experts both within the company and outside the company were established to ensure expertise availability. Organizational processes played a key role in the initial response stage of the incident. Thereafter, organizational processes were also involved in, for example, the development of procedures necessary for resolving the riser incident.

Training for Incident Command

Team leader performance can be improved using techniques such as training the team leader to act as a team facilitator (Tannenbaum, Smith-Jentsch, and Behson, 1998), by which briefing behaviors and post-action reviews are trained. In addition, team dimensional training (Smith-Jentsch, Zeisig, Acton, and McPherson, 1998) focuses on improvements to teamwork-related knowledge and skills, particularly shared mental models, the aim of which is to encourage teams to conduct debriefings in order to support team learning and self-correction, and has been implemented in the military and emergency services. Salas et al. (2002) cite principles for training and assessment of command-and-control teams, for both team members and team leaders. Targeted team leader training to improve overall team performance included training team leaders to act as a team facilitator to develop ways to improve performance. In addition, Salas et al. (2002) describe a team training strategy, Team Adaptation and Coordination Training (TACT) (Serfaty, Entin, and Johnston, 1998), which has been used with military command and control teams. Teams that had undergone TACT training showed improved performance compared to teams who had not undergone TACT training, but, of more relevance for incident command teams, they also performed better during stressor exposure.

As the evidence of this cases illustrates, more emphasis should be placed during ICT training and preparation for recovery, consequence management, and managing the return to normal operations. Exercises should be used to raise awareness of the acute stressors as well as of their effect on individual and team performance (Crichton, Lauche, and Flin, 2005). The main criterion for training is therefore not the degree of technical realism an exercise or simulation achieves, but the extent to which it recreates the experience of making complex decisions under emotional stress and time pressure, and the explicit focus on decision making and communication. This may be achieved by using low-fidelity simulations such as incident scenario descriptions and dynamic interventions from an experienced facilitator. For example, Tactical Decision Games have been used effectively to train team leaders by creating realistic, dynamic scenarios that require decision makers to adapt to changing and increasingly stressful circumstances and communicate their decision regularly (Crichton, Flin, and Rattray, 2000).

Conclusion

This incident illustrated how a macrocognition approach could be applied to understanding which skills are essential during both the response and recovery phases of incident management. During the response phase, knowledge as well as previous experience and practice from participating in emergency exercises proved to be particularly useful for IMT leaders in providing a basic structure and organization. In the recovery phase, a more consultative regime was appropriate as experts were invited to provide necessary knowledge and input to formulate future plans. IMT leaders therefore required the social and cognitive skills to enact the necessary processes in the response phase, and also to anticipate how the event could progress and make the necessary decisions to implement appropriate courses of action.

IMT leaders who participated in both the exercise and the actual event commented that although the earlier full-scale exercise provided knowledge and understanding of the processes and procedures of the incident management system, as well as roles and responsibilities, there was little opportunity to experience making decisions under stress, particularly pressures of time, uncertainty, and event novelty. In contrast, these pressures and challenges were experienced for many of the decisions that arose during the actual event, particularly during the recovery phase, which involved more knowledge-based decision making. No specific training in command skills, for example, how to make decisions under pressure, had typically been received, nor does this form part of the debrief process.

Recommendations

A number of general recommendations for IMT leaders based on this case study can be drawn:

1. All IMTs should hold "what if" sessions to consider how they would manage the recovery stage of potential incidents.
2. Directed practice and training in incident command skills, necessary for effective performance when managing an incident, should be provided to all IMT leaders. Guidance for training and practice in incident command skills and organizational processes for incident management should be based on domain-specific incident command skills (Crichton, Lauche, and Flin, 2005). Moreover, training should be designed to include periods of heightened stress (for example, exerting time

pressure on participants to make a decision), so that participants can experience the effects of stressors on cognitive skills, that is, situation awareness and decision making. Stress has been shown to affect attention, working memory capacity, result in strategy shifts, and cause physiological effects such as arousal that then compete for cognitive resources (Orasanu, 1997).

3. IMT leaders should receive personal individual feedback and debriefing on their performance in emergency exercises, specifically addressing incident command skills (Salas, Cannon-Bowers, and Weaver, 2002; Tannenbaum, Smith-Jentsch, and Behson, 1998).

4. Supplementary experiential training and guided debriefing, based on authentic and realistic scenarios, can be used to enhance incident command skills for tactical and strategic-level IMT leaders to overcome this training gap (Pliske, McCloskey, and Klein, 2001). Techniques that are primarily aimed at improving incident command skills include low-fidelity Tactical Decision Games (Crichton, Flin, and Rattray, 2000; Schmitt and Klein, 1996), and higher-fidelity synthetic training environments (Crego and Harris, 2002; Schiflett et al., 2004). These techniques are adaptable in presenting scenarios that can target the recovery and consequence-management phase of an incident.

Acknowledgements

The views expressed here are those of the authors and do not necessarily represent the views of the case study organization. We would like to take this opportunity to thank all those who assisted with this research.

References

Borodzicz, E. and van Haperen, K. (2002). "Individual and Group Learning in Crisis Simulations." *Journal of Contingencies and Crisis Management*, 10(3): 139–46.

Breakwell, G., Hammond, S., and Fife-Schaw, C. (2000). *Research Methods in Psychology* (2nd edn). London: Sage.

Cacciabue, P.C. and Hollnagel, E. (1995). "Simulation of Cognition: Applications." In J.M. Hoc, P.C. Cacciabue, and E. Hollnagel (eds), *Expertise and Technology: Cognition and Human-Computer Cooperation*. Hillsdale, NJ: Lawrence Erlbaum Associates (pp. 55–73).

Cannon-Bowers, J. and Salas, E. (1998). "Individual and Team Decision Making under Stress: Theoretical Underpinnings." In J. Cannon-Bowers and E. Salas (eds), *Making Decisions under Stress: Implications for Individual and Team Training*. Washington, DC: APA (pp. 39–59).

Crandall, B., Klein, G., and Hoffman, R.R. (2006). *Working Minds: A Practitioner's Guide to Cognitive Task Analysis*. Cambridge, MA:Bradford Books/MIT Press.

Crego, J. and Harris, C. (2002). "Training Decision-Making by Team Based Simulation." In R. Flin and K. Arbuthnot (eds), *Incident Command: Tales from the Hot Seat*. Aldershot: Ashgate (pp. 258–69).

Crichton, M., Flin, R., and Rattray, W.A. (2000). "Training Decision Makers—Tactical Decision Games." *Journal of Contingencies and Crisis Management*, 8(4): 208–17.

Crichton, M., Lauche, K., and Flin, R. (2005). "Incident Command Skills in the Management of an Oil Industry Drilling Incident: A Case Study." *Journal of Contingencies and Crisis Management*, 13(3): 116–28.

Cullen, T.H.L. (1990). *The Public Inquiry into the Piper Alpha Disaster*. London: HMSO.

Flin, R. (1996). *Sitting in the Hot Seat: Leaders and Teams for Critical Incidents*. Chichester: Wiley.

Flin, R. and Arbuthnot, K. (2002). "Lessons from the Hot Seat." In R. Flin and K. Arbuthnot (eds), *Incident Command: Tales from the Hot Seat*. Aldershot: Ashgate (pp. 270–93).

Flin, R., Martin, L., Goeters, K., Hoermann, J., Amalberti, R., Valot, C., et al. (2003). "Development of the NOTECHS (Non-Technical Skills) system for assessing pilots' CRM skills." *Human Factors and Aerospace Safety*, 3: 95–117.

Flin, R., O'Connor, P., and Crichton, M. (2008). *Safety at the Sharp End: A Guide to Non-Technical Skills*. Aldershot: Ashgate.

Hoffman, R.R., Crandall, B., and Shadbolt, N. (1998). "Use of the critical decision method to elicit expert knowledge: A case study in the methodology of Cognitive Task Analysis." *Human Factors*, 40(2): 254–76.

Kirton, B., Wulf, G., and Henderson, B. (2004). *Thunder Horse Drilling Riser Break—The Road to Recovery*. Paper presented at the SPE Annual Technical Conference and Exhibition, Houston, TX, (26–29 September).

Klein, G., Orasanu, J., Calderwood, R., and Zsambok, C. (1993). *Decision Making in Action: Models and Methods*. Norwood, NJ.: Ablex Publishing.

Klein, G., Ross, K.G., Moon, B.M., Klein, D.E., Hoffman, R.R., and Hollnagel, E. (2003). "Macrocognition." *IEEE Intelligent Systems*, 18(3): 81–5.

Lardner, R. (1996). *Effective Shift Handover. A Literature Review.* (No. OTO 96 003): Health and Safety Executive.

Omodei, M.M., Taranto, P., and Wearing, A.J. (2003). *Networked Fire Chief* (Version 1.33). Melbourne: La Trobe University.

Orasanu, J. (1997). "Stress and naturalistic decision making: Strengthening the weak links." In R. Flin et al. (eds), *Decision Making Under Stress. Emerging Themes and Applications*. Aldershot: Ashgate (pp. 43–66).

Pliske, R.M., McCloskey, M.J., and Klein, G. (2001). "Decision Skills Training: Facilitating Learning From Experience." In E. Salas and G. Klein (eds), *Linking Expertise and Naturalistic Decision Making*. Mahwah, NJ: Lawrence Erlbaum Associates (pp. 37–53).

Salas, E., Cannon-Bowers, J., and Weaver, J. (2002). "Command and Control Teams: Principles for Training and Assessment." In R. Flin and K. Arbuthnot (eds), *Incident Command: Tales from the Hot Seat*. Aldershot: Ashgate (pp. 239–57).

Schiflett, S.G., Elliot, L., Salas, E., and Coovert, M.D. (eds) (2004). *Scaled Worlds: Development, Validation and Application*. Aldershot: Ashgate.

Schmitt, J. and Klein, G. (1996). "Fighting in the fog: Dealing with battlefield uncertainty." *Marine Corps Gazette*, 80 (August): 62–9.

Serfaty, D., Entin, E.E., and Johnston, J.H. (1998). "Team coordination training." In J. Cannon-Bowers and E. Salas (eds), *Making Decisions Under Stress: Implications for Training and Simulation*. Washington, DC: American Psychological Association (pp. 221–35).

Smith-Jentsch, K.A., Zeisig, R.L., Acton, B., and McPherson, J.A. (1998). "Team Dimensional Training: A Strategy for Guided Team Self-Correction." In J. Cannon-Bowers and E. Salas (eds), *Making Decisions Under Stress: Implications for Individual and Team Training*. Washington, DC: APA (pp. 271–97).

Tannenbaum, S.I., Smith-Jentsch, K.A., and Behson, S.J. (1998). "Training Team Leaders to Facilitate Team Learning and Performance." In J. Cannon-Bowers and E. Salas (eds), *Making Decisions Under Stress: Implications for Individual and Team Training*. Washington, DC: APA (pp. 247–70).

Zsambok, C.E. (1997). "Naturalistic decision making research and improving team decision making." In C.E. Zsambok and G. Klein (eds), *Naturalistic Decision Making*. Mahwah, NJ: Lawrence Erlbaum Associates (pp. 111–20).

Chapter 7

Making Sense of Human Behavior: Explaining How Police Officers Assess Danger During Traffic Stops

Laura A. Zimmerman

The decisions made by law enforcement officers in response to daily crises involve making high-stakes, time-pressured decisions when little information is available. Police commanders and critical response teams (for example, SWAT) decide how to handle hostage situations, kidnappings, bomb threats, riots, and so on. Police officers on patrol handle dynamic situations such as robberies-in-progress, street fights, domestic disturbances, and suicide threats. They also make decisions about when to pull over vehicles, when to stop and search suspicious people, and when to use varying degrees of force. When officers encounter these situations, they must quickly make sense of what is going on. They do this by interpreting cues from the environment and by reading the nonverbal and verbal behaviors of suspects, victims, and witnesses.

The first step in obtaining understanding, or awareness, of a situation is to actively explore the environment and seek out information (Endsley, 1995). This conscious attempt by decision makers to understand the current situation has been termed "sensemaking" (Weick, 1995). Sensemaking is one component of a larger set of macrocognitive functions that decision makers utilize in real-world settings. A person will engage in sensemaking if they do not have an adequate understanding of the situation (Klein et al., 2007). The situations encountered by police officers often require that they quickly and accurately perceive and process relevant information. Similar to other naturalistic decision making domains, policing situations often lack vital information, change quickly, and provide little time for assessment. If police officers incorrectly interpret unfolding situations, or take too long to comprehend the true nature of the situation, people could lose their lives.

Klein et al. (2007), developed the Data/Frame model of sensemaking to describe the activities decision makers engage in while they attempt to make sense of situations. According to this model, decision makers seek data from the environment and apply that data to internal frames. Klein et al. describes frames as scripts or scenarios constructed by decision makers from previous experiences. When decision makers encounter situations that resemble one of their existing frames, they attempt to match that frame to the external data. When mismatches occur, decision makers engage in sensemaking activities in an attempt to match data to a frame. People create frames from their experiences and those who have access to a large selection of frames are more likely to find a match to external situational data. Experienced decision makers are better able to match incoming data to their preexisting frames, allowing them to rapidly sort through information, recognize the problem, and develop a solution (Klein, 1998). Decision makers with less experience in a given domain do not have access to as many frames, making sensemaking more difficult.

Some of the decision-making requirements found in law enforcement environments have not been explored fully in other areas of decision-making research. In policing situations, the level of uncertainty is extremely high, and officers must quickly learn to function in such environments (Zimmerman, 2006). In order to understand ambiguous policing situations, officers must draw inferences based on the observable behaviors of other people. For example, a police officer encounters a belligerent drunk person. From the drunken person's behaviors, the officer must derive the internal thoughts, motivations, plans, and future actions of that person. The officer's course of action depends on this interpretation. For instance, if the officer decides the drunk person is about to become violent, he will quickly attempt to subdue that person, briefly increasing the risk that someone will get hurt. If the officer thinks he can talk the drunk person into calming down, he will use verbal skills to deescalate the situation; however, human behavior is unpredictable, thus the calmed-down drunk person may suddenly erupt in violence. The data/frame model of sensemaking provides a structure for studying how police assess behavioral data to make sense of situations and take effective actions.

This chapter presents results from a study that examined police officer sensemaking. Police officers viewed a videotape of an actual traffic stop. While watching the video, I stopped the tape at three predetermined points and asked officers questions about the scene. This methodology allowed multiple participants to comment on the same situation, so I could make comparisons across groups of participants. In this study, I compared the responses given by experienced officers to the responses of novice officers. Laboratory research tends to show that people, in general, are not very skilled at accurately making inferences from human behavior (Vrij and Mann, 2005), yet officers rely on behavioral cues when determining danger. This chapter focuses on the sensemaking aspects of police decision making, and pays particular attention to the behavioral cues they use to understand situations. In the naturalistic decision-making literature, not much research has focused on how behavioral cues assist decision makers make sense of situations.

The following introduction addresses three main concepts, starting with a discussion of sensemaking and a description of the data/frame model of sensemaking (Klein et al., 2007). The second section focuses on how domain experience contributes to effective sensemaking and discusses some general findings from the expertise literature. The third section discusses the use of behavioral cues to detect deception and presents general findings from this research area.

Sensemaking and the Data/Frame Model

Pioneered by Weick (1995) in organizational contexts, sensemaking is a deliberate process decision makers use to actively search for information in order to understand an ambiguous situations. In high-stakes, time-pressured, dynamic situations, the decision maker will attempt to detect and identify problems, form explanations, anticipate future events, and find leverage points, all in an effort to develop solutions and make decisions (Klein et al., 2007). The data/frame model of sensemaking provides an explanation of the activities decision makers engage in as they begin the decision making process. *Data* are the elements present in a situation that contribute to finding and building an applicable frame. *Frames* are the internal mental representations that sensemakers use to interpret and understand the data (Phillips and Battaglia, 2003). In attempt to make sense of a situation, a person will gather more data, compare observations to inferences, check assumptions, generate expectancies, reconcile elements that are inconsistent with prior experiences, and put forth explanations (Klein et al., 2007).

The frame a decision maker enters a situation with may be inadequate upon closer inspection of the data. Thus, decision makers will search for data that is relevant to their frame. If they cannot find confirming data, they may adjust or disregard that frame. This discribes an active exploration of the environment to find a fit between the data and the frame. Klein et al. (2007) discusses the similarities between frames and the traditional concepts of scripts and schema. Scripts provide people with an outline of how situations typically unfold. They use scripts to navigate through the common and uncommon situations they encounter in everyday life. Schemas are cognitive structures that provide people with descriptions of domain or situation-specific situations. People use schemas to search for, filter through, and organize incoming information (see Klein et al.). In creating the data/frame model, Klein et al. concluded that the subtle differences between these terms are not enough to warrant a distinction. Instead, they chose the term "frame" and acknowledged that this term encompasses the essence of scripts and schemas.

The data/frame model of sensemaking is not the only model that attempts to describe how decision makers seek to understand a situation. Cohen, Freeman, and Wolf (1996) developed the Recognition/Metacognition model to describe how decision makers recognize a situation and then use metacognition to "critique and correct" their understanding of the situation. Endsley (1995) developed a model of Situation Awareness in dynamic settings and described three steps decision makers take to achieve situation awareness. The first step, "perceptions of the elements in the environment," corresponds with sensemaking. The recognition-primed decision model also describes levels of decision making (Klein, 1998). In this model, Level 2 focuses on situation assessment. Decision makers engage in situation assessment when the situation is not easily recognized or matched to internal representations. These sensemaking models provide valuable insight about decision makers' initial processing of ambiguous situations, however the data/frame model provides the most comprehensive explanation of the *activities* used by decision makers to make sense of situations, thus providing the most applicable model for the purposes of this study.

Klein et al. (2007) identified the six sensemaking activities that decision makers engage in while connecting data from the environment to a frame[1] (see Figure 7.1).

The six sensemaking activities are:

---

1   For a full description of each activity, see Klein et al. (2007).

1. *Elaborating the frame*: The sensemaker elaborates the frame by using data presented in the current situation to fill in missing details, and extend the current story.
2. *Questioning the frame*: When the data does not fit the frame, the sensemaker questions if the frame is mismatched to the situation or the data is incomplete or inaccurate.
3. *Preserving the frame*: Preservation of the frame occurs when the sensemaker's inquiry leads him or her to conclude that the frame is adequate. It also occurs when he or she explains away or ignores inconsistent data, or becomes "fixated" on a frame. This can lead to inefficient decision making.
4. *Comparing frames*: This occurs when the sensemaker attempts to fit two contradictory frames with the data. Through continuous evaluation, one frame eventually offers the most compelling explanation.
5. *Seeking an alternative frame*: When the sensemaker deems the current frame inadequate, he or she seeks another frame that more adequately explains the data.
6. *Reframing*: This involves taking a different approach to understanding the problem, reincorporating recently presented data and reevaluating less-favored frames.

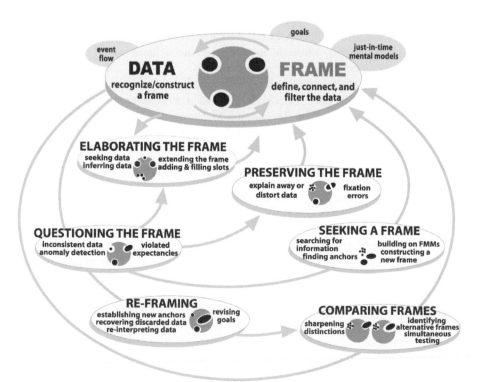

Figure 7.1   Data/frame model of sensemaking

According to the data/frame model, decision makers select the initial frame prior to beginning any conscious sensemaking activities. Once sensemaking begins, the sensemaker elaborates, revises, or discards the frame in response to the incoming data. The methodology used in this study provides opportunity to explore how these initial frames drive subsequent sensemaking activities. Klein et al. (2007) developed the data/frame model through analysis of previous research investigating the sensemaking activities of domain-specific decision makers. The research presented in this chapter provides a application of the data/frame model to newly gathered data, with the purpose of fitting this model outside the research literature that constructed the model.

Expertise and Sensemaking

Much research has examined how expertise develops (Ericsson and Charness, 1994; Ericsson, Krampe, and Tesch-Römer, 1993; Ericsson and Smith, 1991). According to Ericsson and colleagues, a person achieves outstanding performance when they have obtained domain-specific skills. These skills are not natural traits. Instead, these skills develop through repeated and lengthy exposure to a specific skill set. Vicente and Wang (1998) reviewed the expertise literature and concluded that experts acquire skills by adapting to environmental constraints. They use opportunities provided by the environment to achieve goals. When processing information, experts attend to relevant information while filtering out irrelevant information. Based on findings in the expertise research literature, Ericsson and Charness (1994) suggest that expert decision makers instantly form mental representations of the problem, allowing them to make quick and accurate decisions. Studies show that, compared to their novice counterparts, domain experts are better able to perceive and process information, recognize pertinent data, and match that data to previous experiences in a manner that facilitates successful action (Ericsson and Charness, 1994; Goodrich, Stirling, and Boer, 2000).

Dreyfus and Dreyfus (1986) proposed a five-stage model of skill acquisition to describe how decision making changes as experience increases. At the *novice* level of this model, the decision maker recognizes objective facts about the situation and applies "context-free" rules to the task. As decision makers progress through the *advanced beginner*, *competent*, and *proficient* stages, they begin to recognize meaningful elements of specific situations and begin to incorporate those elements into decisions. As expertise develops, decision makers begin to perceive situations as a whole, rules become less important, and decision makers become more flexible and react faster to incoming information. In the final, *expert*, stage, decision makers intuitively recognize situations and match them to previous experiences. They use mental simulation to predict events and action outcomes, and they deal with uncertainty by story-building and actively seeking information (Dreyfus and Dreyfus, 1986; Ross et al., 2005). This model provides a template for how sensemaking changes as decision makers acquire experience.

Decisions About Behavior

Police officers make sense of situations by reading the behavior of other people (suspects, bystanders). This is different from decision making done in many other naturalistic domains, where decision makers focus on equipment failure (airplanes, oilrigs), fire patterns, and weather patterns. Decision makers in military situations also assess human behavior patterns, but, until recently, these assessments were generally of the enemy as an entity, as in battlefield or enemy warship and aircraft movement (see Flin, 1996; Pliske and Klein, 2003, for summaries of these works). The ability to assess and predict human behavior is more complex than predicting the behavior of non-human entities. People interact socially and interpret other people's intentions and actions based on their own previous experiences, their emotions before and during the event, their mental state, personality traits, cultural knowledge, and previous knowledge or misinformation (Forsythe, 2004; Hunter, Hart, and Forsythe, 2000).

Shanteau (1992) made the differentiation between making "decisions about things" versus making "decisions about behavior." He found that people who made decisions about things (weather forecasters, test pilots, chess masters) performed well, whereas people who made decisions about behavior often exhibited poor performance (clinical psychologists, court judges, parole officers). He found that specific task characteristics determined good or poor performance. If the tasks were dynamic, unique, unpredictable, and if decision aids and feedback were unavailable, then decision performance was poor. Many domains, such as weather forecasting and aviation, have developed technology and decision aids to reduce the unpredictable aspects of these environments. Decision makers who rely on human behavior to make sense of situations have not yet been provided with such tools.

The large body of research grounded in understanding peoples' ability to read human behavior focuses on human ability to detect when someone is lying or telling the truth. A great deal of this research has "unpacked" the verbal and nonverbal cues that accurate and inaccurate lie detectors utilize when making their assessments. Studies that have evaluated lie-detection accuracy generally show accuracy rates around 50 percent, or chance levels (Vrij and Mann, 2005). This finding generally holds for police officers, although some studies have reported police accuracy rates significantly above chance levels (Ekman and O'Sullivan, 1991; Ekman, O'Sullivan and Frank, 1999; Mann, Vrij, and Bull, 2004).

In deception studies, researchers usually present video- or audio-taped statements of people who are purposely lying or telling the truth to participants who then decide whether that person is being deceptive. In these studies, researchers focus on one dichotomous outcome measure: "Is this person lying, yes or no?" This static task does not provide participants a full range of contextual cues or allow them the opportunity to seek information and read unfolding events. These researchers do not collect data about the process participants use to read deceptive behavior and incorporate their interpretations into their understanding of a situation. When police are participants in deceptions research, they likely use the behavior reading skills they use in real-world situations. This difference may help explain why laboratory researchers often discredit the use of behavioral cues to detect deception but police officers explicitly trust these cues to help them navigate through life-threatening situations. In a recent meta-analysis of deception studies, DePaulo et al. (2003) suggested that laboratory research does not display the full range of self-conscious behaviors exhibited by liars in high-stakes field situations. They concluded that this methodological weakness might produce results that inadequately demonstrate the degree to which behavioral cues indicate deception.

In this meta-analysis, DePaulo et al. (2003) analyzed 116 studies on deception and found that some behavioral cues, such as higher voice pitch, voice tension, and pupil dilation, were indicative of deception. Liars also were less cooperative than truth-tellers, made negative comments, and had less pleasant facial expressions. Many of the typical cues people commonly associate with deception, such as increased fidgeting and averted gaze (Strömwall and Granhag, 2003), were not associated with deception. In fact, this meta-analysis showed a significant decrease in foot and leg movement associated with deception, but only when liars had high motivation to deceive ($d = -0.13$). A small effect was found for a decrease in eye contact of highly motivated liars ($d = -0.15$). Verbal cues such as logical structure and plausibility of content decreased with deception. Deceivers used fewer gestures to illustrate their speech and less often provided spontaneous corrections to erroneous information. There is some evidence that with instruction and practice, people can improve their deceit detection capabilities. Vrij et al. (2004) trained participants to watch for reliable cues to deception and make judgments based on only those cues. Accuracy was 74 percent with this training.

In the current study, I presented participants with a video recording of an actual traffic-stop incident. At three designated points in the video, I stopped the tape and asked participants a series of questions about their sensemaking activities. By presenting the video in segments, participants provided information about their sensemaking as it progressed. This methodology is different from many naturalistic decision making studies, which often use retrospective accounts of real incidents or simulations (Crandall, Klein and Hoffman, 2006). By stopping a videotaped incident at set points, I was able capture their sensemaking as it unfolded, instead of relying on stories presented with hindsight. If the incoming data did not fit into participants' current traffic-stop frame, they would need to search for more data or adjust their frame to fit the new and novel data. I could track these activities by using the methodology presented in this study.

Method

Participants and Design

Thirty law enforcement officers (twenty-eight males, two females) from the city and county of El Paso, TX participated in this study. Twenty-seven officers were from the El Paso County Sheriff's Office and three officers were from the El Paso Police Department. Twenty-two participants identified themselves as Hispanic, seven as non-Hispanic White, and one as African-American. For the novice group, the average number of years as a police officer was three (range one–seven years), and the average for the experienced group was 13 years (range seven–thirty years).

Both the El Paso County Sheriff's Office and El Paso Police Department utilize a certification process, which calculates years of experience, hours of law enforcement education, and hours of academic education.[2] The number of participants in each certification level was: Basic = eleven, Intermediate = four, Advanced = ten, Master = five. For this study, the novice group consisted of officers with Basic and Intermediate certifications and the experienced group consisted of officers with advanced or master certificates. Assignment of officers into the novice or experienced group took place after the interviews were completed. The interviewer did not know officers' experience levels during the interview, however, officers frequently made remarks that revealed their experience levels.

Materials

Video   Participants viewed a videotape of an actual traffic stop. This video was part of a series of training videos that use real incidents to teach police officers safety lessons (Barber and March, 1995). Participants viewed the video on a large-screen television. The scene began as a routine traffic stop, with an officer pulling over a motorist, but escalated to the point where weapons were drawn and the officer critically wounded the motorist (see Appendix 7.1 for a summary of the scenario).

Interview   I conducted all interviews. The interview questions focused on participants' interpretations of the unfolding event, the cues they were attending to, their predictions about what would happen next, the types of errors that might occur, and what action-choice they would make at each point (see Appendix 7.2). I derived this procedure from the research presented by Dominguez (2001) and Dominguez et al., (2004).

Data collection   Interviews took place in a private room at the El Paso County Sheriff's Academy during officers' regular shifts and were recorded on digital voice recorders. The experimenter transferred the interviews to a computer and transcribed them for data processing.

Procedure

Participants sat in front of the video monitor as the experimenter explained that they would be watching a video in four segments. The experimenter encouraged them to think-aloud during each segment by saying what they were thinking as the scene progressed. At three predetermined points in the video, the researcher stopped the tape and participants answered questions about the events witnessed thus far. After the entire incident ended, participants answered the same questions as in the previous segments and then provided an overall evaluation of the scene and answered follow-up

---

2   Certification levels are: Basic (1 yr experience/400 hrs education); Intermediate (2 yrs experience/2,400 hrs education through 8 yrs experience/400 hrs education); Advanced (6 yrs experience/2,400 hrs education through 12 yrs experience/800 hrs education), and Master (10 yrs experience/4,000 hrs education through 20 yrs experience/1,200 hrs education).

questions based on their earlier responses. I adopted this procedure from the research presented by Dominguez (2001) and Dominguez et al. (2004).

Background questionnaire   At the end of each interview session, the participants completed a background questionnaire. This questionnaire asked participants to provide information about their certification level, their position, years on the force, and hours of education. From this information, participants were assigned to their experimental condition (novice or experienced). In addition, officers provided standard demographic information (age, ethnicity, and so on).

Analysis   From the interview data, I created seven main categories. Two coders read each interview transcript and recorded each time participants made comments that fit into these categories. The seven categories were alternative action, evaluative comment, general procedure, nonverbal cues, verbal cues, predictive comment, and safety. The items listed in the first four categories were further broken down. For example, alternative actions were classified into categories such as: call backup, question driver, secure (with handcuffs), and issue citation. We tallied coded categories separately for each segment of the video, allowing for the assessment of differences across segments. The coders resolved coding discrepancies by discussing the rationale for their coding until they reached agreement.

Interview length   The average interview length for the novice group was 34 minutes (range 26–56 minutes), and for experienced group was 37 minutes (range 24–56 minutes).

Results and Discussion

This first portion of the results section focuses on the participant information obtained after Segment 1. The comments made during this portion of the video reflected the participants' initial sensemaking and more readily revealed differences between novice and experienced participants. Because participants answered the same questions after every video segment, this first segment is the only portion of the interview where participants did not know they type of information they would need to provide. The second section incorporates data from Segments 2 and 3 into the analysis of sensemaking activities. Segment 4 of the video presented a significant change in action, thus I did not analyze this data in the context of the previous three segments. The third section presents a brief summary of Segment 4.

Initial Frame

After the first segment, participants discussed the cues that led to them to believe the traffic stop was dangerous. In answer to the first question, "What do you think is going on here?", all participants focused on either officer safety concerns and/or the behavioral cues displayed by the driver. Novice and experienced participants tended to focus on the same types of cues, but differed in how they elaborated on their observations. This finding is similar to previous research showing that more and less experienced decision makers use similar sensemaking strategies, but experts show a deeper understanding of the situation and provide richer explanations (Klein et al., 2002). When discussing the driver's behavior, participants stated that the driver was overly nervous. Novice participants explained that this behavior made them think "Something is not right" about the situation. Experienced participants usually followed their comments about the driver's behaviors with action suggestions. For example, a novice participant stated:

> Well, he [the driver] keeps crossing his arms, and he keeps moving around. I'm always aware of my surroundings, but I never get too complacent either. He looks like he's nervous 'cause he keeps moving around, he's nervous about something. I wouldn't know exactly, there's several things. He might be wanted or something.

This participant provided a vague explanation for the driver's behavior, and clearly indicated that he did not have enough information to make sense of the situation. While discussing the driver's behavior, this participant remained focus on procedural safety issues by discussing the importance of vigilance while making traffic stops. In contrast, an experienced participant stated: "He's giving me very good nervous indicators that he's got something in there that he doesn't want me to see. So I'd start questioning him more, maybe I'd even put my hand on the trunk and see what he does. See if he just stares at the trunk, stares at my hand." This participant made specific interpretations of the driver's nervous behaviors and followed immediately with the types of information-seeking actions he/she would take.

Comments about officer safety issues followed a similar pattern. When discussing safety issues, participants commented on actions made by the officer that decreased his safety. Novices focused on proper traffic-stop procedures and suggested what the officer should do differently, such as moving to the side of the road and not turning his back to the driver. Experienced participants also mentioned the officer's unsafe actions; however they tended to follow with interpretations of why the officer was behaving in such a manner. For instance, they made statements like "That is probably his department's procedure," or "He is trying to put the driver at ease by acting casual." Eight experienced participants started the interview by stating that the officer's true motivation for making the traffic stop was to investigate drug trafficking, not because of a traffic violation. None of the novice officers offered such a hypothesis. This demonstrates the ability of more experienced participants to synthesize data quickly and construct a reasonable story about the situation.

Experienced participants relied on their prior experiences to frame the situation, make predictions, and discuss typical courses of action (general procedures). Experienced participants made comments such as "Typically, what we do is…" "What I would do at this point is…" "That means he is thinking about…" The frame for novices focused around procedural issues, such as the proper way to approach a vehicle and the correct positioning when talking to drivers. Their comments started out with statements like "I would never get that close…" "He [the officer] is standing incorrectly…" "He is not being safe…" This indicates that the novices' frames centered on understanding how to conduct safe traffic stops according to their academy lessons. By relying on textbook models, they display characteristics found in the less skilled stages of the 5-stage model of skill acquisition. Experienced participants, on the other hand, displayed advanced expertise skills by incorporating a large amount of data and quickly drawing reasonable inferences, which fits into the later stages of the 5-stage model of skill acquisition (Dreyfus and Dreyfus, 1986; Ross et al., 2005).

Data Leading to Initial Frame

Nineteen participants stated the driver's behavior indicated the traffic stop was not typical. Participants relied on nonverbal cues to apply their initial frame. They focused on the driver's hand movements, such as fidgeting and hand-wringing, and body language, such as feet shuffling and looking around (see Table 7.1). According to the deceit detection literature, specific cues, such as fidgeting hands and agreeableness are not reliable indicators of deception in laboratory studies (DePaulo et al., 2003). However, overall nervousness correlates positively with deception. Overall nervousness was defined by DePaulo et al. as "the speaker seems

Table 7.1   Critical cue inventory: Initial detection of danger in traffic stops

Cue	Descriptors
Fidgeting hands	Hands in and out of pockets, wringing hands, moving hands around, picking fingernails, clenching fists, fidgeting with warning ticket, repeatedly putting warning ticket in pocket and removing it
Overall body movements	Pacing, feet shuffling, crossing and uncrossing arms, keeps moving around
Eye and head movements	Looking at feet, looking at car trunk, looking at ticket book, looking up and down highway, looking at hands, repeatedly reading warning ticket
Verbal responses	Answering a question with a question, indirect answers, evasive answers, contradictory answers, shaky, hesitant, over-polite, inconsistent stories
Rental car	Driver not the renter, driver doesn't know who rented the car, renter not in car, car not supposed to be taken out of state
Passenger	Fidgeting in front seat, passenger not the renter, refuses to roll down his window
Vehicle	Turn signal left on, driver looking at trunk, officer looked in back seat

nervous, tense; speaker makes body movements that seem nervous." In the verbal cue category, participants said that the driver's statement about not being the renter of the rental car led them to be suspicious that something more was going on beyond a typical traffic stop. They relied on the content of the driver's statements rather than tone of voice, voice tension, or statement constructions (plausibility, logical structure, negative comments). It is these latter cues that research shows are reliable cues to deception (DePaulo et al. 2003).

Most participants made statements that indicated they need more data to understand the situation fully. This indicates that their frames were not adequate to fit the data. Most participants stated that they were not sure what exactly the driver was hiding, or could not provide a firm prediction about what the driver might do next.

Sensemaking Activities

As the video progressed, participants elaborated their frames and adjusted their stories to fit incoming information. During Segment 2, the officer left the driver standing on the roadside and went to talk to the passenger in the car. Experienced participants increased their discussion about officer safety, stating concern about how the officer was handling two motorists by himself. Nine experienced participants continued to discuss the nervous behaviors of the driver, whereas only three novice participants mentioned the driver. The experienced participants seemed to be able to consider data from multiple sources (driver and passenger), whereas the novice participants narrowed their information search to the interaction between the officer and passenger. Novices determined from the passenger's behavior (for example, he would not roll down his window, he had rolling papers) that he was hiding drugs or "something." Experienced participants became less concerned with what exactly was going on and discussed the actions they would take to gather information and build their case for a legal arrest.

By Segment 3, most of the experienced participants (N=11) were concerned about the officer's safety as he gathered information from the passenger and driver. They were concerned that the officer had not secured the driver and they were concerned about the officer's position in front of the passenger window. Only six novice participants mentioned these safety concerns. This finding raises a possible training issue. In Segment 1, the majority of novice participants discussed the correct procedures officers should take when they first approach and interact with a driver. When the traffic stop began to deviate from typical traffic stops, novices did not notice or did not continue to mention safety or procedural issues. Perhaps they did not have textbook procedures in their mental storehouses. The experienced officers found these safety issues a real concern and did have frames for dealing with the events in the video. If novices are lacking knowledge about how to remain safe and vigilant through an entire traffic stop, training courses should seek to transfer the knowledge experienced officers have gained from conducting these stops.

As participants attempted to make sense of the situation, they engaged in the six sensemaking activities defined in the data/frame model of sensemaking. To clarify these levels and put them in context of a policing situation, below are some examples of participants' progression through the sensemaking activities.

Elaborate the frame   Participants who stated in Segment 1 that "something is not right" elaborated the frame by incorporating incoming information and applying this new data to better fitting frames. For instance, because of the passenger's behavior these participants elaborated their original interpretation and suggested that the driver and passenger were transporting drugs, money, and/or people, or they had stolen the car. Participants who suggested methods for gathering information, such as questioning the driver, searching the car, and calling their dispatcher to get additional driver and vehicle data, were actively elaborating the frame.

Questioning the frame   Participants in the first segment who judged that the stop was routine, or determined the driver's nervousness was normal, now questioned that frame as they received inconsistent data, such as the passenger's uncooperativeness. After Segment 1, an experienced participant stated, "There's not much information right there in the video, but the guy's [the driver's] body language was OK. When he thought he was going to get a ticket, he crossed his arms. He doesn't like what the officer was telling him. But once he found out it wasn't going to be a citation, he relaxed again." In Segment 2, this participant elaborates and starts questioning the frame, "There may be drugs in the car, with the rolling papers and the fact the passenger is not cooperating. To me, the driver is acting suspicious, his body language is saying that he's worried. I guess that would be typical of anybody that is pulled over by the police…" In Segment 3, this participant is continuing to question the frame as he/she compares inconsistent data, "The passenger and the driver do not have their story straight. I still think there is a strong possibility there are drugs in the car…I keep looking at the driver to see if he's really nervous, but I just don't see him being overly anxious yet. There are signs of nervousness, but to show there are drugs in the car, that I just don't see, I don't know."

Preserving the frame   Participants who initially determined that the driver was transporting illegal cargo used new information to confirm their initial assessments. For instance, they incorporated incoming data about the passenger into the previous information about driver and made stronger conclusions about potential illegal activities. One experienced participant's first statement was, "He's [the officer] on a fishing expedition to find drugs. He's running some sort of interdiction stop. He's probably working a drug corridor and he's using minor violation to gain probable cause to make traffic stops on people." In Segment 2, this participant stated, "He's seeing evidence that they do have something with them when he spotted the rolling papers." At the end of Segment 3, this participant continued, "He's building clues to articulate his probable cause. He is thinking they might be transporting drugs."

Comparing frames   Some participants made the interpretation that the driver's nervousness could be due to a minor crime, such as an outstanding traffic ticket or taking the rental car across a state line, or to a major crime such as transporting narcotics or stealing the rental car. As participants gathered more data, they fit this data into these frames and searched for the best fit. Most officers concluded that the driver's nervousness was extensive and eventually discarded the frame that the driver was hiding a minor crime. For instance, after Segment 1, a novice participant stated,

"It is just a traffic stop for an improper lane change…no concerns, the thing about it is, he's [the officer] too close, he's leaving himself unprotected." At the beginning of Segment 2, this participant stated, "At this time, I think something is not right…it's a rental vehicle, well, that's not necessarily bad, but if the guy is acting all nervous then he's hiding something…maybe it's nothing but that's what makes you decide to interview these guys." By the third segment, this participant was reframing and comparing frames: "They might be transporting something illegal. First the rolling papers, then they have conflicting stories. I kept thinking they might be hiding something, especially, the guy was nervous because he was repeating questions, people do it when they are nervous."

Seeking an alternate frame   When the officer first mentioned seeing rolling papers in the car, many participants stated that this was evidence that drugs were in the car. Other participants stated that some people might use rolling papers for cigarette tobacco, or non-legal amounts of marijuana. They were seeking an alternative frame and testing its fit to the data by considering that perhaps the motorists were not carrying a large load of drugs. When participants attempted to explain the officer's behavior by saying that he was acting according to his department's policies, or that he was trying to act nonchalant, they were seeking alternative frames to explain why the officer was behaving in an unsafe manner.

Reframing   A few participants stated after the first segment that what they saw was a normal traffic stop, attributing the driver's behavior to normal nervousness. As the video revealed more information, they had to discard this earlier frame and seek an alternative frame, usually by focusing on the passenger's behavior, which provided clearer information about the situation than did the driver's behavior. This experienced participant was seeking to elaborate the frame after Segment 1, "He [the driver] said that the person who rented the car is not in the car. I'd like to know why he has the car. I'd like to know that it is in fact a rental and who rented it…when it was rented, and from where." In Segment 2, the participant introduces a new frame: "I think they have been smoking [marijuana] and my concern now is that it's not going to be just a traffic violation or warning. Now my concern is going up because I'm probably going to have some contraband in the car." In Segment 3, this participant has enough information to stop seeking a frame, recognizing that the situation as unsafe enough to act even though the situation remained ambiguous: "I'm not going to be too worried about the car anymore, if it's stolen or not or if there is narcotics in there. It's the securing of these two guys that is important."

Segment 4

By Segment 4, all participants reported being reasonably sure the driver was hiding something illegal and that he was looking for either a way to escape or a way to harm the officer. In this segment of the video, the passenger pulled a gun, which resulted in the officer shooting the passenger. Throughout this segment, the officer and a backup officer were acting in direct response to the life-threatening actions of the passenger. Participants mostly made evaluative comments regarding the officers' actions and both novice and experienced officers provided assessment of the driver's actions in this segment. Most participants reported that the driver's actions indicated that he was thinking about grabbing the backup officer's gun. This segment was highly emotional for many participants, who sensed how mortally high the danger was for the backup officer, thus this was the focus of their attention.

Conclusion

The experienced/novice differences found in this study are consistent with the findings of other expertise research, but this work adds contribution by using the data/frame model to demonstrate the activities these decision makers engage in during sensemaking. Similar to prior research, the experienced participants in this study provided descriptions of the event that were more elaborate; they provided and interpreted more cues, and made more interpretations and predictions about what was, or would, occur (Klein et al., 2002; Phillips et al., 2004). This research used a methodology that allowed for analysis of the activities participants used to formulate these interpretations and reach decisions, and demonstrated how more and less experienced participants made sense of a high-stakes situation as it unfolded. Analysis revealed the types of cues police officers attend to during traffic-stop situations and revealed a variety of ways they interpret these situations.

Most naturalistic decision-making research focuses on how decision makers assess situations containing non-human cues. This study identified the behavioral cues decision makers use to form decisions. This study extends prior research on deceit detection by evaluating how police officers utilize behavioral cues within a contextually rich real-life scenario. This has important implications for deception researchers. Regardless of which cues are reliable indicators of deception according to laboratory research, police are using these behavioral cues to make sense of evolving situations and they take action based on the behaviors they observe. Participants reported that behavioral cues were important indicators of criminal activity and they relied on these cues to determine the internal state and motives of the both the driver and passenger. If deception researchers utilize naturalistic decision-making methodologies, perhaps they will gain a more complete understanding of human ability to detect deception.

The results presented here highlight the need for training in how to read human behavior in high-stakes situations. Vrij et al. (2004) demonstrated that it is possible to improve ability to detect deception by teaching people to focus on reliable deception cues. The study presented in this chapter demonstrated experience differences in how participants interpret behavioral cues to obtain situational awareness and determine courses of action. These findings indicate that training should go beyond instruction about reliable deception cues and incorporate real-world conditions that account for the entire decision-making process.

Finally, this study used a methodology that allowed for a demonstration of the data/frame model of sensemaking by showing videotaped situations to obtain decision-maker knowledge. Dominquez and colleagues (2001; 2004) first used this method to demonstrate expert/novice differences in surgeons. Using video instead of simulations provides a low-cost method for collecting decision-making data from domain experts and allows researchers to gather information from a variety of decision makers about the same situation. All the police officers in this study conduct traffic stops, thus they all had preexisting frames built from their traffic-stop experiences. The traffic-stop videotape provided each officer with data they could use to select and adjust their frames. The questions they answered provided insight into which sensemaking activities they engaged in. Videotapes of actual incidents provide domain experts with a typical situation that is rich in cues, ambiguity, and uncertainty, where they can apply their knowledge and provide researchers with a broad understanding of sensemaking in that domain.

References

Barber, R. and Marsh, D. (producers) (September 1995). *Georgia State Trooper* [Video tape] (Available from "In the Line of Fire," 1-800-462-5232).

Cohen, M.S., Freeman, J.T., and Wolf, S. (1996). "Metarecognition in time-stressed decision making: Recognizing, critiquing and correcting." *Human Factors*, 38: 206–19.

Crandall, B., Klein, G., and Hoffman, R.R. (2006). *Working Minds: A Practitioner's Guide to Cognitive Task Analysis*. Cambridge, MA: The MIT Press.

DePaulo, B.M., Lindsay, J.J., Malone, B.E., Muhlenbruck, L., Charlton, K., and Cooper, H. (2003). "Cues to deception." *Psychological Bulletin*, 129: 74–118.

Dominguez, C.O. (2001). "Expertise in laparoscopic surgery: Anticipation and affordances." In E. Salas and G. Klein (eds), *Linking Expertise and Naturalistic Decision Making*. Mahwah, NJ: Lawrence Erlbaum Associates (pp. 287–301).

Dominguez, C.O., Flach, J.M., McDermott, P.L., McKellar, D.M., and Dunn, M. (2004). "The conversion decision in laparoscopic surgery: Knowing your limits and limiting your risks." In K. Smith, J. Shanteau, and P. Johnson (eds), *Psychological Investigations of Competence in Decision Making*. New York: Cambridge University Press (pp. 7–39).

Dreyfus, H.L. and Dreyfus, S.E. (1986). *Mind Over Machine: The Power of Human Intuition and Expertise in the Era of the Computer*. New York, NY: The Free Press.

Ekman, P. and O'Sullivan, M. (1991). "Who can catch a liar?" *American Psychologist*, 46: 913–20.

Ekman, P., O'Sullivan, M., and Frank, M.G. (1999). "A few can catch a liar." *Psychological Science*, 10: 263–6.

Endsley, M.R. (1995). "Toward a theory of situation awareness in dynamic systems." *Human Factors*, 37: 32–64.

Ericsson, K.A. and Charness, N. (1994). "Expert Performance." *American Psychologist*, 49: 725–47.

Ericsson, K.A., Krampe, R.T., and Tesch-Römer, C. (1993). "The role of deliberate practice in the acquisition of expert performance." *Psychological Review*, 100: 363–406.

Ericsson, K.A. and Smith, J. (1991). *Toward a General Theory of Expertise*. New York: Cambridge University Press.

Flin, R. (1996). *Sitting in the Hot Seat.* West Sussex, England: John Wiley and Sons, Ltd.

Forsythe, C. (2004). "The future of simulation technology for law enforcement." *F.B.I. Law Enforcement Bulletin,* 73 (1 January): 19–23.

Goodrich, M.A., Sterling, W.C., and Boer, E.R. (2000). "Satisficing revisited." *Minds and Machines,* 10: 79–110.

Hunter, K.O., Hart, W.E., and Forsythe, C. (2000). *A Naturalistic Decision Making Model for Simulated Human Combatants* (Report No. SAND2000-0974). Albuquerque, NM: Sandia National Laboratories.

Klein, G. (1998). *Sources of Power.* Cambridge, MA: The MIT Press.

Klein, G., Phillips, J.K., Battaglia, D.A., Wiggins, S.L., and Ross, K.G. (2002). "Focus: A model of sensemaking." (Interim Report—Year 1. Prepared under Contract 1435-01-01-CT-31161 for the US Army Research Institute for the Behavioral and Social Sciences, Alexandria, VA). Fairborn, OH: Klein Associates.

Klein, G., Phillips, J.K., Rall, E., and Peluso, D.A. (2007). "A data/frame theory of sensemaking." In R.R. Hoffman (ed.), *Expertise Out of Context: Proceedings of the 6th International Conference on Naturalistic Decision Making.* Mahwah, NJ: Lawrence Erlbaum Associates.

Mann, S., Vrij, A., and Bull, R. (2004). "Detecting True Lies: Police officers' ability to detect suspects' lies." *Journal of Applied Psychology,* 89: 137–49.

Phillips, J.K. and Battaglia, D.A. (2003). "Instructional methods for training sensemaking skills." *Proceedings of the Interservice/Industry Training, Simulation, and Education Conference 2003.*

Phillips, J.K., Klein, G., and Sieck, W.R. (2004). "Expertise in judgment and decision making: A case for training intuitive decision skills." In D.J. Koehler and N. Harvey (eds), *Blackwell Handbook of Judgment and Decision Making.* Victoria, Australia: Blackwell Publishing (pp. 297–315).

Pliske, R. and Klein, G. (2003). "The naturalistic decision-making perspective." In S.L. Schneider and J. Shanteau (eds), *Emerging Perspectives on Judgment and Decision Research.* New York: Cambridge University Press (pp. 559–85).

Randel, J.M. and Pugh, H.L. (1996). "Differences in expert and novice situation awareness in naturalistic decision making." *International Journal of Human-Computer Study,* 45: 579–97.

Ross, K.G., Phillips, J.K., Klein, G., and Cohn, J. (2005). *Creating Expertise: A Framework to Guide Technology-based Training* (Technical Report Contract M67854-04-C-8035, Office of Naval Research).

Shanteau, J. (1992). "Competence in experts: The role of task characteristics." *Organizational Behavior and Human Decision Processes,* 53: 252–66.

Strömwall, L.A. and Granhag, P.A. (2003). "How to detect deception? Arresting the beliefs of police officers, prosecutors and judges." *Psychology, Crime and Law,* 9: 19–36.

Vicente, K.J. and Wang, J.H. (1998). "An ecological theory of expertise effects in memory recall." *Psychological Review,* 105: 33–57.

Vrij, A., Evans, H., Akehurst, L., and Mann, S. (2004). "Rapid judgments in assessing verbal and nonverbal cues: Their potential for deception researchers and lie detection." *Applied Cognitive Psychology,* 18: 283–96.

Vrij, A. and Mann, S. (2005). "Police use of nonverbal behavior as indicators of deception." In R.E. Raggio and R.S. Feldman (eds), *Applications of Nonverbal Communication*. Mahwah, NJ: Lawrence Erlbaum Associates (pp. 63–94).

Weick, K.E. (2005). *Sensemaking in Organizations*. Thousand Oaks, CA: Sage Publications.

Zimmerman, L.A. (2006). "Law enforcement decision making during critical incidents: A three-pronged approach to understanding and enhancing law enforcement decision processes." Unpublished doctoral dissertation, University of Texas at El Paso.

Appendix 7.1

Video Description

Segment 1 (1:07 min)   A Georgia State Trooper pulls over a driver on a busy interstate highway. He walks up to the driver's side window and asks the driver to step out of the car. The driver follows the officer as they walk to an area between the driver's vehicle and the officer's patrol car (in front of the camera). The officer asks the driver for his driver's license and places it on his notebook to read it. The officer asks the driver if the car is his and the driver states that the vehicle is a rental and that the person who rented the car is not in the car. The officer explains that the driver made an illegal lane change and cut in front of the officer. The driver asks if he is going to get a ticket. The officer tells the driver "No," that he is only going to write him a warning. The officer stands squarely in front of the driver and looks down multiple times. The driver crosses and uncrosses his arms many times and takes his hands in and out of his pockets. A passenger sits in the front seat of the car throughout this segment.

Segment 2 (0:41 min)   The officer asks the driver, "You haven't got a gun on you or anything have you?" The driver says "No." The officer tells the driver to stay standing where he is while the officer goes and talks to the passenger. The officer asks the passenger to roll down his window and the passenger refuses, instead asking "Why?" The officer responds, "Because I'd like to talk to you. Let me see the rental agreement." As the passenger searches for the rental agreement, the officer leans on the doorframe of the car, looking inside the car and notices rolling papers. The passenger denies they are his, claiming that he does not smoke. The passenger then hands the rental car papers to the officer through a narrow opening in the window. While the officer talks to the passenger, the driver fidgets with a piece of paper in his hand, and alternates between watching the officer and passenger and reading the piece of paper.

Segment 3 (0:44 min)   The officer returns to the driver with the rental agreement. According to the narrative, the driver states that he thought the passenger rented the car. The narrative also states that the officer discovers that the car was not supposed to be taken out of Florida and that he has a strong suspicion that drugs are aboard.

The officer questions the driver, then goes back, and questions the passenger about where they are going. Their stories conflict about where they are going, but the stated purpose of their trip is consistent. While the officer talks to the passenger, the driver puts the piece of paper in his pocket, takes it out, reads it, puts it back, and takes it out again.

Segment 4 (2:33 min) Between the third and fourth segment the officer calls for backup. This segment starts when the backup officer arrives. The officer tells the backup officer to watch the driver while he goes to talk to the passenger. When he approaches the passenger's side window, the passenger informs the officer that he has a gun. The officer instantly pulls his gun and aims it at the passenger telling the passenger multiple times to "Drop the gun." The officer threatens to shoot the passenger if he does not drop the gun and get out of the car. After repeating the commands multiple times, the officer opens the passenger door and continues with his commands. The officer then switches his gun to his left hand and reaches in with his right hand to pull the passenger out. The video does not allow the viewer to see what goes on inside the car, but as the officer is reaching in his gun goes off and the passenger is shot. The passenger falls out of the car groaning as the officer tells the backup officer to call for an ambulance. While this was going on, the backup officer has the driver put his hands on the hood of the patrol car and spreads his legs. When the officer pulls his gun, so does the backup officer. The backup officer leaves the driver's side and walks to the driver's side window, opposite the passenger. He then walks back toward the driver, alternately raising and lowering his gun. He goes back and forth multiple times. The driver watches the officer and the backup, occasionally taking his hands of the hood of the car. The officer instructs the backup officer to call for more backup. The backup officer pulled out his radio, with his gun still in his hand, while standing next to the driver. He then walked back toward the driver's car. After the passenger is shot, the backup officer walks back to the driver, holsters his gun, and puts handcuffs on the driver.

Appendix 7.2

Questions Asked at Each Decision Point

1. What do you think is going on here? What specific factors (or cues) are leading to this interpretation?

2. Do you have any concerns at this time? What are they?

3. What errors would inexperienced officers be likely to make in this situation? Are there any cues they might miss?

4. Can you give me a numerical rating, from 1 to 7, of your comfort level with continuing as this officer is continuing, using the anchored scale shown here?

5. If I told you that the officer decided to call for a roll-by at this time, would you think that was a reasonable course of action? Why or why not?

6. Given that your overall goal is to complete this call without incident, what are your short-term objectives at this time?

7. Are there any alternative courses of action that might work? Would you do anything different than this officer?

8. Are there any other cues you see that are influencing your actions that you have not mentioned yet? Are there cues that you expect to see that are not present? As a supervising officer, would you be satisfied that all actions taken thus far are acceptable?

Chapter 8

Cultural Variations in Mental Models of Collaborative Decision Making

Anna P. McHugh, Jennifer L. Smith, and Winston R. Sieck

Introduction

This chapter describes an initial research effort to characterize mental models of collaborative decision making across diverse culture groups. The purpose of this work is to highlight the areas of collaborative decision making where members of multinational teams commonly have disconnects, so that interventions for developing "hybrid" cultures in multinational teams can be appropriately targeted. This research comes at a time when we are experiencing a strong trend toward using multinational teams to tackle highly complex problems, particularly as globalization gives way to international strategic alliances in both commercial and governmental settings (Earley and Gibson, 2002; Maznevski and Peterson, 1997; Peterson et al., 2003; Shapiro, Von Glinow, and Cheng, 2005). Coupled with this trend is an accumulating recognition that cross-cultural differences in norms, values, and cognitive patterns influence the decision-making process (for example, Choi, Choi, and Norenzayan, 2004; Peterson et al., 2003; Yates and Lee, 1996), and that these differences may influence the effectiveness of multinational collaborative decision making (Granrose and Oskamp, 1997; Ilgen, LePine, and Hollenback, 1997; Klein and McHugh, 2005). Given such developments, understanding the dynamics of multinational team decision making and striving to enhance the performance of these teams has taken on increased significance throughout the global community.

Multinational decision-making teams are frequently assembled due to their expected value and competitive edge. Scholars and organizational leaders often purport that members of different nationalities bring a variety of perspectives and skills that can enhance creativity and lead to a broader array of solutions than would be possible in a culturally homogenous team (Cox, 1993; Joshi, Labianca, and Caligiuri, 2002; Maznevski, 1994). Yet, this benefit is frequently not realized (Thomas, 1999), creating a "diversity paradox." The diverse knowledge and perspectives that can yield creative ideas and solutions can also contribute to a unique set of challenges for collaborative decision making. In many cases, without any intervention or anticipation of these challenges, the decision making and ultimate performance of the team falls far below expectations (Distefano and Maznevski, 2000; Thomas, 1999).

In attempting to manage the "diversity paradox" and facilitate the achievement of high levels of multinational decision-making team performance, practitioners might consider an array of approaches or interventions. One approach is to turn to prescriptive models of effective teamwork for guidance about the types of teamwork attitudes and behaviors that organizations should foster in their teams. Several models have been developed that incorporate key behaviors or team "competencies" that have been found to contribute to effective team performance. For example, Salas and his colleagues have developed a set of core team competencies known as the "Big Five" (Salas, Sims, and Burke, 2005), which include attitudes and behaviors such as Leadership, Back-up Behavior, Performance Monitoring, and others. Fleischman and Zaccaro (1992) have developed a similar taxonomy of key team competencies that offers an expansive set of behaviors and attitudes central to successful teamwork. Such models could provide a starting point for supporting and training essential team behaviors in multinational decision-making teams.

The primary attraction to this approach is that the prescriptive teamwork models and taxonomies alluded to above are grounded in over two decades of research on team effectiveness. The researchers have based their taxonomies on studies of a variety of teams within diverse contexts. This approach also has drawbacks, however. A key disadvantage is that the approach is not sensitive to cultural differences in behavioral tendencies and conceptions of teamwork (Klein and McHugh, 2002). The behaviors and attitudes prescribed in these models are based primarily on research with Western cultural groups. Given this reality, it is quite possible that the attitudes and behaviors prescribed in these taxonomies may be incompatible with a given team member's cultural norms, values, and cognitive styles.

A second approach that might be considered for enhancing the effectiveness of multinational decision-making teams is one that sensitizes team members to a small set of cultural dimensions to increase their awareness of each others' cultural tendencies across a variety of situations. This approach is grounded in the theoretical premise that one can make reasonable predictions about differences in cognition and behavior in a wide range of specific contexts (for example, teamwork), based on an understanding of a few domain-general cultural values. Programs developed from this theoretical premise rely heavily on the work of Hofstede (1980; 2001), who in a seminal research study, identified a core set of dimensions for describing differences across cultures. These dimensions include Individualism–Collectivism, Power Distance, Uncertainty Avoidance, and Masculinity–Femininity. A benefit to a dimension-based approach is that it provides collaborators with increased sensitivity about some of their key similarities and differences due to cultural background. When specific personal information about an individual or group is not available, it can provide a "best guess" to some potential ways in which the people might differ in the way they think and behave. A risk of this approach, however, lies in moving from general national differences to more specific contexts (Atran, Medin, and Ross, 2005), such as collaborative decision making. There is some evidence that cultural dimensions, such as those of Hofstede (1980; 1981), may not be as useful as one might expect to predict cognitive or social patterns within the context of a specific situation (Sieck, Smith, and McHugh, 2006; Tinsley and Brett, 2001). When considering the cultural dimension of independence–interdependence (Markus and Kitayama, 1991), for example, one might expect that higher levels of interdependence would predict higher levels of "team orientation," as defined by Salas, Sims, and Burke (2005). Instead, Sieck et al. (2006) found that participants from high-independence cultures placed greater value on team orientation than those from high-interdependent cultures. Sieck et al. also found evidence of movement toward convergence on teamwork values across cultures. Such convergence also suggests a reduction in the value of using domain-general cultural dimensions to predict teamwork and behaviors in any given team situation. Furthermore, the domain-general approach is too far removed from the practical question that most multinationals face of "How are we going to function to accomplish our objectives on this team?"

A promising alternative to the two approaches described above is to help multinational decision-making teams improve their collaboration through facilitating development of a hybrid team culture (Earley and Mosakowski, 2000). A hybrid culture is a shared and emergent culture that occurs when highly diverse teams develop and enact a new set of patterns, shared meanings, norms for operations, and expectations about team processes (ibid., 2000). The assumption is that members of a newly formed team determine their own set of patterns and processes for accomplishing the work within the specific context in which they are working. The co-created "hybrid" culture serves as a basis for facilitating team-member interaction and communication that should lead to improved collaborative decision-making performance.

The development of a hybrid culture depends, at least in part, on team members resolving disconnects in their mental models of collaboration and converging on a shared mental model. Mental models are explanations about how things work that enable people to form expectations and understanding (Gentner and Stevens, 1983; Rouse and Morris, 1986). Klein et al. (2003) identified mental models as macrocognitive processes that are critical for supporting the full spectrum of macrocognitive functions, including decision making. Although Klein et al.'s focus was presumably on mental models of physical domain knowledge that supports decision making, people also possess mental models about the nature of psychological processes (for example, Gopnik and Wellman, 1994; Van Boven and Thompson, 2003). In particular, people possess mental models about the critical macrocognitive function of decision making. Mental models are naturally domain specific since they pertain to the workings of particular artifacts and natural processes. Furthermore, mental models can vary across cultures in ways that are constrained only by the domain itself and any cognitive universals that ground shared understanding across humanity (Hirschfield and Gelman, 1994).

Given the inherent linkage between convergence of mental models and development of a hybrid team culture, a first step towards accelerating the development of hybrid cultures is to characterize cultural differences in mental models of collaboration. Though there are certainly a variety of mental models related to collaboration, the mental models of interest in the current study are those related to the collaborative decision-making process. The specific aim of the current research was, thus, to uncover the salient disconnects in mental models of collaborative decision making among people from diverse cultures.

The remainder of this chapter will describe a study that explored the common points of disconnect in the way members of a variety of cultures understand collaborative decision making. Through in-depth interviews, we collected fragments of mental models and experiences from individuals from a diverse set of cultures.

We analyzed the qualitative responses in order to uncover aspects of respondents' mental models of collaborative decision making where cultural disconnects arise. We did not initiate this study with specific hypotheses. Instead, we sought to explore concepts of collaborative decision making across cultures and lay the foundation for a set of hypotheses and additional research.

The various themes that surfaced in the data are captured in a comprehensive and cultural-general mental model of collaborative decision making, as shown in Figure 8.1. The model suggests a process with aspects that are shared widely across cultures, including divergence, convergence, deciding, gaining commitment to the decision, executing the decision, and adapting to change. Yet, as will be seen, the ways in which these aspects of the process are expected to unfold vary considerably across cultures. Thus, the elements contained in the general model of collaborative decision making in Figure 8.1 represent the points at which incompatible ideas would need to be addressed within a multinational team in order to build a hybrid team culture.

As Figure 8.1 shows, *divergence-convergence* in the comprehensive mental model of collaborative decision making refers to the process by which collaborators present alternative ideas and viewpoints for consideration by the team and then winnow those ideas down to a smaller number of possibilities. The *decision point* refers to the point at which commitment to an actual course of action (or set of actions) is stated by the key decision maker(s), and *commitment* refers to the process by which the broader team members come to endorse that decision. *Execution* refers to the process by which team members implement the chosen course of action. Finally, the *opportunity for change* refers to the occasion in which the team members face new information or unexpected circumstances that may lead them to re-examine and/or modify their decision.

In addition to the core aspects of the collaborative decision-making model, there is a set of key social-context variables that emerged from these data and from previous work as important to consider in collaborative decision making and that differ across

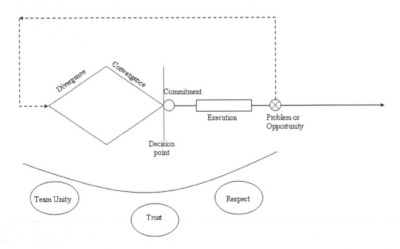

Figure 8.1   Comprehensive mental model of collaborative decision making

cultures. These variables are team unity, trust, and respect. Team unity refers to the degree to which the team members feel a sense of oneness with their teammates and see themselves as a collective unit, rather than as a group of individuals. Trust refers to the sense of confidence that team members have in one another due to close personal bonds. And respect refers to the extent to which team members endeavor to uphold one another's sense of honor and dignity. The data we have gathered, along with related prior literature, will be described together in the results section.

Method

Our research team conducted interviews with 61 people from a variety of nations. The nations represented in the study included the US, China, Korea, India, Japan, Taiwan, Chile, Brazil, and Italy. Given the small sample of interviewees from the Latin countries (Chile, Brazil, and Italy), the data presented will primarily represent descriptions of US, East Asian (China, Japan, Korea, and Taiwan), and Asian Indian teams.

The interviewees represented a mix of vocational backgrounds. Some worked for large corporations, others for small, privately owned companies. Some were graduate students in business or healthcare. Others were members of their country's military services. The interviews ranged from one-and-a-half to two hours in length, and the questions posed consisted of topics ranging from how decisions are made in a team, to how conflict is addressed, to the role that the leader plays, to how the teams deal with change. From each of the main categories of questions, we sought to extract mental models of collaborative decision making by eliciting specific real-lived examples of team experiences within the interviewees' home culture, and their beliefs about how practices actually occur within teams in their home culture. In some of the interviews, the interviewees also offered their perspective on differences in teamwork in other cultures based on their experiences in multinational settings. In some cases, these data are presented to illustrate the contrast between perceived differences and actual differences in mental models of collaborative decision making.

Results

As described above, a general model of the collaborative decision-making process emerged from the data. Highlighted in the model are core elements of the decision-making process that were shared across all the cultures interviewed, but that vary in the means by which they are accomplished. These areas of variation are further supported by the extant literature on collaborative decision making and cultural differences. Although variations within cultures certainly exist, it is clear that certain patterns are more common in some cultures than in others. The areas highlighted in the model represent points of likely disconnect and/or misunderstanding in multinational teams. Thus, they should be targeted as areas of negotiation among multinational team members in the beginning phases of the team's formation as part of developing a hybrid team culture. Variations along the core points in the collaborative decision-making process are described below. There are certainly overlaps among the data presented. This is to be expected given the qualitative nature of the data, and given the interrelatedness among the various elements of the collaborative decision-making process. With that in mind, each section presents interviewee quotes that we believe most effectively illustrate cultural variations in each core stage within the collaborative decision-making process.

Divergence/Convergence

Both conflict management and cross-cultural scholars have consistently indicated a pattern of conflict avoidance and maintenance of harmony in East Asian cultures, due to a concern for preserving relationships with others (Obhuchi and Takahasi, 1994; Obuchi, Fukushima, and Tedeschi, 1999; Ting-Toomey, 1994; Tjosvold, Hui and Sun, 2004). Upon considering this pattern within the context of collaborative decision making, one might infer that some cultural groups altogether avoid a divergence–convergence phase of collaborative decision-making, and thereby refrain from arguing and expressing uncommon or unpopular points of view. As indicated in our initial set of findings, however, respondents from all the cultures represented in our data set do conceptualize divergence and convergence processes as part of their mental models of collaborative decision making. It is the way in which the phase of divergence–convergence is realized that can be quite distinct, depending on the values and cultural norms that are in play.

Our initial data indicate that a key difference in the way teams in different cultures experience divergence and convergence is the extent to which the process is transparent. It is not uncommon for team members in Western cultures, such as the US, to express divergent opinions and hash out conflicting ideas during public forums such as face-to-face team meetings, teleconferences, or via collaborative technologies. Expressing disagreement, even if sometimes heated, is often viewed as a productive means of reaching an optimal decision and attempting to achieve consensus (Jehn, 1995; Jehn, Northcraft, and Neale, 1999). A US interviewee supported this stance as he explained: "We keep debating and hashing things out until it comes more clear which direction we should take…everyone should feel heard. The group process should be open and transparent to everyone—no backroom power deals. Team members should be aware of where their loyalties lay within the team."

This pattern of public divergence, however, does not appear to be shared among cultures who place a high premium on "saving face" and upholding the honor of oneself and others. Members of East Asian and Asian Indian cultures, for example, tend to avoid the public conflict, disagreement, or criticism that may be considered hallmarks of the divergence phase of collaborative decision making in the US. It is believed that such open conflict can undermine oneself, one's superiors, or fellow team members and cause these individuals to feel devalued in front of their colleagues. It is an uncomfortable and uncommon process to debate and challenge one another's ideas in public. As a Chinese interviewee expressed: "In our teams in China, we're more likely to talk about different ideas in private. We don't like conflict in public. Even though it's important to get out the counter ideas in a team, many people in

China don't feel comfortable with that." Similarly, an Indian interviewee stated: "If I disagree with someone on my team, I express my disagreement. But I don't do it at the actual meeting. I disagree privately."

As an alternative to public disagreement and heated discussion, our initial data indicate that divergence in East Asian and Asian Indian cultures proceeds in a private and non-transparent manner. Private conversations take the place of public discussion, ensuring that alternative viewpoints and perspectives of individual team members are taken into consideration, while remaining anonymous to fellow team members. This enables individuals who might have viewpoints that are unaligned with their fellow team members or with their manager's ultimate decision to maintain face. As stated by a Chinese interviewee: "As team leader, I like everyone to speak up. But I know most will not. So I interview them privately to learn more about their opinions...people think that if they speak up in front of the group there may be some conflict." Similarly, an Indian interviewee explained: "We [the team members] present our ideas to the leader before the meeting, and the leader then presents the ideas during the meeting as 'These are two team ideas, what do you all think?' The leader doesn't attach names to the ideas..."

These different patterns in divergence–convergence have the potential to create friction and misunderstanding in culturally heterogeneous teams. Collaborators who are accustomed to a more private process might find open divergence disconcerting and possibly insensitive. On the other hand, those accustomed to a more private process may perceive some cultural groups as restricting an open exchange of ideas and opinions. Such a perception is exemplified in the statement made by an American officer who worked with Koreans and Thais: "When we were working together, neither the Koreans nor the Thais had the free-flowing exchange of opinions that the Americans had." Yet, as interviews with several non-US interviewees suggest, this may not have been the complete story. The actual difference may not lie in whether or not divergent ideas and opinions are exchanged, but the manner in which these divergent perspectives are exchanged.

The process of divergence–convergence is particularly unique within Japanese collaborative decision making. It is well-documented that when decisions need to be made or new plans developed, Japanese teams and organizations follow a semi-structured consensus-building process known as *Nemawashi* (Fetters, 1995; Nishiyama, 1999). In this process, input is sought from all parties who will be affected by the decision. The intent is to elicit different perspectives and conflicting ideas, and to iron out those differences and ultimately reach consensus prior to announcing the decision in a formal and public meeting. Due to the goal of obtaining input from all affected parties and ultimately achieving consensus, *nemawashi* can be quite time-consuming (Fetters, 1995). Where Western teams may find themselves taking time to iron out differences during decision execution, Japanese teams take time to sort out these differences and achieve a level of detail and precision ahead of time. That way, once a decision is reached, the course of action can be executed rapidly and smoothly (Lazer, Murata, and Kosaka, 1985). Clearly, such markedly different approaches to divergence and convergence can lead to points of disconnect in multinational settings. One example of such differences is described by a Japanese military officer:

If we skip *nemawashi*, we can't proceed with the project. We have to take time to decide. This was a problem when the US helicopter recently crashed into Okinawa University. Luckily nobody was killed. But it was decided that the airport needed to be moved so things like that wouldn't happen again. But we felt like we weren't being given enough time by the US to make the decision about where the new air base would go. These decisions take time.

Decision Point

The cultural groups involved in this research hold a shared conception that the process of divergence and convergence ultimately steers teams towards a decision point so that they can move forward in implementing the chosen course of action. Though a certain level of cross-cultural variability exists, our interview data indicate some important commonalities in this phase as well. In nearly every case, the team manager is ultimately the decision maker for his/her team, regardless of the level of input sought from team members up to the decision point. As a Chinese interviewee explained: "The leader of our team is expected to listen to every team member's opinion; but it is his responsibility to make the final decision." Similarly, an interviewee from India purported: "In India, everyone has to give input on the decision. But ultimately the leader has the final say."

This perspective is not much different from how Westerners think about the decision point, as exemplified by a US interviewee: "The leader listens to all the team members' ideas, but ultimately he brings all these ideas to a conclusion." Acknowledgement of this similarity is sometimes neglected when both researchers and practitioners describe differences in decision making across cultures. It is not uncommon to hear Westerners refer to their own team decision-making process as "democratic" while referring to other cultures as highly "autocratic." For example, an American who observed a team of Jordanians—who exhibit some similarities in team process to Asians—stated: "When a decision was made, it was never made by the team, never by consensus. It was always autocratic. The leader would say how it's supposed to be done. He/she never engaged the team members to make decisions."

Yet our data from this study indicate that this American view of non-Western teamwork may be an oversimplification. It is not surprising that Westerners would interpret the decisions in some cultures as autocratic, given that the input and exchange of ideas prior to the decision point usually takes place in private settings rather than in open forums. Westerners working on a multinational team might witness a leader make a decision, without any visible involvement from other members of the team. Yet, the individuals we interviewed from cultures that are often described as autocratic (Korea, Japan, China, and India) described a more collaborative approach. The team manager may ultimately make the decision in the team, but his decision is informed by private discussions in which he seeks input from the various team members.

A subtle difference, however, is that due to the transparent process leading up to the decision point in Western cultures, team members are generally aware of which other team members most strongly influenced the leader's decision, and generally have a good sense of who the leader's decision aligned with or contradicted. This is far less likely in East Asian cultures, where the leader makes the decision privately after soliciting input from the team members. He then uses a public forum such as a team meeting as an opportunity to announce the decision.

Variation among cultures also exists in the tempo of the decision making. In the US, being "decisive" and moving to action as rapidly as possible is generally valued and encouraged (Lewis, 1999). The decision in a US team might be reached sooner, relative to an East Asian team. As a US interviewee explained: "We don't want to spend too much time making a decision, because it's probably going to change." In other cultures, however, great importance is placed on investing time to build consensus among team members prior to making the decision—such as in the Japanese process of *nemawashi*. This may result in a slower decision-making tempo, or a longer period of time to reach the decision point.

Commitment

A shared belief across all the cultural groups we interviewed is that commitment among team members is needed in order to move forward with implementation once a decision is made. In some cases, there may be an explicit process used to gain commitment to a decision. For example, a Taiwanese interviewee explained:
"I as the team leader make the decision. I then have a meeting with the team for them to endorse what I do, to get agreement and consensus on my decisions, to avoid conflict later."

But our data suggest that in many Asian teams, commitment of all parties to a decision is assumed. This may be partly due to the deliberative process that occurs during the divergence and convergence phases, giving all key stakeholders the opportunity to express their perspectives and ideas. Given that there is a great investment in building consensus up-front, there is a smaller likelihood that team members will be surprised or have major criticisms or aversions to the plan or decision. Furthermore, given the hierarchical nature of certain cultures that rate high on Power Distance (Hofstede, 1980; 2001), such as the Asian cultures in our sample, team members are simply expected to be committed to the leader's final decision, and to work vigorously toward its implementation: "In India, even if the boss is incorrect, you don't say anything. You just go with it. It's a more authoritative relationship. The leader ultimately has the final say. Even if people don't agree, they sometimes hold back and don't make that known." Another Indian interviewee claimed: "[Once a decision is announced]…It's expected that you obey the project driver." Similarly, a Chinese interviewee explained: "As a team member, you must accept and support the leader's decision, even if you do not agree with it and even if it leads to a bad outcome. Once the leader makes his final decision, all members must follow his idea, even if it seems wrong."

In some cultures such as the US, however, commitment to a decision may not be assumed. In fact, the team may move into implementation knowing full well that particular members of the team may not be on board or committed to the plan. This lack of commitment to a decision can result in more time spent during implementation as team members try to iron out some disagreements and explicitly work to get buy-in to the decided course of action.

Execution

Another culturally common aspect of mental models of collaborative decision making is the execution of a decision. As mentioned previously, certain cultural groups such as the Japanese tend to invest quite a bit of time and effort in the divergence and convergence stages, addressing the various ideas and viewpoints of the team members, managing details, and attempting to gain buy-in from all key stakeholders. Thus, once the decision is made, they are prepared to move swiftly to action (Lazer et al., 1985). In contrast, Western teams have the tendency to move to a decision point more quickly—often before every one on the team has "buy in" (Lewis, 1999). It is not uncommon in Western cultures to be encouraged to decide *something*, move to execution, and then use what is learned from the execution process to define or redefine the goals and plans along the way (Klein, 2005). This theory rests on an assumption that goals tend not only to be ill-defined, but undefinable in principle prior to execution. The different emphasis on reaching a decision point and moving to execution quickly can lead to a certain degree of resistance during execution as issues are raised and ironed out by those who have not yet fully bought in to the decision. As described by a US interviewee: "In that team, I didn't have everyone on board with my plan up-front. That caused a lot of headaches when it came time to actually getting the work done. We went through a bunch of fits and starts before we were able to make any real progress."

It appears that there is also substantial variation across cultures in mental models of how decision execution proceeds. One variable that surfaced in our data is the level of involvement and situational awareness of the leader. Though much of this is certainly dependent upon the personality of the leader, our data indicate that there may be a general tendency for leaders of US-based teams to delegate and take a more "hands-off" approach during the execution phase. Check-ins at various points throughout implementation can be enough to satisfy the leader's requirements for situation awareness. As an interviewee from the US explained: "I would delegate the work, and just check in from time-to-time with my subordinates. If I really wanted to be more involved, I'd just do the task myself."

This serves as a contrast to some non-Western cultures, however, whose leaders may maintain a much greater level of involvement and a higher degree of situation awareness of the team's progress during implementation. A Taiwanese interviewee indicated: "The Taiwanese chairman I worked for was very careful about details. He wanted to know everything about every project. I don't see many people like him here in the US who want a hand in everything." Similarly, a Chinese interviewee explained his experience working on a multinational team in the US, led by a Chinese manager. He describes the conflict that can occur if a member of another culture is not accustomed to the close level of monitoring by a team leader: "The Chinese manager would call the US all the time to request details about his work. The US guy interpreted this as distrusting and so he quit. Chinese leaders like lots of information about the details of the team members' work."

Though observational in nature, an American noted a Korean leader's performance monitoring in a multinational military setting: "If there was work to be done, the Korean NCO would stand there to make sure it was done. The American NCO would expect it to be done, and would provide periodic checkups. The Koreans were more hierarchical. They did lots more checking and double-checking."

Wherein some cultures, the leadership tends to conduct a high degree of performance monitoring of the team, in other cultures it is the team members themselves who are suggested to play a more active role in monitoring the performance and progress of the team toward its goal. Team researchers such as Salas et al. (2005), for example, argue that this behavior is critical to effective team performance. The monitoring serves the purpose of checking in on fellow collaborators' progress and identifying gaps and weaknesses so that resources can be shifted appropriately. Our data indicate that performance monitoring in some cultures might serve the additional function of gauging one's performance relative to other team members. Several interviewees from East Asian and Indian cultures acknowledged monitoring others as a function of competition that exists among team members. The competition is generally covert and unacknowledged within the team interactions, but is certainly perceptible to team members. Several of the Chinese and Indian interviewees described this. An excerpt from a Chinese interviewee follows:

> Team members in China compete with each other. But this competition is under the table. People don't talk about it. But it's in everyone's head. If you talk about it out loud, it would offend the honor of the other person. You also have to make it look natural. If you stand out too much, you get targeted. Everyone wants to do better than the other people on the team. So we may check in to see if we're doing better.

Indian interviewees expressed a similar tendency: "In India, team members are very competitive. But it's very covert. People find out from each other things like how far they moved along on their goals. If I ask someone that, they don't give me the true picture of where they are though. And they know I don't give the true picture of where I am either."

The mental models concerning how resources are shifted, in the event of any identified gaps and weaknesses, may also vary across cultures. Our data indicate that in some parts of the world, such as East Asia, the pervasive belief is that the leader typically is the one who identifies the need and establishes who on the team might assist another team member or take over a given task. In other cultures, such as in the US, there tends to be an explicit expectation that team members will assist another teammate or take over a task if needed. Compensating for team members does not have the same meaning in all cultures, however. In cultures such as the Middle East, where a high level of value is placed on maintaining face and preserving one's honor (for example, Feghali, 1997), stepping in for someone and filling their role might be viewed as harming one's honor (Klein and McHugh, 2005).

Opportunity for Change

Inevitably, every team faces situations in which new information is acquired, circumstances change, and the original plan or decision needs to be reexamined. The way in which team members consider how to address such circumstances, may be culturally dependent. Given that certain cultures tend to spend a high level of effort up-front in the divergence and convergence phases, they may have a higher threshold for change given that they have much invested in the ultimate decision made or plan developed. Thus, in the face of new information or changing circumstances, any change in the decision or plan will require moving thoughtfully and cautiously, and will require the investment of additional time and consideration. If such time and resources are not available, maintenance of the existing plan may be the best way forward. For example, recall the description of the Japanese process of *nemawashi* above. Such a deliberate and time-consuming process can mean that changing a plan is not a simple matter. This aspect of a collaborative decision-making mental model is described by a Japanese interviewee:

> We (the Japanese) are not very flexible…we need to keep our course of action once we've decided on it. During our exercise with the US military, the US tended to be very flexible. They would discuss a new COA. We would say "no, we have to take the COA that we already discussed and decided on." The US would say "no, we need to take a new COA." We often couldn't find a compromise. In Japan, if a decision changes too quickly, the lower-level staff can't respond quickly enough. If we do change, we need a lot of time. We don't have the flexibility like the US troops.

Our initial data indicate that some cultures tend to have a higher level of comfort with change and a quicker pace in adapting. This is also supported by previous research (for example, Klein, Klein, and Mumaw, 2001; Klein and McHugh, 2005). This variability may be associated with a smaller investment of time that some cultures make in the initial divergence–convergence phases of the team decision making process. Cultures such as the US tend to spend less time up front and expect that changes will need to be made once the decision is executed. Thus, there is some evidence that they feel more comfortable in adapting plans as new information is acquired or circumstances change. For example, as a Taiwanese interviewee expressed: "Americans will feel more comfortable to change a plan. In my team in the US, if I point out that something is wrong, everyone will say OK and they'll change the plan. In Taiwan if I point out something wrong, it would be difficult to get people to change."

In addition to the relationship between adapting and investing time and resources, there appears to be another variable that affects certain cultural groups' comfort with change. A common theme that was represented in most of our interviews with East Asian cultures was the importance of honor and preserving "face." Such values have been described above as they relate to other aspects of teamwork. This theme was also interwoven in interviewees' discussions about adapting. Some of the interviewees expressed a resistance to change due to concern that it could cause members of the team (particularly the team leader) to lose face, particularly if the change somehow suggested that the original decision or plan was flawed. For example, a Chinese interviewee stated: "A leader would lose face if he reverses a decision. The leader should try to get the best outcome, but he may lose face in the process."

Social Context of Decision Making: Team Unity, Trust, and Respect

The mental models of collaborative decision-making process that have been described throughout this paper are influenced by a variety of contextual factors that can impinge on how the process unfolds. From our interviews, a set of key factors emerged as particularly important to the collaborative decision-making process in certain cultures. These factors constitute unity, trust, and respect.

A sense of trust and unity have been described as important aspects of high-performing teams (for example, Earley and Mosakowski, 2000; Salas et al., 2005). Yet what constitutes a sense of unity and trust in one's co-workers—and the degree of importance placed on these aspects of collaboration—appears to vary across cultures. Knowing collaborators on a personal level is considered as the cornerstone of team unity and trust and viewed as essential to effective collaboration in some cultures. In East Asian, Latin, and Middle Eastern cultures, team members tend to attach a prominent level of importance to developing strong personal bonds with teammates, with the expectation that it will improve the cohesiveness in the team and lead to higher levels of productivity (Gibson and Zellmer-Bruhn, 2001). Such a perspective was clear from the Asians in our sample. As an interviewee from the Korean military explained: "It is critical that each soldier get to know the other soldiers in order to make it through the work. We want our soldiers to become friends so that they can rely on them in times of trouble." Similarly, an interviewee from India described the importance of personal bonds among collaborators, and the barriers that can exist in circumstances where such relationship-building, trust, and unity is not fostered:

> Currently there's a lot of discontent with computer firms that put teams together that don't really know each other. Two of my friends work for IT firms in the US, and have been very upset because they're expected to work on teams with certain people, and they don't really know them. There's much less of a personal relationship than they're used to. It makes it harder to work together.

The manner in which team members from some cultures develop unity and establish trust among one another is often through spending time together outside of the work environment. Teams and their leaders use group dinners, conversations over drinks, and other opportunities for social interaction to enable co-workers to learn about one another on a personal level. A Korean interviewee described the types of social interactions that are typical in his teams: "It is important to them that [team members] get to know the members of their team. They hold lots of informal meetings to get to learn about people, their families, their hobbies…It is not uncommon for them to stay in the office late into the evening talking. They like to drink wine together…" Similarly, a Chinese interviewee explained:

> We always throw a party at the beginning of a team project. We like to eat dinner and drink lots of wine together and get to know one another. Good relationships are important to have in case you have a problem and then your teammates can help. We develop relationships by eating out together, playing games, going out together on weekends. Our families stay together. We make the whole team like a family.

An Indian interviewee also expressed a similar pattern of interaction: "When you work in teams, there are lots of family interactions. You have dinners and drinks together with families. My dad knew everyone I worked with in my HR dept in India."

Whereas Westerners tend to promote a certain degree of separation between personal and professional life, no clear lines exist between professional and personal relationships in other cultures. As stated by a Chinese interviewee:

> There's not so much difference between professional and private life. If you are teammates or colleagues, chances are you're really good friends outside of work. US people tend to show very obvious differences between professional life and private life. There's more focus on relationships in China than there is in the US. In China we expect to know personal things about our teammates…

Team members in Asian cultures expect to be aware of many aspects of one another's life that other cultures, such as the US, might typically view as too personal or private, such as family issues or financial status. For example, a Japanese interviewee explained: "In Japan, the leader needs to know everything about his team. He needs to know about their family, their girlfriends, their hobbies, money issues…If a staff member is having trouble with money, the leader will keep track of his bank card." A Chinese interviewee similarly explained:

> You need to know a lot about the family background of your co-workers. For example, you should know the parents of the members and the members' birthdays. In China, birthday celebrations are very important within the family. On that occasion, it's good if you toast the parents and congratulate them, especially the father. This builds the individual relationships that are important.

A final factor that emerged as particularly salient in the mental models of collaborative decision making in Asian cultures is respect. The importance of respect and enabling oneself and one's co-workers to maintain honor and preserve "face" is a factor that has been woven throughout nearly all aspects of the collaborative decision-making process described earlier. It is an aspect of collaboration that is clearly important to understand when working in multinational teams. It has implications for how ideas and perspective are shared, how disagreements are voiced, whether and how criticism is provided, and how conflict is managed.

Conclusion

The effort described here was designed to tap into participants' mental models of collaborative decision making. Identifying common points of disconnect in culturally diverse team members' mental models provides insight into the areas in which members of multinational teams are likely to suffer from misunderstandings and process losses. It also highlights areas where interventions to facilitate hybrid culture development might be appropriately targeted. The data collected in this effort provide an initial indication of differences in mental models of collaborative decision making between the US and Asian cultures. The research is not without limitations, however. In particular, the data collection and analysis were likely affected by the Western bias of the team of US researchers. Yet while more research is certainly necessary to support and deepen on these findings, the evidence thus far suggests some important trends.

Our findings indicate that despite some important differences in mental models of collaborative decision making across cultural groups, the basic set of core elements comprising these mental models seems to be shared. It is the means by which these various elements are carried out that varies. Failure to understand these differences may lead military leaders, business managers, and members of multinational teams to experience varying degrees of friction and misunderstandings as multinational teams seek to accomplish critical decision making tasks. From an East Asian perspective, for example, one might see US collaborators as rash in their decision making, disrespectful, and generally too quick to cede a developed plan. From a US perspective, one might assume that East Asian collaborative decision making is autocratic, with little opportunity for the exchange of different ideas and viewpoints, or generally resistant to change. Such stories, however, are overly simplistic. The differences in collaborative decision making between Asians and the US are far more complex and nuanced than they may appear on the surface, with values of respect, consensus, and preservation of face playing crucial roles in how the process unfolds, including its degree of transparency.

When multinational teams are temporary and short-lived, as so many of them are in organizational settings today, they must find ways to work together productively early in their existence. Creation of a shared set of mental models of collaborative norms and processes—or a "hybrid" culture—is a recommended means for doing so (for example, Earley and Mosokowsi, 2000). To facilitate the development of a hybrid culture, members of multinational teams must be aware of common points of disconnect in the way they think about the process of collaborative decision making so that they can focus on reconciling discrepant conceptions and building a set of processes and norms to employ in that particular team. Interventions that foster the convergence of mental models and development of a "hybrid" culture will enable culturally heterogeneous teams to position themselves for smoother interactions and higher levels of productivity earlier in their life cycle.

References

Atran, S., Medin, D.L. and Ross, N.O. (2005). "The cultural mind: Environmental decision making and cultural modeling within and across populations." *Psychological Review*, 112(4): 744–76.

Choi, I., Choi, J., and Norenzayan, A. (2004). "Culture and decisions." In D. Koehler, and N. Harvey (eds), *Blackwell Handbook of Judgment and Decision Making*. Malden, MA: Blackwell (pp. 504–24).

Cox, T. (1993). *Cultural Diversity in Organizations: Theory, Research, and Practice*. San Francisco, CA: Berrett-Koehler.

Distefano, J.J. and Maznevski, M.L. (2000). "Creating value with diverse teams in global management." *Organizational Dynamics*, 29(1): 45–63.

Earley, C. and Gibson, C. (2002). *Multinational Work Teams: A New Perspective.* Mahwah, NJ: Lawrence Erlbaum Associates.

Earley, P.C. and Mosakowski, E. (2000). "Creating hybrid team cultures: An empirical test of transnational team functioning." *Academy of Management Journal*, 43: 26–49.

Feghali, E. (1997). "Arab cultural communication patterns." *International Journal of Intercultural Relations*, 21(3): 345–78.

Fetters, M. (1995). "Nemawashi essential for conducting research in Japan." *Social Science and Medicine*, 41(3): 375–81.

Fleishman, E.A. and Zaccaro, S.J. (1992). "Toward a taxonomy of team performance functions." In R.W. Swezey and E. Salas (eds), *Teams: Their Training and Performance*. Norwood, NJ: Ablex (pp. 31–56).

Gentner, D. and Stevens, A.L. (eds) (1983). *Mental Models*. Mahwah, NJ: Lawrence Erlbaum Associates.

Gibson, C. and Zellmer-Bruhn, M. (2001). "Metaphors and meaning: An intercultural analysis of the concepts of teamwork." *Administrative Science Quarterly*, 46(2): 274–303.

Gopnik, A. and Wellman, H.M. (1994). "The theory theory." In L.A. Hirschfield and S.A. Gelman (eds), *Mapping the Mind: Domain Specificity in Cognition and Culture*. Cambridge: Cambridge University Press (pp. 257–93).

Granrose, C., and Oskamp, S. (1997). *Cross-cultural Work Groups*. Thousand Oaks, CA: Sage.

Hirschfield, L. and Gelman, S. (eds) (1994). *Mapping the Mind: Domain Specificity in Cognition and Culture*. New York: Cambridge University Press.

Hofstede, G. (1980). *Culture's Consequences*. Beverly Hills, CA: Sage.

—— (2001). *Culture's Consequences: Comparing Values, Behaviors, Institutions, and Organizations Across Nations* (2nd edn). Beverly Hills, CA: Sage.

Ilgen, D., LePine, J., and Hollenbeck, J. (1997). "Effective decision making in multinational teams." In P. Earley and M. Erez (eds), *New Perspectives on International Industrial/Organizational Psychology*. San Francisco, CA: Jossey-Bass (pp. 377–409).

Jehn, K.A. (1995). "A multimethod examination of the benefits and detriments of intragroup conflict." *Administrative Science Quarterly*, 40: 256–82.

Jehhn, K.A., Northcraft, G., and Neale, M. (1999). "Why differences make a difference: A field study of diversity, conflict and performance in workgroups." *Administrative Science Quarterly*, 44: 741–63.

Joshi, A., Labianca, G., and Caligiuri, P. (2002). "Getting along long distance: Understanding conflict in a multinational team through network analysis." *Journal of World Business*, 37(4): 277–84.

Klein, G., Ross, K.G., Moon, B.M., Klein, D.E., Hoffman, R.R., and Hollnagel, E. (2003). "Macrocognition." *IEEE Intelligent Systems*, 18(3): 81–5.

Klein, H.A. (2004). "Cognition in natural settings: The cultural lens model." In M. Kaplan (ed.), *Cultural Ergonomics: Advances in Human Performance and Cognitive Engineering Research*, 4. Oxford: Elsevier (pp. 249–80).

Klein, H.A., Klein, G., and Mumaw, R.J. (2001). *A Review of Cultural Dimensions Relevant to Aviation Safety* (Final Report Prepared for Boeing Company under General Consultant Agreement 6-1111-10A-0112). Fairborn, OH: Wright State University.

Klein, H.A., and McHugh, A. (2005). "National differences in teamwork." In W. Rouse and K. Boff (eds), *Organizational Simulation*. Hoboken, NJ: Wiley-Interscience (pp. 229–51).

Lazer, W., Murata, S., and Kosaka, H. (1985). "Japanese marketing: Towards a better understanding." *Journal of Marketing*, 49(2): 69–81.

Lewis, R. (1999). *When Cultures Collide: Managing Successfully Across Cultures, a Major New Edition of the Global Guide*. London: Nicholas Brealey.

Maznevski, M.L. (1994). "Understanding our differences: Performance in decision-making groups with diverse members." *Human Relations*, 47(5): 531–52.

Maznevski, M.L. and Peterson, M.F. (1997). "Societal values, social interpretation, and multinational teams." In C.S. Granrose and S. Oskamp (eds), *Cross-cultural Work Groups*. Thousand Oaks, CA: Sage (pp. 61–89).

Nishiyama, K. (1999). *Doing Business with Japan: Successful Strategies for Intercultural Communication*. Honolulu, HI: University of Hawaii.

Nydell, M.K. (2002). *Understanding Arabs: A Guide for Westerners* (3rd edn). Yarmouth, ME: Intercultural Press, Inc.

Obhuchi, K. and Takahasi, Y. (1994). "Cultural styles of conflict management in Japanese and Americans: Passivity, covertness, and effectiveness of strategies." *Journal of Applied Social Psychology*, 24(15): 1345–66.

Obuchi, K., Fukushima, O., and Tedeschi, J. (1999). "Cultural values in conflict management." *Journal of Cross-cultural Psychology*, 30(1): 51–71.

Patai, R. (2002). *The Arab Mind*. Long Island City, NY: Hatherleigh Press.

Peterson, M.F., Miranda, S.M., Smith, P.B., and Haskell, V.M. (2003). "The sociocultural contexts of decision making in organizations." In S.L. Schneider and J. Shanteau (eds), *Emerging Perspective on Judgment and Decision Research*. New York: Cambridge University Press (pp. 512–58).

Rouse, W.B. and Morris, N.M. (1986). "On looking into the black box: Prospects and limits on the search for mental models." *Psychological Bulletin*, 100(3): 349–63.

Salas, E., Guthrie, J.W., Wilson-Donnelly, K.A., Priest, H.A. and Burke, C.S. (2005). "Modeling team performance: The basic ingredients and research needs." In W.B. Rouse and K.R. Boff (eds), *Organizational Simulation*. Hoboken, NJ: John Wiley and Sons (pp. 185–228).

Salas, E., Sims, D.E., and Burke, C.S. (2005). "Is there a 'Big Five' in teamwork." *Small Group Research*, 36(5): 555–99.

Shapiro, D.L., Von Glinow, M.A., and Cheng, J.L. (2005). *Managing Multinational Teams: Global Perspectives*. Oxford: Elsevier/JAI Press.

Sieck, W.R., Smith, J.L., and McHugh, A.P. (2005). *Team Competencies in Multinational Collaboration* (Final Report prepared under Cooperative Agreement DAAD19-01-2-0009 and Subcontract No. 05-R2-SP1-RT1 for the US Army Research Laboratory). Fairborn, OH: Klein Associates.

Thomas, D. (1999). "Cultural diversity and work group effectiveness: An experimental study." *Journal of Cross-Cultural Psychology*, 30: 242–63.

Ting-Toomey, S. (1994). "Managing intercultural conflicts effectively." In L. Samovar and R. Porter (eds), *Intercultural Communication*. Belmont, CA: Wadsworth (pp. 360–72).

Tinsley, C. and Brett, J. (2001). "Managing work place conflict in the US and Hong Kong." *Organizational Behavior and Human Decision Processes*, 85(2): 360–81.

Tjosvold, D., Hui, C.H., and Sun, H. (2004). "Can Chinese discuss conflicts openly? Field and experimental studies of face dynamics in China." *Group Decision and Negotiation*, 13(4): 351–73.

Van Boven, L. and Thompson, L. (2003). "A look into the mind of the negotiator: Mental models in negotiation." *Group Processes and Interpersonal Relations*, 6: 387–404.

Yates, F. and Lee, J. (1996). "Chinese decision making." In Bond, M.H. (ed.), *Handbook of Chinese Psychology* (pp. 338–51). Hong Kong: Oxford University Press.

Chapter 9

Athletes, Murderers, and a Chinese Farmer: Cultural Perspectives on Sensemaking

Mei-Hua Lin and Helen Altman Klein

Introduction

In naturalistic contexts, people often need to make sense of complex and sometimes contradictory information to prepare for effective decision making. This presents a problem for teams. If three people work together in a complex and dynamic situation, each may identify a different "sense" of the situation. This is because sensemaking depends on past experiences and current goals as well as on how each person attends to, selects, categorizes, and integrates available information. These processes are sensitive to individual but also to national differences in cognition.

The increased internationalization of transportation, commerce, communication, and technology has dramatically increased global interdependence. The expansion of social, technical, and economic systems across national borders has made it more common for multicultural teams to work together on planning, coordination, decision making and other complex tasks. Professionals are more likely to be asked to make predictions about what allies, competitors, or adversaries might do. This means professionals need good information. Advances in technology are expanding the availability of information. Finding information on the Internet and elsewhere is not a problem; facing the confusion and contradictions in this information can be overwhelming. These trends contribute to the urgency of understanding national differences in cognition that influence information use and sensemaking in complex and dynamic environments.

Our research group has collected naturalistic data in a variety of international settings. In samples of multinational peacekeeping personnel, commercial pilots, and international students in US universities, we have identified cognitive differences in how people from different groups work together. We have learned that people from different groups sometimes have dramatically different views about how teams should function with regard to monitoring, team orientation, leadership, trust, and communication (Klein and McHugh, 2005). This chapter addresses national differences in cognition that may influence how people search for information and select from the information available; how they handle contradictory and/or changing information, and how they organize and categorize this information to make sense of new situations. Differences in performing macrocognitive tasks, including sensemaking, make team coordination difficult, common ground evasive, and prediction error-prone. Without an understanding of the impact of national differences on sensemaking, multinational interactions will remain haphazard.

We make the case that sensemaking, like other macrocognitive processes, is shaped by cognitive differences that vary over national groups. Understanding these national differences in cognition can improve team functioning during multinational interchanges and allow people to better anticipate how people of other national groups will act. Moreover, an understanding of national differences can do more than support multinational interchanges and improve predictions. Currently, it is primarily Western researchers using Western research paradigms with Western participants who undertake research on macrocognition. Extending our study to national variations in cognition can fundamentally alter and extend our understanding of sensemaking and other macrocognitive functions including planning, coordination, and common ground. This can make the macrocognitive framework more universal and add testable hypotheses to our research programs.

In this chapter, we will explore four national differences in cognition: Attention, Causal Attribution, Tolerance for Contradiction, and Perception of Change. These have been documented in laboratory studies, which may shape the way people use information and understand the natural world. Naturalistic and laboratory researchers have sometimes been considered to be warring armies. We believe this is a mistake. To understand why we have borrowed laboratory data to address sensemaking, a macrocognitive process, we need to look at the goals of naturalistic decision making. Naturalistic decision making emerged, in part, to capture the richness of real-world contexts and in part as a response to the limitations of parametric laboratory research. The existence of limitations in field research does not mean that we must choose between the two. Rather, just as traditional experimental researchers might enrich their science by considering the outcomes of naturalistic studies to formulate better laboratory questions, so too would NDM researchers do well to use laboratory outcomes to identify new dimensions and approaches to understanding cognition in the world. It is in this spirit that we turned to laboratory research on analytic and holistic thinking to suggest national differences important during sensemaking in naturalistic settings.

Before we begin this examination of culture and sensemaking, a few caveats are needed. First, most research on culture and cognition has compared Westerners, people from English-speaking nations and northern Europe, with East Asians, principally people from Japan, Korea, and China. These comparisons probably do not capture the variations in cognition found worldwide and may even distort some groups. A full review will have to wait until we have a broader representation of nations in the research literature. Second, while comparisons among national groups have shown clear cognitive tendencies in cognition for specific groups, we make no claim that these tendencies describe every person. To do so would ignore the power of individual differences. Third, cultures are dynamic systems that emerge from a particular setting and that change with modernization and contact with other groups. We cannot assume that patterns identified in a nation's rural communities will describe patterns among urban people or that patterns identified in the past are true today. The patterns we identify are only meant to be useful starting points for understanding a particular group. Finally, national groups differ in many ways:

customs, behavior, values, ideologies, and social roles. The emphasis in this chapter is on cognition and not on the other important ways in which national groups differ.

In this chapter, we define sensemaking within the macrocognitive framework and outline some of the demands that make it vulnerable to national differences in cognition. We present the four cognitive differences identified in laboratory settings, describe how each is linked to cognition in natural settings, and suggest how each might influence the course of sensemaking. Finally, we describe the implications of these differences for sensemaking and more generally for macrocognition.

Sensemaking

> You are driving to an important meeting in a GPS-equipped rental car in an unfamiliar city. You have the directions from your host. You need to attend to the road, stay alert for unexpected driving patterns, and watch for your next turn. If there is a conflict between the GPS and your directions, you will have to decide what to do; if you make an error, you will need to figure out how best to recover. As you approach your destination, you will need to watch for parking places.

In such complex situations, we may face an unpredictable stream of information —the trajectories of other vehicles, obscure or hidden street signs, confusing or contradictory directions, and unfamiliar driving customs. While much of the available information may be ignored, some of it demands immediate explanation, interpretation, and action: "Why did that happen?" "What does this mean?" "What should we do now?" We must organize, interpret, and use the information to make sense of continual changes by differentiating what is relevant from irrelevant (Choo, 1998). People use their past experiences to interpret and reinterpret situations and to form possible explanations as new information becomes available. *Sensemaking* is the process that people use to identify compelling problems, construct meaning, select frameworks for new information, and provide causal explanations (Weick, 1995). It builds on other macrocognitive processes, such as problem detection and problem identification, and it triggers and guides adaptive planning and decision making. If successful, it can guide us to our important meeting on time.

Sensemaking begins when a person becomes aware of a compelling problem: a change, anomaly, or surprise in the stream of information (Thomas, Clarke, and Gioia, 1993; Weick, Sutcliffe, and Obstfeld, 2005). The recognition of any of these can initiate the gathering of additional information (for example, Thomas, Gioia, and Ketchen, 1997). Categorizing information allows us to verify findings by comparing similar cases, past and present. This may help suggest plausible explanations or propose effective actions. Making sense of situations in this way can lead to effective action by individuals and by organizations (Daft and Weick, 1984). Actions, in turn, produce changes in the environment that reflect on the effectiveness of one's sensemaking. This can allow closure, propel additional action, or suggest reinterpretations (Thomas, Shankster, and Mathieu, 1994). In this way, sensemaking is a dynamic and ongoing process.

Klein et al. (2007) suggest a data/frame model of sensemaking to describe the deliberate effort undertaken to understand events. They propose that incoming information can suggest frames—mental models for organizing and understanding. These frames are similar to Minsky's (1975) notion of frames as structures for representing known situations. They may include information about dynamic relationships among components, expectations for the future, and appropriate actions. The frame helps to delineate what counts as data and guides the search for additional information. Contradictions and inconsistencies may provide cues for elaborating the frame or reconsidering previously discarded data. If people cannot explain events by elaborating their initial accounts, they must question their frame, perhaps rejecting it and shifting to another in their repertoire. People differ in how hard they work to preserve their original frame. They may, for example, be lured down a "garden path" to explain away inconsistent data by deciding that the data is unreliable. Some people may track several frames and, as events develop, compare these frames to find one that fits best. Both Weick's and Klein et al.'s frameworks identify crucial sensemaking processes. Their formulations suggest several questions about sensemaking and the variation it exhibits, the answers to which may vary across national groups:

- *What is an anomaly or problem?* Identifying a problem is important but what is considered a problem may differ from one person to another and from one national group to another. This is especially true when problems are ill-defined. When problem identification differs, the subsequent sensemaking will also differ.
- *What frames are available to provide an initial sense of the situation and how is one selected in a particular case?* People use their own experiences to suggest frame(s) for a specific situation and these frames guide further exploration. When people have had different experiences, they are likely to use different frames. To the extent that national settings provide different experiences, these experiences can shape the pool of available frames for the group.
- *What is the range and content of information considered?* A compelling problem may call for additional investigation and action. Information is considered based on its relevance to the current sense of the situation and the explanatory frames that are being used. Because people use differing frames, the information they select can differ. Even when there is a common frame for understanding, people may vary in the scope and the types of information to which they attend. Because national groups differ in the range and amount of information considered relevant, selection and the sensemaking that follows is likely to vary.
- *How is material categorized or otherwise organized?* In order to deal with the large store of available material, people categorize information (Hamilton and Trolier, 1986). Differences in categorization can reduce information overload but at the same time can change subsequent information gathering and lead to differences in sensemaking.
- What counts as an explanation? Sensemaking helps explain ongoing events in a way that allows for future actions. When people differ in causal beliefs, they will accept different causal explanations. This can generate different action plans.

• *How open are people to new or contradictory information?* In dynamic situations, people often face contradictory information and this has been exacerbated with the advent of the Internet. There are individual differences in openness to new and to contradictory information: what it takes to change frames during sensemaking. If there are national differences in how people deal with contradictory information, these can also influence sensemaking.

These questions about sensemaking are part of our ongoing research agenda. When people from different nations work together on complex problems, the processes underlying sensemaking sometimes show considerable variability. National groups may bring their own distinct cognitive styles to teamwork. Predicting the actions of those from other nations is more difficult if there are cultural variations in cognition.

In the next section, we review the origins and nature of four dimensions of Analytic-Holistic Thinking: Attention, Causal Attribution, Tolerance for Contradiction, and Perception of Change. We describe how these cultural differences may influence what is considered a problem, the information deemed necessary to address a problem, and the way it is attended to, selected, categorized, interpreted, and used to make sense of complex situations. Each of the differences can shape the course of sensemaking in natural settings and can introduce confusion in multinational settings.

Cultural Differences in Cognition

People from Western nations, including Americans, tend to be analytic thinkers and focus on objects and dispositions. Those from Eastern Asia show more holistic thinking and focus on relationships. Analytic thinkers view the world as composed of separate elements that can be understood independently, while holistic thinkers focus on the relationships among different elements (Nisbett, 2003). Choi, Koo, and Choi (2007) propose four dimensions of analytic-holistic thinking. These dimensions provide the conceptual basis for the present analysis. The dimension of *Attention* defines the scope of information considered or needed: analytic thinkers focus on central features in the field while holistic thinkers attend to the field as a whole. *Causal attribution* directs the search for explanations to situational or dispositional causes. Analytic thinkers target dispositional causes while holistic thinkers include situational causes. *Tolerance for contradiction* describes the difference between analytic differentiation, polarizing goals and options to define the most important, on one hand, and holistic naïve dialecticism, merging goals and options by synthesis, on the other. *Perception of change* describes beliefs about change, whether phenomena are viewed as linear by analytic thinkers or as cyclical, non-static patterns by holistic thinkers. These national differences in cognition can make sensemaking vulnerable to cultural differences.

Analytic and holistic differences can be understood from two perspectives. First, the Ecocultural Model provides a framework for understanding how ecological constraints are related to perceptual and cognitive differences leading people from different ecological settings to see the world differently. The work was based on field work with 21 traditional groups, from North America, Africa, Australia, and Europe engaged in a variety of subsistence patterns including farming, herding, hunting and gathering. Berry (1976) reported political stratification, social stratification, and family organization as they related to cognitive and perceptual functioning. He reported consistent relationships between ecological demands on the one hand and perception and cognition on the other. These ecocultural constraints provide a lens through which members of a group see the world (Klein, 2004).

Patterns of cognition are relatively enduring because they grow out of the socialization practices within a particular culture, and they have survival value for people in that culture. Groups who, in recent generations, engaged in hunting and gathering, for example, are more likely to exhibit field independent perception while those who have, in recent generation, engaged in farming are more likely to exhibit field dependent perception (Berry, 1986). Adults who have these adaptive skills are likely to be more successful in their culture.

Cognitive patterns appear to be perpetuated by social structure and childrearing patterns. Caregivers model and reinforce the patterns successful in and characteristic of the culture. The role of caregivers in shaping cognition is seen in the difference between the childrearing of Japanese and Americans. Fernald and Morikawa (1993) report how Japanese mothers use few labeling nouns and many more verbs when playing with their babies. This serves to focus the infant's attention on relationships and the context of objects, for example, "See the mother feeding the little girl." In contrast, American mothers label many objects and focus on categories of objects when playing with their babies, for example, "Let's put all the red blocks in this box." These lessons from the early years of life are consistent with later cognition. The Japanese adult looks for relationships while the American looks for distinctions.

A second perspective for understanding analytic-holistic differences comes from Richard Nisbett's recent work. Nisbett (2003) asserts that the analytic and holistic systems of thought originated from two ancient philosophic traditions: Greek Aristotelian thought and Chinese Confucian thought, respectively. The Greeks saw power as located in the individual's sense of personal agency. Their analytic thinking is seen in the tradition of debate (Cromer, 1993), the rule-based understanding of the world (Lloyd, 1991), and the speculative creation of causal models about the nature of the objects and events. In contrast, the Chinese tradition held a sense of reciprocal social obligation or collective agency. Individuals felt that they were a part of a large and complex system where behavior should be guided by the expectation of the group. In ancient China, debate was not generally encouraged (Cromer, 1993). Understanding of the natural world proceeded by intuition and empiricism rather than formal models (Lloyd, 1991). The social system focused attention on the larger broader picture and cultivated holistic thinking.

The four dimensions—attention, causal attribution, tolerance for contradiction and perception of change—considered in this chapter stem from the analytic-holistic distinction (for example, Nisbett, 2003; Nisbett et al., 2001). These dimensions are further explored below.

Attention

Our attention allows us to limit the information available for sensemaking in a complex environment. Consistent with early mother-child communication patterns, holistic thinkers, including East Asians, attend to the relationships among objects and context. They see the whole picture, emphasizing relationships and interconnections, a more field-interdependent view, at the expense of a focal object. Also consistent with early communication patterns, analytic thinkers, including most Westerners, look to individual objects and items rather than to the field as a whole. The Westerners pay more attention to individual parts and they are more field independent. How might this look in the cockpit of a commercial plane?

> There can be a lot going on in the cockpit during an emergency—multiple conversations with multiple people talking to the pilot—it is impossible to process them all. Flight instructors and check pilots report a common trend: Middle Eastern and Chinese pilots have more trouble with prioritizing information during overload. They are less able to "turn off" the low-priority conversations. They work to attend to all of them and often miss the most important input. It is as if they are afraid to miss anything because they give it all equal importance. [Klein, Klein, and Mumaw, 2001]

These observations illustrate differences in how different cultural groups see and attend to their auditory world. Westerners typically focus on focal information, even to the expense of contextual information. Other groups appear to share attention to focal information with more contextual information including conversations and routine functions. The strength is that this information may be more available for later use. The cost is less attention to immediate, focused demands.

Masuda and Nisbett (2006) looked at these same phenomena in the laboratory. They showed Americans, presumably analytic, and Japanese, presumably holistic, a set of video clips depicting an aircraft in flight over a crowded airfield. A large aircraft, a medium-sized plane, and a helicopter are in the foreground. Peripheral objects such as the control tower and additional planes appeared in the background. In the animated clips, there were changes related to the plane in flight and the large aircraft on the ground, the focal objects, and in the peripheral objects and context. Participants were asked to report changes from the first to the last frame. The set of video clips include many changes in the physical environment so the viewer can only attend to some of them. The goal was to capture attention difference between the American and Japanese samples. What did they see? The more analytic Americans noticed more changes in properties such as color, shape, and number of the aircraft in flight. They noticed, for example, changes in the position of the flight aircraft's wheel. They were less likely to notice changes in the background or the distance between the helicopter and planes. The Japanese participants noticed more changes in background, for example, the control tower in the background and the changes in the distance between the helicopter and the planes indicating attention to relationship between two objects. They noticed fewer changes in the plane, the focal object.

Differences in attention are not limited to the physical environment but have also been shown using social stimuli in a simple but elegant experiment. Masuda et al. (2008) showed Americans and Japanese participants 56 stimuli each consisting of a central cartoon figure together with four smaller background cartoon figures. The facial expression of the central figure was depicted as Happy, Neutral, or Sad. The smaller figures for each stimulus were depicted as Happy, Neutral, or Sad. When Americans rated the emotion of the central figure, their judgment focused on the central figure, uninfluenced by the emotions of the background figures. In short, they performed as requested. In contrast, the Japanese participants modulated their judgments to reflect the emotions of the small surrounding figures. If a neutral figure was surrounded by happy figures, they rated the figure as more happy. This study found no consistent gender by culture interaction. In a later study, Americans, Japanese, Koreans and Taiwanese were asked to make this same judgment. The Americans attended to the central figure while participants from the three East Asian groups modulated their judgments of the central figure by the emotions of the background figures (Klein et al., 2006). They attended to the social context of the cartoon figure, not just the figure alone. These holistic thinkers are likely to have more peripheral social information available for later use.

East Asians were also found to attend to both background information and focal information in solving complex decision-making tasks. Strohschneider and Guss (1999) gave Asian Indians and Germans an interactive computer simulation of a small semi-nomadic tribe. They were asked to take the role of a developmental aide worker and work to improve the conditions of the tribe over time. To do this, they could ask for information they thought necessary. Even though Indian participants generally asked for less information than did the German participants, they asked for more background and context information, such as the social conditions, in their effort to accomplish their task.

In a similar study, Gelfand et al. (2000), extended the role of attention to judgments of information usefulness. Participants were given relational or individuating information about a target person they were to interact with across four situations. They were asked to rate the usefulness of the information and how confident they thought they were in predicting their own and the target's behavior across the four situations. Chinese students saw relational information—social groups, family, social class—as more useful for predicting their own and other person's behavior, whereas US students favored individual information such as personal accomplishments (Gelfand et al., 2000). Because of this difference, it is likely that analytic and holistic thinkers would have different information available at the beginning and during sensemaking.

Sensemaking starts with awareness of an anomaly or problem that focuses attention. It continues to the selection and evaluation of a frame to provide a sense of the situation. When national groups differ in their initial attention to the visual field, to social context, and to problem space, they are likely to notice different anomalies. When people vary in the range of information they consider relevant, they will have different examples and counter-examples for use in sensemaking. Holistic thinkers may use information more intuitively because they have more information available to consider. Analytic thinkers may favor rule or cost-benefits analysis because they attend to a narrow set of information. Taken together, attention appears to direct problem identification and set the stage for subsequent sensemaking.

Causal Attribution

Attribution describes how people assign cause (Heider, 1958) and so directs the selection and use of information. Dispositional attribution identifies internal causes such as competence, personality, and values as most explanatory. Situational attribution looks also to external causal factors such as task demands, environment barriers, and surrounding people. Analytic thinkers, including Westerners, typically attribute behavior to the actor's disposition (Gilbert, Pelham, and Krull, 1988) while ignoring situational causes (for example, Gilbert and Malone, 1995). Holistic thinkers, including East Asians, use both situational and dispositional factors to identify the driving forces for behavior and events (for example, Choi et al., 1999; Morris and Peng, 1994). Because both dispositional and situational factors are included, sensemaking is likely to be systemic in nature.

To study the differences in attribution as reflected in press coverage, Markus et al. (2006), reviewed Japanese and American media accounts from the 2000 and 2002 Olympics for explanations of Olympic performances. Coverage from 77 Japanese athletes and 265 Americans athletes were coded and analyzed. The analysis provided contrasting responses to a journalist's question, "How did you succeed?" as recorded in a respected newspaper from their native nation. Would it be dispositional, that is, hard work and discipline, or situational, namely, the support received from teammates and a good night's sleep the night before? The Japanese accounts included more categories describing athletes' positive and negative personal attributes, background, and social and emotional experience. American media accounts included fewer categories and emphasized positive personal characteristics and features of the competition. Capturing this difference, one Japanese athlete responded, "Here is the best coach in the world, the best manager in the world, and all the people who support me—all of these things were getting together and became a gold medal. So I think I didn't get it alone, not only by myself." In contrast, an American responded, "I think I just stayed focused. It was time to show the world what I could do. I am just glad I was able to do it. I knew I could beat Suzy O'Neil, deep down in my heart I believed it, and I know this whole week the doubts kept creeping in, they were with me on the blocks, but I just said, 'No, this is my night.'"

In a follow-up laboratory study, participants reviewed the explanations from both Japanese and American media and chose the most relevant information about the Olympic athletes (Markus et al., 2006). Responses mirrored that of each national press: the Americans favored dispositional explanations while the Japanese used more categories and found the situational components more compelling.

Morris and Peng (1994) address the question of attribution, looking at media treatment of two well-covered mass murders. One of the crimes was committed by a Chinese graduate student and the other by an Irish-American postal worker. The researchers reviewed, coded, and analyzed dispositional and situational attributions provided by the articles related to the crime over a two-month period published in New York by the *New York Times* and the Chinese-language *World Journal*. The English-language newspaper accounts reflected significantly more dispositional attributions. They describe the graduate student as having a "very bad temper," and "psychological problems with being challenged," and the postal worker as a "man was mentally unstable," and "had repeatedly threatened violence." The Chinese-language newspaper provided more situational explanations for the graduate student—"did not get along with his advisor," and was "isolated from the Chinese community," and for the postal worker—"gunman has been recently fired" and "followed the example of a recent mass slaying in Texas."

The researchers then asked American and Chinese graduate students to rate probable causes and their importance as well as things that might have averted the tragedies. Their judgments were consistent with the journalistic report: the American students included more dispositional causes and rated them as more important, while the Chinese students included both dispositional and situational causes and rated both as important. Taken together, this research confirms attribution differences.

Choi et al. (2003) used Morris and Peng's (1994) murder incident to confirm the attributional differences between groups. They asked participants to read a scenario describing the murder: "Suppose that you are the police officer in charge of a case involving a graduate student who murdered a professor (the dead professor was the graduate student's adviser). Why would the graduate student possibly murder his or her adviser? As a police officer, you must establish the motive."

After reading this scenario, participants were given a list of 97 potentially useful facts for making sense of the murder. They were asked to indicate which of the facts they considered irrelevant. For example:

- Whether or not the graduate student was unhelpful.
- Whether or not the professor was religious.
- Whether or not the graduate student was far away from his/her hometown.
- Whether or not the graduate student liked rock music.

Americans, with their analytic thinking, excluded more information as irrelevant than did a sample of Koreans when they make sense of the scenario (Choi et al., 2003). The same difference was found when Americans were compared to samples of Japanese, Korean, and Taiwanese students (Klein et al., 2006), all presumably holistic thinkers. Overall, Westerners attended to a more focused range of information while holistic thinkers attend to information that is more diverse (Choi, Choi, and Norenzayan, 2004).

An early laboratory study provided a parallel indication of the influence of attribution on sensemaking. Miller (1984) presented Americans and Hindu Indians with this scenario describing a motorcycle accident:

This concerns a motorcycle accident. The back wheel burst on the motorcycle. The passenger sitting in the rear jumped. The moment the passenger fell, he struck his head on the pavement. The driver of the motorcycle—who is an attorney—as he was on his way to court for some work, just took the passenger to a local hospital and went on and attended to his court work. I personally feel the motorcycle driver did the wrong thing. The driver left the passenger there without consulting the doctor concerning the seriousness of the injury—the gravity of the situation—whether the passenger should be shifted immediately—and he went on to the court.

Participants were asked why the driver left the passenger at the hospital without staying to consult about the seriousness of the passenger's injury. While both Americans and Indians mentioned the state of the driver at the time of the accident as a reason for his leaving, the Americans were more likely to attribute the behavior to the disposition of the driver, such as irresponsibility or drive to succeed at work. The Indians, in contrast, were more likely to also mention situational attributions, such as responsibilities and obligations at work. One accident, different attributions.

Research into causal differences also suggests the power of information. Choi and Nisbett (1998) manipulated information saliency in a scenario to determine causal attribution outcome. When situational information was limited, both Koreans and Americans used dispositional attribution to explain outcomes. However, when situational information was salient, Americans ignored the information that did not fit their dispositional frame while Koreans were likely to change frames. In making sense of an organizational scenario, Lin (2004) presented Malaysians and Americans with scenarios consisting of both dispositional and situational information. They were then given recognition tests of the information presented in the scenarios and attribution assessments of the scenarios. Participants' holistic tendencies were also measured. The study found Malaysians to be more holistic in their thinking than Americans. They remembered significantly more situational information and identified both dispositional and situational explanations, while Americans rated situational causes as less likely. The different value placed on situational information would be expected to lead to different explanations during sensemaking.

Causal attribution and categorization appear to be linked. People who favor dispositional explanations appear to group objects and concepts using objective traits, while those who favor situational explanations group by relational characteristics. We can see this difference in a simple task: "What goes with the cow? Chicken or grass?" If you categorize based on dispositions, you will categorize the cow with the chicken because they are both animals. If you categorize based on situational attributions, you will group the cow with the grass because cows eat grass. Norenzayan et al. (2002) found that Japanese participants typically chose relationship-based categorization with the most similar attributes grouped while US participants were more likely to use single attributes and formal rules. Similarly Ji and colleagues (Ji, Zhang, and Nisbett, 2004), showed participants sets of three words, for example: "Cow—Milk—Pig" or "Foot—Shoe—Hand." When asked to identify the pair that belonged together, East Asians categorized based on relationships (that is, foot and shoe) while Americans favored dispositional categorization (that is, cow and pig). Faced with a complex task, people may seek a frame that uses a category based on past experience. If we experience stomach distress, we may categorize it with our last experience with stomach distress. If that was appendicitis, we might rush to the hospital while if it was a minor flu, we might drink hot tea. Classification can support sensemaking and decision making by guiding the selection of comparison cases. Classification may also make it easier to access information later to revise the sense of the situation.

Because people from different cultures begin with distinctive causal explanations, they may attend to, value, and accept different information. They categorize information using different dimensions, generate different explanations for situations, and make different predictions for future events (Choi et al., 1999; Choi and Nisbett, 1998; Ji, Nisbett, and Su, 2001). These differences shape their sense of the problem space and direct their planning and decision making. This can create problems in settings where exchanges of information are important. Differences in attribution can mean preference for different information-management processes, one that is rich with related or connected information and the other, focused but detailed information. When multinational team members seek and retain different information, this can take them in different directions.

Tolerance for Contradiction

Tolerance for contradiction describes how people typically manage divergent information. Analytic thinkers avoid contradictions whenever possible (Peng and Nisbett, 1999). Information, goals, and options are polarized to identify the most important feature of a situation. Consistent with analytic logic, each statement, philosophy, or technique must be true or false but not both. In contrast, holistic thinkers tend to use naïve dialecticism. They deal with contradiction by searching for the "middle way" between opposing propositions, retaining and synthesizing basic elements of each. Holistic thinkers tolerate contradiction because they see truth in opposing views (Peng and Nisbett, 1999).

Peng and Nisbett (1999) compared cultural folk wisdom reflected in everyday language using proverbs from traditional Chinese and American sources. A dialectical proverb contains contradiction. For example, "beware of your friends not your enemy" and "too humble is half proud." A nondialectical proverb contains no such contradiction. For example, "one against all is certain to fall" and "for example is no proof." They found four times as many dialectical proverbs in the Chinese source as compared to the American, suggesting contradiction plays a larger role in the Chinese folk wisdom. They then asked American and Chinese research participants to evaluate the proverbs. American participants preferred nondialectical to dialectical American proverbs and the Chinese participants preferred dialectical to nondialectical Chinese proverbs. To control for familiarity of proverbs, Yiddish proverbs were used. Again, the result reflects a preference for dialectical proverbs by the Chinese than the American participants.

Two other domains capture differences in tolerance for contradiction. For the Western world, a religious system is seen as having integrity in its representation and expression of truth. A person can be a Christian or a Moslem but not both. In contrast, the pattern of Japanese life may include Shinto, Buddhist, and even Christian expressions. Chinese religious expression may incorporate both Buddhist and Taoist elements. For many people in Peru, Catholicism and traditional pagan worship stand side by side. These are not seen as conflicting but rather as capturing a broader reality.

Similarly, the Western world has generally adopted a biological model for healing. We demand medications and treatments based on sound science and research. While there is recent interest in holistic and alternative therapeutic approaches, mainstream medicine remains "science" based. In contrast, in holistic cultures, the use of state-of-the-art science is not seen as incompatible with a wide variety of traditional healing practices, ranging from herbs to rituals evoking the aid of higher powers. In these groups, shamans using rituals to counter the possibility of a curse as the source of illness are not seen as contradictory to modern medicine.

Tolerance for contradiction influences openness to new and contradictory information during sensemaking. Dialectical thinkers seek potential truth in divergent positions while differentiation thinkers seek the correct explanation and explain away other options. Choi and Nisbett (2000) looked at cultural differences in the judgment of contradiction. In the Good Samaritan story below, taken from Darley and Batson's Good Samaritan study (Darley and Batson, 1973), the character is described as religious, generous, and helpful but also under time pressure and stress:

> John, a seminary student, is very religious, generous, and helpful. He is taking a sermon course and has to give a practice sermon as a course requirement. On the day he was supposed to give the sermon he was 10 minutes late for class. The professor was known to be harsh with students for being late. While John was rushing to class, he saw a man lying on the ground needing help. If John stayed to help he would not make it to class.

Participants were assigned to one of three conditions. They were told John helped the man, he did not help the man, or they were told nothing about the outcome. They were then asked about their expectations and reactions to alternative outcomes.

Choi and Nisbett (2000) expected that Koreans and Americans would react differently when given contradictory information and they did. The Koreans showed more hindsight bias than did the Americans in the "no help" condition, accommodating this new information. They saw the conclusion opposite to their initial one as also plausible and were less surprised by it. Americans were more surprised when they were told John did not help the man, contradicting their expectation. The same finding was confirmed with a story where the target was described negatively but ended up doing a positive action. The Americans here also were more surprised by the new information. The Koreans were less influenced by the "mismatch."

To study this same distinction, Peng and Nisbett (1999) used pairs of statements to present possible conflicts to participants from China and the US. For example:

> A) A social psychologist studies young adults and asserted that those who feel close to their families have more satisfying social relationships.

> B) A developmental psychologist studies adolescent children and asserted that those children who were less dependent on their parents and had weaker family ties were generally more mature.

They queried East Asian and Western participants to see if statements A) and B) were seen as contradictory or if they could both be true? The East Asians, thinking dialectically, understood the two statements as noncontradictory and parts of a whole rather than as dichotomous descriptions. They accepted the seeming contradiction as multiple perspectives of a single truth (Chu, Spires, and Sueyoshi, 1999; Nakamura, 1985). In contrast, differentiating reasoning which is typical of Westerners seeks constancy. Because contradictory propositions are unacceptable by formal logic, they believe that contradictory statements cannot both be true. Differentiation thinkers polarized the contradiction, decided which position was correct, and explained away the other position. They considered the propositions in a restricted context rather than embedded in a broader context (Cromer, 1993). Hence, Westerners saw statements A) and B) as opposing, and decided which one was correct.

During sensemaking, differences in tolerance for contradiction influence the way information is selected and retained. These differences provide conflicting paths in complex situations. Differentiation thinkers seek the best goal while dialectical thinkers seek a harmonious, intermediate goal. In choosing the best goal, the differentiation thinkers may reduce cognitive dissonance by avoiding or dismissing contradicting information. They may, however, exclude information needed as the situation unfolds, new information emerges, and existing frames must be reexamined or changed. Differential thinkers may also be more likely to seek out confirming rather than disconfirming information. They simplify information to reduce overload (O'Reilly, 1980; Weick, 1979). In contrast, dialectical thinkers may experience more information overload and have difficulty in settling on a course of action. At the same time, they may see more information as related (Choi et al., 2003) and so may experience little dissonance and be more prepared for change and surprise.

During sensemaking, information can reduce ambiguity (Weick, 1995). However, people who avoid contradictory information may err in not considering alternate positions and valid objections. People who are comfortable with contradiction remain open to new information longer. They may track several frames simultaneously so that they can merge frames or modify them. Their readiness to change their sense of the situation and their decisions may depend on how much they can prolong the sensemaking process. These differences can hinder collaboration when high-tolerance people are more flexible in decision making and view the low-tolerance people as rigid and/or if low-tolerance people seek to complete work and view high-tolerance people as indecisive.

Finally, tolerance for contradiction extends to criticism and self-appraisal. Differentiation thinkers are more likely to be threatened by information that questions their choices and are more likely to disregard unfavorable information. In contrast, dialectical thinkers may welcome unfavorable information and not necessarily evaluate it as threatening. Their awareness of situational constraints allows them to see the external pressures on their choices relieving a sense of failure or dissonance (Choi et al., 1999; Hiniker, 1969). Criticism may be seen as information that can lead to improvement.

Overall, differences in tolerance for contradiction mean that dialectical and differential thinkers may use conflicting information in different ways which may lead to differences in selection and retention of information. Together, these differences are expected to generate different problems, approaches, and goals for sensemaking and also for planning, decision making, and team interaction.

Perception of Change

Perception of change describes the belief about the nature of change. Is change viewed as a linear, monotonic pattern, or as a cyclical, dynamic pattern? Holistic thinkers appear to see the world as cyclical and so seek to understand one point in time by reference to long-term cyclic patterns of change (Nakamura, 1985). An ancient Chinese folk tale illustrates this cultural belief about change:

One day an old Chinese farmer's horse ran away. His neighbors came to comfort him, but he said, "How can you know it isn't a good thing?" A few days later, his horse came back, bringing a wild horse with it. His neighbors came to congratulate the old man, who said, "How can you know it isn't a bad thing?" A few weeks later, the old man's son tried to ride the new horse and fell off, breaking his leg. Again, the neighbors came to comfort the old man, who said, "How can you know it isn't a good thing?" Some months later, a war broke out, and all the young men in the region were recruited for the war. The old man's son was spared because of his broken leg. [adapted from Ji, Nisbett, and Su, 2001]

In contrast, analytic thinkers have a linear view of change (Ji, Nisbett, and Su, 2001; Peng and Nisbett, 1999). Western folk tales reflect this perception of change. The hero or heroine is a wonderful person in an unfortunate situation but all comes out well in the end. Grimms' *Tales for Young and Old* capture this difference (Grimm and Grimm, 1977).

Sleeping Beauty (Brier Rose in the German version) tells us that at the celebration given to mark her birth, one of 13 Wise Women of the realm was not invited and so placed a curse on the baby. Nevertheless, Sleeping Beauty "grew to be so beautiful, so modest, so sweet-tempered and wise that no one who saw her could help loving her." While things were bad for a time—a pinprick and a hundred years of sleep—in the end, her prince came and all turned out well: "The prince and Brier Rose were married in splendor, and they lived happily to the end of their lives." The lives of Snow White and Cinderella (Ashputtle in the German version) follow similar paths.

Holistic thinkers, like the old Chinese farmer, believe that reality, as a process, is always in flux and expect that ups and downs will alternate cyclically (Peng and Nisbett, 1999). They see everything in the world as connected with complex interactions among elements. In contrast, Westerners with their linear perspective, expect stability over time with few dramatic changes. The neighbor in the old farmer story represents this view as do the "happily ever after" Grimm's fairy tales endings.

Predictions about change are part of how we make sense of the world. Ji, Nisbett, and Su (2001), asked Chinese and Americans to make predictions about the future patterns of 12 natural events including economic growth rates and global cancer death rates. They reviewed 12 graphs showing three points to indicate the development of the rate across three periods of time. For example, the points on the graph of global economy growth rates showed annual percentage change in real GDP, of 3.2 percent, 2.8 percent, and 2.0 percent for 1995, 1997, and 1999, respectively. Participants were asked to predict the probability for the trend to go up, down, and remain the same. They also indicated what they thought the next two points would be. The predictions could show growth or decline and the rate of change could accelerate or decelerate. Americans made more predictions consistent with the given trend whereas Chinese were more likely to deviate from trends making predictions in the opposite direction.

In a second study, Ji et al. (2001) asked people to predict the course of their own life happiness. American and Chinese participants were presented with eighteen trends: six linear and 12 nonlinear trends, four were parabolic nonlinear trends. They were asked to select the trend that best represented their expectations for happiness throughout their lifetime. The Chinese participants were likely to predict nonlinear rates of change or directions of movement, and parabolic nonlinear change for both rate and direction of change. In contrast, Americans selected the linear trend, moving in one direction, to represent their life happiness.

Both studies found Chinese participants more likely than American participants to assume that upcoming events would deviate from the current trend and even reverse direction. Americans made predictions in the direction consistent with the current trends with more stability. The Chinese participants also reported greater confidence about their prediction than did Americans. The studies suggest that Americans may be more likely to respond to immediate information while the Chinese look at things holistically and from a long-term perspective.

A concept of change provides a frame for making sense of an ongoing situation and for formulating expectations about an unfolding situation. Differences in beliefs about change and the resulting differences in expectations mean that different events are considered to be anomalous. When the view of the world is stable, there may be less need to review a frame. When the world is seen as in constant change, a person may be constantly reviewing the frame as well as anticipating change. This difference is important as a particular sense of how a situation might change guides planning and action.

Discussion

Implications for Sensemaking and Macrocognition

National differences in cognition shape the way people approach macrocognitive tasks. The Cultural Lens model describes the emergence and mechanisms by which cognition can vary for different national groups (Klein, 2004). As described by the Cultural Lens model, people see the world through different lenses. Each culture provides schemas for information management and sensemaking that tend to generate different interpretations (Bhagat, Harveston, and Triandis, 2002). Sensemaking is sensitive to cultural variations because it reflects the complex lives of people and their cognitive plasticity. The four dimensions described in this chapter illustrate how specific differences can shape the nature of sensemaking and other macrocognitive functions. A few examples illustrate this:

- *What is an anomaly or problem?* Cultural differences in attention guide the detection of anomalies. For holistic thinkers, an anomaly or problem can be very broad encompassing the context and including situational, interactional, and systemic factors. In contrast, the analytic thinker focuses on features central to the task at hand. Cognitive notions about the nature of change also guide the identification of anomaly. A change in a trend is anomalous for analytic thinkers but not necessarily for holistic thinkers.
- *What frames are used to integrate information and provide a sense of the situation?* For holistic thinkers, the solution to a problem is to find the middle way—a compromise position. Holistic thinkers may have more complex

frames with interconnected causal factors. In contrast, for analytic thinkers, only the best option will do. Concepts of change also contribute because people with a cyclical rather than linear view about events may be more willing and ready to anticipate shifts in the flow of event and to incorporate this in their sense of a situation.

• *What is the range and content of information considered?* Once an anomaly is identified, differences in attention shape the range of awareness and the search for sense. While analytic thinkers may discard information if it does not fit into the dominant frame, holistic thinkers may try to accommodate dissenting information by extending frames. They may be less likely to discard information as irrelevant because things are viewed as interconnected. Those who use a holistic approach may attempt to incorporate information that supports seemingly contradicting goals. Because holistic thinkers, with their situational attribution, incorporate a wide range of information as potentially relevant, they are more likely to experience information overload. In contrast, analytic thinkers, using more dispositional attributions, focus on key information and so may lack contextual information when problems shift.

• *How is material categorized or otherwise organized?* Causal attribution helps define salient categories for organizational schemas. Cultural differences in the nature of causal attribution will affect the categories and classification schemes employed by sensemakers.

• *What counts as an explanation?* Cultural differences in the nature of causal attribution will determine what counts as an acceptable explanation. Some people want a single underlying cause while others look for a nuanced set of factors. The nature of the explanation is a key step towards sensemaking.

• *Openness to new or contradictory information.* People with broad attributions, both dispositional and situational, may be more open to contradictory information because they believe all information is part of a meaningful whole. People from cultures that tend to avoid contradiction may ignore or explain away potential contradictions. A rejection of contradictory information can lead to a cleaner sense but the sense may not reflect reality. Embracing contradictory information can lead to a more comprehensive and realistic solution but may overload information-management capacity.

Implications for Team and Organizational Processes

Cultural differences in cognition can impose constraints on multinational teams and on organizations. Team sensemaking shares all the complexities of NDM because goals may be ill-defined and consensus lacking, information may be incomplete and ambiguous, decision makers may be pursuing multiple or competing objectives, and real constraints limit the time available to manage the information at hand. National differences in cognition further complicate teamwork by shaping information sharing, and affecting communication patterns during collaboration. People from different groups may seek and transmit different information, assign different causes, and use different frames for sensemaking. Together these cognitive differences can lead to different senses of the situation, resulting in different decisions about how to act. When the professionals who make up multinational teams encounter anomalies, unexpected events, and discrepancies with implications for their organizations, they may differ radically in their subsequent sensemaking, problem solving, and decision making (see also McHugh, this volume). Differences in cognition can have benefits by providing different views and diverse solutions. Understanding cultural differences may enable multinational teams to share information more effectively in order to arrive at a common understanding that is richer than the interpretation forged from any single cultural perspective.

The difficulties are even more pronounced in multinational organizations where cultural differences in cognition point to different problems and frames, and lead to different assessments of information and problem resolution (Bhagat, Harveston, and Triandis, 2002). This is an important issue because organizational psychologists describe the importance for organizations to identifying information needs; acquiring, organizing, and storing information; developing information products and services; distributing information, and using information in order to continually learn (Choo, 1998; Cohen and Levinthal, 1990). The goal is to harness organizations' resources and capabilities to enable the organization to make sense of and adapt to a changing environment. These processes are critical in complex domains such as international commerce, communication, military operations and aviation. However, Choo (1998) emphasized that information management must address the social and situational context of information use. In view of this, we propose that the theories of information management should include variation in cultural cognition. By understanding cultural differences, organizations have the opportunity to shape communication based on group's information needs and to be aware of the additional demands for consensus building.

Next Steps

While this chapter suggests several relationships between culture and sensemaking, it also reveals gaps in the research literature and suggests an agenda for future work. We believe high priority needs to be given to several areas of research.

Additional Research

While we can describe conceptual links between cultural differences and sensemaking, there is little data available for directly testing these predicted links. We need both naturalistic observation studies and microworld experimentation to identify culture-related patterns of sensemaking in different groups. One reason for this void has been the dearth of paradigms for measuring sensemaking. New efforts are now underway to assess these complex processes (for example, Klein et al., 2006) which may provide tools for exploring the impact of national differences on sensemaking. Our laboratory is exploring methods to effectively describe and measure sensemaking patterns.

Additional Dimensions

While this analysis has explored four specific cognitive dimensions with implications for sensemaking, several other dimensions may also contribute to sensemaking differences across national groups and deserve research attention. *Tolerance for uncertainty* describes reactions to uncertainty (Hofstede, 1980). This may influence comfort with incomplete information and dynamic change. *Hypothetical-concrete thinking* describes the distinction between thinking that is based on abstract speculation and thinking that is grounded in the reality of past cases (Markus and Kitayama, 1991). Hypothetical thinkers would select frames and attend to information that allows hypothesis-driven sensemaking while concrete thinkers would seek past cases to understand current anomalies and suggest solutions. National groups also differ in their *time horizon*. Some look to the distant future while others look to the days or weeks ahead (Kluckhohn and Strodtbeck. 1961). This is reflected in the scope of sensemaking and the information considered. *Mastery-fatalism* describes the efficacy people feel for taking action and making changes (Kluckhohn and Strodtbeck, 1961). It may influence reactions to problems and the kinds of solutions seen as plausible. The frames available to people would reflect their expectations for efficacy. In organizational settings, social and interpersonal differences may also be powerful forces. *Power distance* describes the extent to which members of a group expect and accept the uneven distribution of power (Hofstede, 1980). During team sensemaking, high power distance can reduce input from lower-status group members hastening decision making while reducing the engagement and input of knowledgeable group members (see also McHugh, this volume).

Managing Information

In a myriad of domains including military, transportation, and business, information management is both important and vulnerable during sensemaking. Blunders in these domains can compromise judgment and decision making (Choo, 1998). Information management is vulnerable because the *supply* of information is immense and varied in content, reliability, and accuracy. People must have enough data to identify the problem but excessive information can make it difficult to "see" what is relevant. Overload, not scarcity may be the bigger challenge, because information is often managed under time pressure and stress (Choo, 1998). An important direction for future research will be to describe the role of cultural cognition on the mechanisms underlying information management. Research underway in our laboratory is exploring the role of culture-linked cognition on information management.

Additional Regions

Current research has focused on comparisons between East Asian and Western groups. This is an important start but leaves a number of important regions almost untouched. These include South America, sub-Saharan Africa, Southeast Asia, Eastern Europe, the Arab Middle East, and the Indian subcontinent. We need a better understanding of these regions for two reasons. First, the dimensions of difference we study are ones seen between East Asia and the West. To understand human variation, we need to look at a broader range of civilizations. Second, among the less studied regions are ones with great importance for commerce and conflict resolution. Our ongoing research includes samples from both Southeast Asia and India extending the available information in these critical regions (Klein et al., 2006).

Conclusion

As long as macrocognition is grounded primarily in research with Westerners, we can only claim an understanding of Western thinking, not human thinking. Sensemaking is a good starting point for describing the effect of cultural differences on macrocognition. These processes are sensitive to cultural variations because they are embedded in the complex lives of people who vary in upbringing, ecocultural constraints, and social pressures. Our understanding of macrocognition will gain as we appreciate the variations possible for pursuing each of these functions. The inclusion of cognitive variations in our descriptions of sensemaking may help to describe this complex, dynamic process, and the variations it exhibits around the world. Future efforts should extend this work to other complex processes including decision making and planning/replanning.

References

Berry, J.W. (1976). *Human Ecology and Cognitive Style: Comparative Studies in Cultural and Psychological Adaptation*. New York: Sage.

—— (1986). "The comparative study of cognitive abilities: A summary." In S.E. Newstead, S.H. Irvine, and P.L. Dann (eds), *Human Assessment: Cognition and Motivation*. Dordrecht: Martinus Nijholt (pp. 57–74).

Bhagat, R.S., Harveston, P.D., and Triandis, H.C. (2002). "Cultural variations in the cross-border transfer of organizational knowledge: An integrative framework." *Academy of Management Review*, 27: 204–21.

Choi, I., Choi, J.A., and Norenzayan, A. (2004). "Culture and decisions." In D.J. Koehler and N. Harvey (eds), *Blackwell Handbook of Judgment and Decision Making*. Malden, MA: Blackwell (pp. 504–24).

Choi, I., Dalal, R., Kim-Prieto, C., and Park, H. (2003). "Culture and judgment of causal relevance." *Journal of Personality and Social Psychology*, 84: 46–59.

Choi, I., Koo, M., and Choi, J. (2007). "Individual differences in analytic versus holistic thinking." *Personality and Social Psychology Bulletin*, 33: 691–705

Choi, I. and Nisbett, R.E. (1998). "Situational salience and cultural differences in the correspondence bias and actor–observer bias." *Personality and Social Psychology Bulletin*, 24: 949–60.

—— (2000). "Cultural psychology of surprise: Holistic theories and recognition of contradiction." *Journal of Personality and Social Psychology*, 79: 890–905.

Choi, I., Nisbett, R.E., and Norenzayan, A. (1999). "Causal attribution across cultures: Variation and universality." *Psychological Bulletin*, 125: 47–63.

Choo, C.W. (1998). *The Knowing Organization: How Organizations Use Information to Construct Meaning, Create Knowledge, and Make Decisions.* New York: Oxford University Press.

Chu, P., Spires, E., and Sueyoshi, T. (1999). "Cross-cultural differences in choice behavior and use of decision aids: A comparison of Japan and the United States." *Organizational Behavior and Human Decision Processes*, 77: 147–70.

Cohen, W.M. and Levinthal, A. (1990). "Absorptive capacity: A new perspective on learning and innovation." *Administrative Science Quarterly*, 35: 128–52.

Cromer, A. (1993). *Uncommon Sense: The Heretical Nnature of Science.* New York: Oxford University Press.

Daft, R.L. and Weick, K.E. (1984). "Toward a model of organizations as interpretation systems." *Academy of Management Review*, 9: 284–95.

Darley, J.M. and Batson, C.D. (1973). "From Jerusalem to Jericho: A study of situational and dispositional variables in helping behavior." *Journal of Personality and Social Psychology*, 27: 100–19.

Fernald, A. and Morikawa, H. (1993). "Common themes and cultural variations in Japanese and Americans mother's speech to infants." *Child Development*, 64: 637–56.

Gelfand, M.J., Sputlock, D., Sniezek, J.A., and Shao, L. (2000). "Culture and social prediction: The role of information in enhancing confidence in social prediction in the United States and China." *Journal of Cross-cultural Psychology*, 31: 498–516.

Gilbert, D.T. and Malone, P.S. (1995). "The correspondence bias." *Psychological Bulletin*, 117: 21–38.

Gilbert, D.T., Pelham, B.W., and Krull, D.S. (1988). "On cognitive business: When person perceivers meet persons perceived." *Journal of Personality and Social Psychology*, 54: 733–40.

Grimm, J. and Grimm, W. (1977). *Grimms' Tales for Young and Old: The Complete Stories* (R. Manheim, trans.). Garden City, NY: Doubleday and Co, Inc. (1st edn 1819).

Hamilton, D.L. and Trolier, T.K. (1986). "Stereotype and stereotyping: an overview of the cognitive approach." In J.F. Dovidio and S.L. Gaertner (eds), *Prejudice, Discrimination, and Racism*. Orlando, FL: Academic Press (pp. 127–63).

Heider, F. (1958). *The Psychology of Interpersonal Relations.* New York: Wiley.

Hiniker, P.J. (1969). "Chinese reactions to forced compliance: Dissonance reduction or national character." *Journal of Social Psychology*, 77: 157–76.

Hofstede, G. (1980). *Culture's consequences: International Differences in Work-related Values.* Newbury Park, CA: Sage.

Ji, L., Nisbett, R.E., and Su, Y. (2001). "Culture, change, and prediction." *Psychological Science*, 12: 450–56.

Ji, L., Zhang, Z., and Nisbett, R.E. (2004). "Is it culture, or is it language? Examination of language effects in cross-cultural research on categorization." *Journal of Personality and Social Psychology*, 87: 57–65.

Klein, G., Phillips, J.K., Rall, E.L., and Peluso, D.A. (2007). "A data/frame theory of sensemaking." In R.R. Hoffman (ed.), *Expertise Out of Context*. Mahwah, NJ: Lawrence Erlbaum Associates (pp. 113–55).

Klein, H.A. (2004). "Cognition in natural settings: The Cultural Lens model." In M. Kaplan (ed.), *Cultural Ergonomics: Advances in Human Performance and Cognitive Engineering*. Oxford: Elsevier (pp. 249–80).

Klein, H.A., Klein, G., and Mumaw, R. (2001). "Culture-Sensitive Aviation Demands: Links to Cultural Dimensions." Technical report completed for the Boeing Company under General Consultant Services Agreement 6-1111-10A-0112.

Klein, H.A., Lin, M.-H., Choi, I., Masuda, T., Lien, Y., Radford, M., et al. (2006). "The Rosetta Project: Measuring National Differences." Technical report prepared for the Air Force Research Laboratory/Human Effectiveness, Dayton, Ohio and AOARD, Tokyo, Japan.

Klein, H.A. and McHugh, A.P. (2005). "National differences in teamwork." In W.B. Rouse and K.R. Boff (eds), *Organizational Simulation*. Hoboken, NJ: Wiley–Interscience (pp. 229–52).

Kluckhohn, F. and Strodtbeck, F.L. (1961). *Variations in Value Orientations*. Evanston, IL: Row Peterson.

Lloyd, G.E.R. (1991). "The invention of nature." In G.E.R. Lloyd (ed.), *Methods and Problems in Greek Science*. Cambridge: Cambridge University Press (pp. 417–34).

Lin, M. (2004). "The role of analytic–holistic thinking on sensemaking." Unpublished master's thesis, Wright State University, Dayton, Ohio.

Markus, H.R. and Kitayama, S. (1991). "Culture and the self: Implications for cognition, emotion, and motivation." *Psychological Review*, 98: 224–53.

Markus, H.R., Uchida, Y., Omoregie, H., Townsend, S.S.M., and Kitayama, S. (2006). "Going for the Gold." *Psychological Science*, 17: 103–12.

Masuda, T., Ellsworth, P.C., Mesquita, B., Leu, J., Tanida, S., and Van de Veerdonk, E. (2008). "Placing the face in context: Cultural differences in the perception of facial emotion." *Journal of Personality and Social Psychology*, 94: 365–81.

Masuda, T. and Nisbett, R.E. (2006). "Culture and change blindness." *Cognitive Science*, 30: 381–99.

Miller, J.G. (1984). "Culture and the development of everyday social explanation." *Journal of Personality and Social Psychology*, 46: 961–78.

Minsky, M. (1975). "A framework for representing knowledge." In P. Winston (ed.), *The Psychology of Computer Vision*. New York: McGraw-Hill (pp. 211–77).

Morris, M. and Peng, K. (1994). "Culture and cause: American and Chinese attributions for social and physical events." *Journal of Personality and Social Psychology*, 67: 949–71.

Nakamura, H. (1985). *Ways of Thinking of Eastern People*. Honolulu: University of Hawaii Press.

Nisbett, R.E. (2003). *The Geography of Thought: How Asians and Westerners Think Differently... and Why*. New York: The Free Press.

Nisbett, R.E., Choi, I., Peng, K., and Norenzayan, A. (2001). "Culture and system of thoughts: Holistic versus analytic cognition." *Psychological Review*, 108: 291–310.

Norenzayan, A., Smith, E.E., Kim, B.J., and Nisbett, R.E. (2002). "Cultural preferences for formal versus intuitive reasoning." *Cognitive Science*, 26: 653–84.

O'Reilly, C.A. (1980). "Individuals and information overload in organizations: Is more necessarily better?" *Academy of Management Journal*, 23: 684–96.

Peng, K. and Nisbett, R. (1999). "Culture, dialectics, and reasoning about contradiction." *American Psychologist*, 54: 741–54.

Strohschneider, S. and Guss, D. (1999). "The fate of the MOROS: A cross-cultural exploration of strategies in complex and dynamic decision making." *International Journal of Psychology*, 34: 235–52.

Thomas, J.B., Clarke, S.M., and Gioia, D.A. (1993). "Strategic sensemaking and organizational performance: Linkages among scanning, interpretation, action and outcomes." *Academy of Management Journal*, 36: 239–70.

Thomas, J.B., Gioia, D.A., and Ketchen, D.J. (1997). "Strategic sense-making: Learning through scanning, interpretation, action and performance." *Advances in Strategic Management*, 14: 299–329.

Thomas, J.B., Shankster, L.J., and Mathieu, J.E. (1994). "Antecedents to organizational issue interpretation: The roles of single-level, cross-level, and content cues." *Academy of Management Journal*, 37: 1252–84.

Weick, K.E. (1979). *The Social Psychology of Organizing* (2nd edn). Reading, MA: Addison–Wesley.

—— (1995). *Sensemaking in Organizations*. Thousand Oaks, CA: Sage Publications.

Weick, K.E., Sutcliffe, K.M., and Obstfeld, D. (2005). "Organizing and the process of sensemaking." *Organization Science*, 16: 409–21.

Chapter 10

Strategies in Naturalistic Decision Making: A Cognitive Task Analysis of Naval Weather Forecasting

Susan Joslyn and David Jones

As explained in the previous chapters of this book, macrocognition is, by definition, cognition in natural contexts. Decades of research on decision making in naturalistic contexts has taught us that cognition is adapted to the constraints of the context in which it occurs (Klein, 1998). In this chapter we examine one such context. The decision makers are US Navy weather forecasters making crucial decisions that affect the safety of naval pilots. This is an extremely complex decision-making process that is adapted to the chaotic environment in which it takes place by those who are limited in time and expertise. And yet, using heuristics and rules of thumb these forecasters are able to supply high-quality forecasts for the adjacent airfield several times each day.

How do weather forecasters arrive at their decisions? What information do they use and how do they combine it to come up with the forecast? In general, weather-forecasting decisions, like other naturalistic decisions, depend heavily on the preliminary stage of situation assessment (Klein and Calderwood, 1991; Wohl, 1981). In weather forecasting, the current state of the atmosphere is determined first by examining large amounts of observational data. When the current state of the atmosphere is ascertained to the degree possible, it is then projected forward in time to obtain the future conditions. Forecasters are aided in this endeavor by numerical weather-prediction models—computer programs that predict weather by taking current weather conditions as input and transforming them algorithmically into future weather conditions according to known principles of atmospheric physics. Computer predictions are fairly accurate for large-scale phenomena over short time periods. However, they still make large errors, especially for small-scale, local phenomena and for forecasts extending further into the future. Thus, the computer predictions cannot be applied directly. Human forecasters must evaluate the computer predictions and adjust them when necessary, based in part on their understanding of local conditions and known weaknesses in the numerical models. It is interesting to note that weather forecasting is one of the few domains in which humans have improved on the decisions made by decision aids such as computer weather-prediction models (Swets, Dawes, and Monohan, 2000). However, computer models are constantly improving with advances in atmospheric science and, as their accuracy increases, the human contribution is reduced (Baars, Mass, and Albright, 2004).

Although little is known about the psychology of forecasters' understanding of computer predictions themselves, previous descriptive research has charted an idealized process by which information is combined to arrive at a weather forecast by expert forecasters. Forecasters are taught to work down the forecast funnel from the global to the local level (Krolak, 2001). They begin by analyzing the large-scale or global weather patterns, referred to as synoptic scale patterns. Then they ascertain the local conditions. At this point, expert forecasters identify a weather problem of the day. This is the weather feature assumed to have the biggest impact on the forecast user. Focusing on an important feature allows forecasters to narrow the scope of information gathering to this critical issue, a necessary step in light of the overwhelming amount of information available to the modern forecaster, much of it on the Internet (Pliske et al., 1997).

From a psychological perspective, forecasters are thought to be creating a mental representation of the current state of the atmosphere that includes both spatial and temporal components. This has been described as a mental model (Hoffman, 1991; Pliske et al., 1997; Trafton et al., 2000, Trafton, 2004). The initial mental model is derived from various information sources, then compared with other information sources and adjusted as necessary. It is at the later comparison stage that expert forecasters first consider the computer predictions, preferring to begin with their own analysis (Pliske et al., 1997). As a result, much of the information gathering takes place at this point in an effort to check the mental model (Trafton et al., 2000). Finally, forecasters project the mental model of the atmosphere forward in time to derive the future weather conditions that result in the forecast (Pliske et al., 1997). Thus, the mental model is actually four-dimensional, including a time component revealing the rate and direction of change over time. This is clearly a computationally intensive process requiring deep concentration and complex understanding of atmospheric physics.

How do the Navy forecasters compare to this idealized process? First, one must consider the context. The Navy forecasters we studied worked in a busy weather-forecasting office at an air station in the Pacific Northwest. Forecasting in this setting is accomplished in parallel with a number of competing duties such as answering questions on the telephone and giving in-person briefings to pilots. We observed one forecaster who was interrupted 13 times while completing a forecast. Figure 10.1 shows a typical scene in a naval weather office where an aircraft crew has stopped by for a weather update prior to a flight. The forecaster (to the right in the picture) must interrupt the forecasting process to search for specific weather information on the Internet to fulfill their request.

An additional challenge for Navy forecasters, compared with their civilian counterparts in the National Weather Service, is that they are often given important forecasting responsibilities after a relatively short training period. Most Navy forecasters have less than one year of formal training. Civilian forecasters receive a four-year undergraduate degree and some go on to graduate studies. Navy forecasters may also have limited local forecasting experience owing to the rapid turnover at any one location.

Yet, as determined by the Navy's own evaluation techniques (for example, pilot post-flight debriefings, weather station monthly verification of Terminal Aerodrome

Figure 10.1 Scene in a naval weather office showing interaction of flight crew and weather forecaster

Forecasts) they routinely provide high-quality forecasts with crucial information for tactical decisions. Indeed, a lens model analysis (Hammond, 1996) of forecasts for the period July–October 2003 revealed that the naval forecasters at the air station we studied[1] added value to the forecast over and above that of the computer predictions (Joslyn and Hunt, unpublished). We compared the human forecasts for three key parameters (wind speed, surface atmospheric pressure, and "ceiling" or the altitude of the cloud base) to the leading computer model (MM5). Correlations were computed between the human predictions and the observed values (criterions) as well as for the computer predictions and the observed values. For all three parameters, the human forecasters added predictive value over and above that provided by the model.

The bottom line is that naval forecasters produce very useful forecasts in a challenging environment with limited training and local experience. We used two methods to determine how they do it. First, we attempted to capture forecasters' thoughts occurring during their decision-making process. We did this by asking several forecasters to think out loud while they produced an operational forecast. One might worry that verbalizing would interfere with the task being studied, but as long as the task itself is nonverbal, such as the forecasting task studied here, interference is minimal (Ericsson and Simon, 1993). The verbalizations were then coded and grouped into well-defined categories. Overall inter-coder reliability was 89–92 percent.

---

1   No specific identifiers were used so it is not possible to determine whether individual forecasters were the same in the two studies.

Although verbal protocol analysis is a labor-intensive approach, it had three main advantages. First, it allowed us to evaluate constructs such as the "mental model" by applying well-defined criteria to verbal descriptions as they unfolded in real time, thus avoiding the loss of information and regularization that sometimes occurs from memory-based self-report. In addition, clear and reliable definitions such as those used here are less susceptible to biases in interpretation by the researcher and permit verification of the results. Finally, this approach allowed us to observe the forecasting process "in the wild" (Hutchins, 1995). This is especially important when studying macrocognition because information-gathering and decision strategies may be altered in response to naturally occurring pressures and advantages. To verify the results of this first study, we conducted a follow-up questionnaire study to evaluate the same issues over a number of different forecasts and weather situations.

A Case Study of the Terminal Aerodrome Forecast

The Navy forecasters produced a Terminal Aerodrome Forecast (TAF) for the think-aloud verbal protocol. A TAF is a 24-hour forecast for several parameters including cloud cover, cloud ceiling, visibility, winds, precipitation, and pressure for the 5-mile radius of the airport. We recorded forecasters' verbalized thoughts as they completed this task, as well as the video signal from the primary computer workstation. Five sessions were recorded over two days. The four Navy weather

Table 10.1   Information source definitions

- *Computer predictions*: A number of different numerical weather prediction models were available, typically via the Internet. These include the University of Washington's implementation of the MM5 model, the Navy's own computer prediction called NOGAPS, and several others including the National Weather Service's ETA and AVN.

- *Satellite imagery*: Satellite images accessed through the Internet or dedicated communication sources.

- *Surface observations*: Standardized reports of the current weather at various locations.

- *Local TAFs*: TAFs produced at other airfields in the region.

- *Radar*: Images that were products of real-time reflectivity and Doppler-derived wind velocity such as rainfall over a kilometer square grid, wind direction, and speed.

- *Weather discussions*: Forecaster discussions of the weather that are made available on the Internet.

- *Knowledge*: Information from forecasters' mental knowledge base. This source was not named directly by the participant but inferred by the coder based on content. Knowledge was broken down into local knowledge and computer model dynamics. Local knowledge was inferred if the statement made reference to the impact of local terrain on the weather, such as impact of the mountain ranges on local wind patterns. Computer prediction knowledge was inferred if the statement made reference to the dynamics or trustworthiness of the computer predictions.

forecasters who participated are referred to as Forecasters A, B, C, and D to preserve anonymity. Forecaster D was recorded twice. Fortuitously, the group spanned a wide range of expertise, from six months' forecasting experience to more than twenty years, allowing comparisons based on expertise. All participants had completed the Navy's course of basic training for weather forecasters.

The auditory recordings were then transcribed and broken down into individual statements in such a way as to isolate units of meaning that could be readily coded in the content areas of interest. This was our raw data. Statements were then categorized in several ways. Because we were interested in the information used in the forecast, statements were coded for the sources of information from which they were derived (see Table 10.1). The specific source was identified because it was mentioned explicitly or because it appeared on the computer screen. We were also curious about forecasters' understanding of the computer models. To illuminate this issue, statements were also coded for knowledge of computer prediction dynamics and uncertainty. Then we identified the goal the forecaster was pursuing at each stage in the process. Each non-overlapping goal subsumed several consecutive statements. The Navy forecasters we observed pursued five major goals overall (see Table 10.2).

Table 10.2   Goal definitions

1. *Analyze synoptic scale weather patterns*: This goal was identified when forecasters discussed large-scale weather patterns such as the movement of high- or low-pressure areas.

2. *Evaluate information sources to determine specific parameter values*: This goal was identified when the forecaster first mentioned specifics of the parameters forecasted, such as timing of the precipitation, cloud heights, visibility, wind speed and direction, and pressure.

3. *Consolidate mental model*: Based on previous research (Hoffman, 1991; Pliske et al., 1997; Trafton et al., 2000), this goal was identified when forecasters described a coherent, qualitative, four-dimensional, causally interrelated, spatial/temporal representation of the weather that was not derived directly to some external information source (involved inference; Trafton, 2004) and incorporated more than one parameter.

4. *Check mental model*: This goal was identified when forecasters compared the above-mentioned representation to other sources of information to evaluate the mental representation.

5. *Write TAF*: This goal was identified when forecasters began to type information into the notepad.

Mental Model versus Rules of Thumb

All forecasters shared three of the five goals. The first goal for all forecasters was the synoptic scale analysis of the weather. All forecasters also conducted an evaluation of specific parameters, such as wind speed or precipitation. Finally, all four forecasters finished by writing the TAF. When we compared forecaster statements to the criteria specified in Table 10.2 for a mental model, we found that only Forecaster B,[2] with the greatest experience, appeared to consolidate a mental model of the atmosphere and check it against other information sources. For him, these were the second and third goals pursued. A complete mental model was not detected in the verbal protocols of the other three Navy forecasters.

Forecaster A, for instance, described a partial mental model of the wind patterns. He said, "Winds will tend to go south of the Olympic Mountains to the southwest into Bremerton and up through the south, southwest…then it becomes south to southeast over us and we typically get a…small craft force over Whidbey Island… that's pretty common with normal systems." However, this is clearly specific to winds alone. In fact, upon closer inspection he appeared to be referring to a "rule of thumb," rather than a mental model. The rule has an if–then format and can be characterized as "If a system approaches the coast, there will be strong winds out of the southeast over the forecast area." Two other forecasters (C and D) refer to this same rule of thumb. A rule of thumb interpretation is further supported by the next line in which Forecaster A said, "These systems aren't normal. It's a weak trough, so it's going to act a little bit different than a normal front coming through." This remark suggests that the situation did not conform perfectly to the "if" portion of the rule, requiring an adjustment. The very fact that a "mismatch" was noted suggests comparison to a pre-formulated rule. If he had been reasoning from a four-dimensional mental model, the forecast would result directly from the manipulations of the causal elements of the mental representation precluding both comparison and mismatch.

Why was this? It may have to do with the working memory demands of creating and maintaining a complex mental representation of the current atmospheric conditions. It has long been known that the amount of information that can be maintained and manipulated in working memory, roughly synonymous with consciousness, is severely limited (Miller, 1956; Baddeley, 1987). Creating a mental model of the current state of the atmosphere means building a dynamic, complex, and detailed spatial representation in working memory and then maintaining it in working memory throughout the forecasting task.

These demands can be circumvented to some extent as expertise develops. As one becomes more familiar with a given domain, features common to similar situations, domain-specific knowledge that is stored in long-term memory, can be used to ease the load on working memory. Then, only the features unique to the specific situation need be maintained in consciousness. In fact, some believe that much of what appears to be the content of an expert's working memory is actually stored in readily available long-term memory organizational structures that have developed for that express purpose (Ericsson and Kintsch, 1995). Prior to the development of such structures, however, the burden on working memory is greater.

Thus, the approach of the majority of the forecasters studied here may have been due to insufficient expertise. Notice that the only forecaster who referred to a mental model was much more experienced than the other three. Perhaps creating and

---

2   Forecaster B was a civilian forecaster, formerly a naval forecaster, with over 20 years' forecasting experience.

maintaining a complex mental representation such as this is a skill that develops with experience and well-developed, domain-specific, long-term knowledge structures to support it.

The bottom line, however, is that much of the information required to create a complete mental model of the current atmospheric situation is not contained in long-term memory, especially among forecasters with less developed expertise. As such, the burden on working memory, or consciousness, is likely heavy. This would be something like visiting a five-bedroom house for the first time and remembering each room and all its furnishings while you analyzed the household budget. You know that there will be such things as window and floor coverings in each room (your knowledge of houses contained in long-term memory), but you must remember the specific style and color of each, the number of windows, and so on (unique qualities that must be maintained in working memory), while you conduct another analysis. It is a pretty demanding task!

Moreover, the Navy forecasters we observed were usually interrupted several times in the process of forecasting. Interruptions are particularly problematic for tasks that rely on working memory because they force forecasters to direct attention away from the forecasting task for several minutes at a time. In this situation, holding the unique features of the mental model in consciousness while directing attention to another task and then back again, may be psychologically overwhelming. Perhaps that is why three of the forecasters studied did not do it.

The three forecasters who did not make reference to a mental model of the atmosphere did refer to what might be described as rules of thumb. Rules of thumb can be memorized and stored in long-term memory because they do not change from one situation to the next. Unlike the long-term memory structures described above, however, rules of thumb do not contain *any* situation specific information. They serve to reduce the strain on conscious-level processing because they are used *instead* of a mental model of the specific atmospheric situation. We speculate that this is a preferable approach in the observed setting. It is similar to approaches taken by novices in other domains (Kirschenbaum, 1992).

Information Sources

The specific information sources forecasters evaluated are shown in Table 10.3. We inferred the degree to which a source was used from the percent of information source statements referring to it. Following the current trend (Doswell, 1992), the naval forecasters relied heavily on the computer predictions. For every forecaster, the largest percentage of information source statements referred to the computer predictions (see Table 10.3). This pattern was significantly different than would be expected by chance. Interestingly, although many computer prediction models are available, all four forecasters relied primarily on one, the University of Washington's MM5, a mesoscale model that takes into account the effects of local terrain. Two forecasters also occasionally used NOGAPS, produced by the US Navy. The latter choice may have been motivated by the NOGAPS convenient interactive interface. Unlike other models, it allowed the forecaster to request pressure, a crucial parameter for the TAF, at a very specific location. This observation is important, not because of the specific models selected, but rather because it suggests that Navy forecasters do not make use of the full range of information available. Moreover, their choices are not necessarily motivated by the quality of the information provided but rather by the need for efficiency.

Table 10.3   Percent of total source statements referring to each source by each forecaster

Source	Forecaster				
	A	B	C	D1	D2**
Computer predictions	55%	40%	70%	57%	79%
Satellite	2%	6%	7%	17%	7%
Radar	8%	1%	5%	0%	0%
Surface	21%	33%	9%	20%	7%
Forecast discussion	3%	0%	0%	0%	0%
Regional TAFs	11%	21%	9%	7%	7%
Chi-square	$\chi^2 (5, N = 242)$ $= 294.78*$	$\chi^2 (5, N = 123)$ $= 107.59*$	$\chi^2 (5, N = 44)$ $= 93.18*$	$\chi^2 (5, N = 60)$ $= 81.60*$	$\chi^2 (5, N = 29)$ $= 82.93*$

* $p < .01$.

** Forecaster D was recorded twice.

Although the computer models offer perhaps the richest source of information for forecasters, very few aspects of the computer prediction were used. The MM5 web interface offers approximately a hundred different charts, graphs, and tables (known as "products") of relevance to the TAF. The forecasters studied here used only 11 of these products overall, focusing most of their time (64 percent) on only five products. Thus, although a huge quantity of information was available, only a small percentage of it was used by the naval forecasters. Even so, they spent a surprising 20 percent of their time navigating between various websites. Perhaps more sources of information were not accessed in part because of the increase in navigation time that would accompany them.

As with the mental model, forecasters' use of information sources varied to a certain extent with expertise. The two least experienced forecasters (C and D) relied most heavily on the computer predictions, while the two most experienced forecasters (A and B) relied least on the computer predictions and more on the surface observations. Moreover, Forecaster B, the most experienced, did not look at the computer predictions until later in the process, as described in the idealized process outlined above, providing indirect evidence that he was relying on a mental model instead. This suggests that, like experts in other domains (Kirschenbaum, 1992), experience allowed Forecaster B to make better use of the raw data and depend less upon the processed information provided by the models. This is especially advantageous when computer predictions are less trustworthy or not available, because experienced forecasters can generate a prediction directly from the raw data if necessary. This approach may also allow them to do a better job of evaluating computer prediction performance in the first place.

Table 10.4   Statements categorized by goal

Goal	Forecaster					
	A	B	C	D1	D2**	Mean %
Goal 1: Synoptic Analysis Source statements	138/346 40%	12/155 8%	32/64 50%	15/94 16%	12/36 33%	31%
Goal 2: Specific Parameter Source statements	178/346 52%	101/155 65%	25/64 40%	6/94 6%	2/36 5%	33.6%
Goal 3: Write TAF Source statements	30/346 9%	5/155 3%	7/64 11%	73/94 78%	22/36 61%	32.4%

** Forecaster D was recorded twice.

When we examined the information source statements by goal, we noticed that most of the forecasters gathered information primarily in the middle of the forecasting process (see Table 10.4), as suggested by previous research (Trafton et al., 2000; Pliske et al., 1997). The approach of Forecaster D was surprisingly different however. He delayed most of the information gathering until the final stage, *after* he began writing his TAF. We speculate that this may have been another attempt to circumvent the limitations of working memory. This approach allowed him to avoid maintaining large quantities of information about the various weather features, across multiple interruptions, and over the entire forecasting process. Instead, he developed an approach in which he did not evaluate most of the information until he needed it in the writing process.

Weather Problem of the Day

We noted also that forecasters appeared to narrow their information search to products relevant to a "weather problem of the day." In the context of the TAF, the weather problem of the day can be inferred from the degree to which forecasters emphasized one weather parameter (for example, precipitation or cloud cover) over another, reflected in the proportion of statements referring to that parameter. Indeed, each forecaster made a larger proportion of statements that referred to a specific parameter (see Table 10.5). Moreover, the emphasized parameter varied somewhat by forecaster, each of whom was working during a different time period.

Table 10.5   Statements referring to forecast parameters

Parameter	Forecaster				
	A	B	C	D1	D2**
Clouds/ visibility/ precipitation	107 (33%)	68 (44%)	34 (52%)	48 (31%)	33 (43%)
Winds	170 (52%)	55 (35%)	21 (32%)	42 (27%)	19 (25%)
Pressure	51 (16%)	34 (22%)	10 (15%)	63 (41%)	24 (32%)
Total statements	328	156	65	153	76
Chi-square	$\chi^2 (2, N = 328)$ $= 64.84*$	$\chi^2 (2, N = 156)$ $= 11.23*$	$\chi^2 (2, N = 65)$ $= 13.32*$	$\chi^2 (2, N = 153)$ $= 4.59$	$\chi^2 (2, N = 76)$ $= 3.97$

*p < .01.
** Forecaster D was recorded twice.

Closer examination of the transcripts revealed that the forecasters were primarily concerned with different stages in the passage of the same strong low-pressure system that was located offshore and heading toward the coast on Monday, 18 February 2002 (see Figure 10.2). Forecaster A wrote his TAF on Sunday evening as a weak low passed through the forecast area preceding the stronger system that was still located offshore. The largest proportion of his statements referred to winds. He was concerned with the timing of the shift in wind direction and the increase in wind speed that would accompany the second front. Six hours later, when Forecaster B's TAF was written, the stronger low-pressure system was closer to the coastline. The largest proportion of his statements concerned cloud cover and precipitation because he was concerned with the onset and duration of the rainfall. He also had some concerns about fog. Forecaster C, writing his TAF 12 hours after Forecaster B, also

**Figure 10.2   GOES 10 Infrared Satellite Picture of 0000 UTC
18 February 2002**

made the largest proportion of statements referring to cloud cover and precipitation. The satellite imagery available at the time showed a break in the shield of clouds and embedded rain that was approaching the forecast area. Consequently, he was concerned with the timing of the beginning and end of two periods of rain before and after the break. Finally, Forecaster D's forecast period included the time during which the low-pressure area passed directly over the forecast region. He made the largest proportion of statements concerning pressure because he was concerned with how far it would drop and when the lowest pressure would occur. He was also concerned with when the pressure would begin to rise again. Thus, both kinds of evidence, the unequal proportions of statements devoted to individual parameters and the close examination of the statements, suggest that each of these forecasters focused on a weather problem of the day.

Understanding of Computer Model Dynamics

We were not surprised to find that the naval forecasters relied heavily on the computer predictions. Computer decision aids are crucial to even highly experienced forecasters in the modern forecasting office. For novice forecasters, such as the majority of those studied here, they were essential. As mentioned earlier, however, computer predictions are error-prone for small-scale phenomena such as the winds in the Strait of Juan de Fuca, or for forecasts more than one or two days in the future. Thus forecasters must evaluate the computer predictions and adjust them when necessary to provide the best forecast. Forecasters' ability to identify situations in which the computer predictions are inaccurate is dependent on their understanding of how the computer models work, referred to here as model dynamics. What did these naval forecasters understand about model dynamics?

To find out, we examined the category of statements called "computer model dynamics," a subset of the knowledge statements (see Table 10.1), for each forecaster. This category was further divided into three subcategories. The first subcategory was "initialization," which refers to the three-dimensional description of the atmosphere with which the computer prediction begins. It includes a blend of actual weather observations (for example, temperature or wind speed) and estimates for those values at locations where no observations are available. Because model errors tend to increase over time, predictions are more accurate over short periods, making a recent initialization better. Statements revealing knowledge about the starting conditions of the model or the recency of the initialization were included in this subcategory. The next subcategory of computer model dynamics was "model bias" and included statements referring to systematic biases associated with particular weather parameters, geographic features, or seasons. For example, computer predictions do not always capture the funneling effect that tends to increase wind speeds in the geographic corridors such as the Strait of Juan de Fuca. Thus, wind speed predictions made by computer models are often too low for such locations. The final subcategory included statements referring to strategies used to check and adjust computer prediction output, such as comparing a past forecast to observations.

We inferred that the naval forecasters had some knowledge of model dynamics because all of them made statements that referred to these issues. Most model dynamics statements were made during the synoptic scale analysis at the beginning of the process. The only exception was Forecaster B, the most experienced, who did not evaluate the models at this stage. The other three forecasters referred to the greatest number of *different* computer predictions during the synoptic scale analysis as well. Examination of the transcripts suggested that both results were due to the fact that most forecasters devoted considerable effort to evaluating overall computer model performance during the synoptic-scale analysis. It is interesting to note, however, that these analyses were largely sequential. In other words, forecasters tended to consider model predictions one at a time. This process may be similar to the sequential consideration of alternatives observed in other naturalistic settings (Klein, 1998), if we assume that each model prediction is regarded as an alternative basic forecast that could be selected as a starting point. Only rarely did forecasters compare the computer predictions directly to one another, a more computationally expensive approach.

We noticed two broad strategies to evaluate computer model performance, an intuitive pattern-matching strategy and an analytic error-estimation strategy. The first strategy involved comparing the spatial and temporal patterns depicted in the computer prediction to those in the observations, usually satellite imagery (see Figure 10.3). This appeared to be a process of mentally lining up patterns (of an

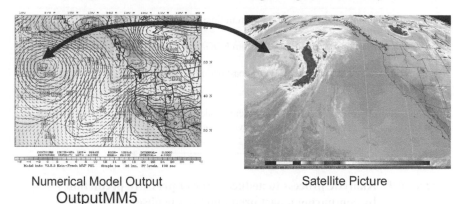

Numerical Model Output	Satellite Picture

OutputMM5

Figure 10.3   Pattern matching process to evaluate model performance

Note: Pattern matching process to evaluate model performance by comparing the position and movement of a weather system depicted in the model (left) and in the satellite imagery (right).

advancing weather system, for instance) in the separate displays and then monitoring parallel changes in prominent features over time. The TAFs of Forecasters A and C, who favored this strategy, suggest that they may have adjusted their forecasts accordingly, although Forecaster A is the only one who announced this intention explicitly. Later in the process, they both focused on individual models and used them to forecast specific parameter values, in some cases with adjustments to timing (for example, when precipitation starts) to adjust for the error observed in the large-scale analysis.

The other two forecasters, B and D, appeared to focus on specific computer model predictions from the start, skipping the earlier multiple-model evaluation stage. However, they did evaluate the predictions of the models they favored, employing the second strategy, analytic error estimation. This involved comparing specific predicted parameter values (for example, pressure) to the observed values. For example, Forecaster D took the pressure predicted for the present time period from an earlier forecast, and subtracted it from the observed pressure to detect a 0.05-inch error. He then adjusted future predicted values by that same amount (see Figure 10.4). The reasoning appears to be if there is 0.05 error now, that same amount of error will propagate into future forecasts. He then pointed out a bias of the particular model—that it tended to under-forecast both dropping and rising pressures—to explain why he thought the error had occurred. These data imply a process in which forecasters first assess the error in the computer prediction and then adjust their own forecast accordingly.

In summary, we observed different approaches among the naval forecasters studied here. To some degree they may be correlated with expertise. Forecaster B, the most experienced, closely followed the idealized process derived from previous work. He was the only forecaster who used a mental model rather than rules of thumb to arrive at his forecast. He employed very few external information sources while creating his mental representation and did his information gathering later in

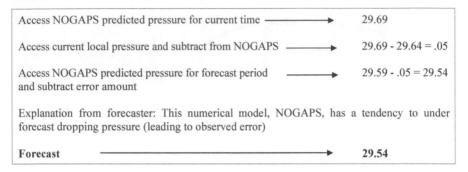

Access NOGAPS predicted pressure for current time ⟶   29.69

Access current local pressure and subtract from NOGAPS ⟶   29.69 - 29.64 = .05

Access NOGAPS predicted pressure for forecast period ⟶   29.59 - .05 = 29.54
and subtract error amount

Explanation from forecaster: This numerical model, NOGAPS, has a tendency to under forecast dropping pressure (leading to observed error)

Forecast ⟶   **29.54**

**Figure 10.4   Analytic process to deduce error in pressure prediction
by comparing a past prediction to the observation and
correcting future predictions by that amount**

the process, in a manner similar to other expert forecasters (Hoffman, 1991; Pliske et al., 1997; Trafton et al., 2000). However, it must be noted that this forecaster was trained prior to widespread use of computer prediction decision aids. Hence, the approach could be a result of the habit and training rather than expertise per se.

Importantly, for the majority of naval forecasters, the forecasting process observed in this real-life setting was different in several respects to the idealized process. The forecasting experience of these forecasters was gained mainly in a busy naval forecasting office such as the one we observed. A new TAF must be posted every six hours. Thus, regardless what other duties are competing for the forecaster's attention, the TAF must be produced on time. The forecasters are interrupted several times during the production of the TAF to answer questions on the phone and to provide in-person briefings to pilots and officers. Our observation suggested that these circumstances induced forecasters to develop a forecasting process to be immune to these pressures. Thus, instead of creating a complex and detailed mental model of the current state of the atmosphere, much of the problem solving was based on if–then rules. This is consistent with the approach observed in non-experts in other studies (Kirschenbaum, 1992; Pliske et al., 1997). Instead of conducting head-to-head comparisons of multiple computer predictions to select the best one upon which to base the forecast, model predictions were compared sequentially or not at all. In fact, half of the forecasters studied here appeared to focus a single computer model rather than comparing models to select the one that was currently performing best. Instead of gathering large quantities of information to bring to bear on the forecast, information-gathering strategies appeared to have been streamlined to use relatively few information sources; many of those were selected for ease of access. In fact, one forecaster delayed gathering information entirely until he had begun to write the TAF.

Because this study was conducted over a two-day period during which particular weather conditions prevailed, it is difficult to tell whether this was a more widespread adaptation or particular to this group of forecasts. To resolve this, we conducted a second study using a very different approach.

TAF Questionnaire Study

In a second study we asked Navy forecasters to fill out anonymous web-based questionnaires immediately after completing several TAFs, when possible, over a four-month period. Four forecasters responded consistently and their responses are analyzed here. They are referred to as Forecasters E, F, G, and H.[3] Because the questionnaire study was undertaken to find out whether the streamlined approach would be observed over several forecasts and weather conditions, the questionnaire asked respondents to indicate the sources of information, weather problem of the day, and model evaluation strategies used for the TAF just completed.

In addition, in light of the well-developed understanding of computer model dynamics observed in the verbal protocol analysis, we wanted to know how forecasters responded when they determined that the computer predictions were doing a poor job of predicting the weather. Perhaps, for instance, forecasters would examine additional or different information sources or use additional model evaluation strategies in an attempt to compensate for the inadequate computer predictions. To that end, we also asked respondents to rate computer model performance, confidence in their own forecast, and the difficulty of the TAF just completed so that we could examine more closely those forecasts for which model guidance was determined to be less trustworthy.

Information Sources

At the end of the four-month period when we examined their responses, we again found evidence for a highly streamlined process. These forecasters also relied on very few information sources, primarily the computer predictions, when producing the TAF. The same two models, the MM5 and NOGAPS, were used repeatedly. Their responses to the inquiry about the weather problem of the day revealed how central the computer predictions were to the process. Recall that this is a common term to refer to the weather feature determined to have the biggest impact. Fog or cloud cover, for instance, have a big impact on the forecasts studied here because these features are crucial for aviation. In fact, respondents cited fog and cloud cover as the weather problem of the day for several TAFs. However, in almost half of the cases, the novice forecasters we studied misinterpreted the question concerning the weather problem of the day, citing the computer predictions as the biggest "problem," for example, "models are underforecasting winds," and "models have a hard time handling strong pressure systems." This demonstrates how crucial the models are to this forecasting group. When the models are deficient, it is their biggest problem.

---

3   Because participants in this study were anonymous we do not have information on their forecasting experience. However, we do know that they all had the Navy's basic training for forecasters.

Forecasting Routine

The forecasters appeared to have a fairly set forecasting routine that they followed for most TAFs. For the most part, they used just two basic strategies. They compared synoptic weather patterns shown in the model initialization to satellite images and they compared predicted with observed parameter values. In about 40 percent of the cases they made model bias adjustments. To a lesser degree, they used more sophisticated model evaluation methods. On fewer than one-third of the TAFs did they compare computer predictions directly with one another. In fewer than 20 percent of TAFS did forecasters evaluate model performance over more than a single observation (trends in model performance). Again, these latter two strategies are more computationally intensive as they require considering multiple comparisons simultaneously to reach a judgment and may have been avoided for that reason.

We anticipated that forecasters would alter their forecasting strategies when model performance rating declined, perhaps using additional evaluation techniques or sources of information to compensate for model prediction deficiencies. However, neither the sources of information nor evaluation techniques correlated with the model performance rating. In fact, there was very little variation in the forecasting routine at all. Two forecasters used virtually the same information sources and model evaluation techniques on *every* forecast. The other two forecasters indicated slightly more variability in the products they used, but the differences were not related to model ratings. Clearly, forecasters had, for the most part, developed a forecasting routine that varied little. This may have been a response to the pressures inherent in the forecasting situation. Making on-line adjustments to information-search and computer-model evaluation techniques in response to the current situation requires conscious-level processing, that is, working memory. On the other hand, using static, tried and true forecasting routines requires little in the way of conscious adjustment. Once memorized, the routine can be stored in long-term memory, retrieved, and enacted automatically, making few demands on working memory resources.

Forecasting from Long-term Memory

Surprisingly, confidence in the forecast did not appear to be related to model performance or the difficulty of the forecast. Instead it appeared to be related to the degree to which the current situation conformed to general principles. The written explanations for forecaster confidence proved revealing in this respect. The explanations fell into two general categories. In the first, level of confidence was explained by the degree to which the observed situation conformed to a rule of thumb, such as, "A building ridge leads to light winds, normally..." The logic seemed to be that when the situation at hand conforms to the "if" part of the conditional, one can be confident that the "then" will follow. In the second category, degree of confidence was explained in terms of model strengths and weaknesses. For example, one forecaster explained lack of confidence in his pressure forecast because "Models have a hard time handling strong pressure systems." Thus, their confidence in the forecast appeared to be related to the degree to which the situation conformed to general principles rather than specifics in the given forecasting situations. That is, confidence was high when the weather situation conformed to a well-known rule of thumb or was a situation known to be handled well by computer predictions.

Thus, Navy forecasters appeared to have some understanding of the uncertainty inherent in computer predictions and made an attempt to evaluate it, albeit using procedures streamlined for efficiency. The forecasters we studied tended to avoid potentially informative but computationally intensive analyses such as direct comparisons of computer predictions and evaluation of trends in model error. Moreover, although we observed a few corrections tailored to the situation, many of their adjustments appeared to be bias corrections. Addressing general well-known model biases allowed them to use the same solutions over and over rather than devising new situation-specific solutions.

In addition, it is important to note that although the use of probabilistic products is of growing interest to the weather forecasting community (National Academies, 1999; National Research Council, 2003) and most major numerical weather prediction centers provide probabilistic forecast products on the Internet, the naval weather forecasters studied here did not report using a single probabilistic product. We speculate that they did not have an understanding of the computer model uncertainty, regarding it instead as error that could be corrected. These Navy forecasters appeared to use their evaluation techniques to make adjustments to the computer prediction, essentially turning forecast uncertainty into a deterministic forecast. This may be why we failed to see a correlation between computer model performance ratings and confidence in the forecast in the questionnaire study. Perhaps the forecasters thought that they had "fixed" any problems with their adjustments rather than regarding these error-prone forecasts as inherently uncertain. Notice that this is a point about their understanding of situation rather than about the forecast they made. We speculate that they would make the same adjustment whether they saw it as error or uncertainty. However, if it were regarded as uncertainty, it would reduce their confidence in the forecast.

General Conclusion

In these studies, we observed a forecasting process that was adapted to the situation in which it was conducted, replete with time pressure, limitations in experience, and interruptions. The process observed here in context is quite different from the idealized process in many respects. Much of the data presented here suggests that situation assessment in the forecasting process was streamlined for efficiency and geared toward routine solutions. Forecasters used very few sources of information overall and relied heavily on the computer prediction decision aids. In many cases they used the computer predictions as a starting point.

We found that most Navy forecasters relied upon rules of thumb rather than elaborate mental models of the current state of the atmosphere to make forecasting decisions. As such, the strain on conscious-level processing was diminished. When the current situation matched the "if" condition in the relevant rule, that same rule could be applied with minimal additional effort.

Moreover, we found that information-gathering and model-evaluation processes had been reduced to a routine. Forecasters relied on a few tried-and-true procedures rather than creating new solutions specific to each situation. Similar success by non-expert decision makers using simple decision rules has been noted in other domains (Leprohon and Patel, 1995). In addition, these forecasters relied on favored models and products, suggesting selection based on habit and ease of access rather than on the weather problem of the day or on superior relative performance of a computer model in the current situation. Although this inflexible approach is unlikely to produce the optimal decision in the abstract, we submit that such strategies may well be a successful adaptation to the circumstances. Instead of an error or a deviation from prescribed analytic procedures, they allowed for complex decision making by relative novices under time pressure and amid constant interruption. This approach provided the forecasters with adequate answers and allowed them to get the job done in a timely fashion.

Acknowledgements

This research was supported by the DOD Multidisciplinary University Research Initiative (MURI) program administered by the Office of Naval Research under Grant N00014-01-10745.

References

Atkinson, R. and Shiffrin, R. (1968). "Human memory: A proposed system and its control processes." In K.W. Spence and J.T. Spence (eds), *The Psychology of Learning and Motivation: Advances in Research and Theory.* New York: Academic Press (Vol. 2, pp. 89–195).

Baars, J., Mass, C., and Albright, M. (2004). "Performance of National Weather Service forecasts versus model output statistics." *Proceedings of the 20th AMS Conference on Weather Analysis and Forecasting/16th Conference on Numerical Weather Prediction, Seattle, WA.* Boston, MA: American Meteorological Society.

Baddeley, A. (1987, c.1986). *Working Memory.* Oxford: Clarendon Press.

Dawes, R., Faust, D., and Meehl, P. (1989). "Clinical vs. actuarial judgment." *Science*, 243: 1668–74.

Doswell, C. (1992). "Forecaster workstation design: Concepts and issues." *Weather and Forecasting*, 7: 398–407.

Ericsson, K.A. and Kintsch, W. (1995). "Long term working memory." *Psychological Review*, 102, 211–45.

Ericsson, K.A. and Simon, H. (1993). *Protocol Analysis: Verbal Reports as Data.* Cambridge, MA: MIT Press.

Hammond, K. (1996). *Human Judgment and Social Policy: Irreducible Uncertainty, Inevitable Error, Unavoidable Injustice.* Oxford: Oxford University Press.

Hoffman, R. (1991). "Human factors psychology in the support of forecasting: the design of advanced meteorological workstations." *Weather and Forecasting*, 6: 98–110.

Hutchins, E. (1995). *Cognition in the Wild.* Cambridge, MA: MIT Press.

Leprohon, J. and Patel, V. (1995). "Decision making strategies for telephone triage in emergency medical services." *Medical Decision Making*, 15: 240–53.

Kirschenbaum, S. (1992). "Influence of experience on information-gathering strategies." *Journal of Applied Psychology*, 77: 343–52.

Klein, G. (1998). *Sources of Power: How People Make Decisions.* Cambridge, MA: MIT Press.

Klein, G. and Calderwood, R. (1991). "Decision models: Some lessons from the field." *IEEE Transactions on Systems, Man, and Cybernetics*, 21: 1018–26.

Krolak, R. (2001). *Aerographer's Mate: Module 5—Basic Meteorology, NAVEDTRA 14312.* Pensacola, FL: US Naval Education and Training Professional Development and Technology Center.

Miller, G. (1956). "The magical number seven, plus or minus two: Some limits on our capacity for processing information." *Psychological Review*, 63: 81–97.

National Academies (1999). *A Vision for the National Weather Service: Road Map for the Future.* Washington, DC: The National Academies Press.

National Research Council (2003). *Environmental Information for Naval Warfare.* Washington, DC: The National Academies Press.

Pliske, R., Klinger, D., Hutton, R., Crandall, B., Knight, B., and Klein, G. (1997). *Understanding Skilled Weather Forecasting: Implications for Training and Design of Forecasting Tools* (Technical Report AL/HR-CR-1997-003). Brooks Air Force Base, TX: Air Force Material Command, Armstrong Laboratory.

Swets J., Dawes, R., and Monahan, J. (2000). "Psychological science can improve diagnostic decisions." *Psychological Sciences in the Public Interest* (a supplement to *Psychological Science*), 1: 1–26.

Trafton, J.G. (2004). "Dynamic mental models in weather forecasting." *Proceedings of the Human Factors and Ergonomics Society 48th Annual Meeting*, 20–24 September 2004, New Orleans, LA. Santa Monica, CA: HFES.

Trafton, J.G., Kirshenbaum, S., Tsui, T., Miyamoto, R., Ballas, J., and Raymond, P. (2000). "Turning pictures into numbers: Extracting and generating information from complex visualizations." *International Journal of Human-Computer Studies*, 53: 827–50.

Wohl, J. (1981). "Force management requirements of Air Force tactical command and control." *IEEE Transactions and Systems, Man, and Cypbernetics*, 9: 618–39.

Chapter 11

Supporting Macrocognition in Health Care: Improving Clinical Reminders

Laura G. Militello, Emily S. Patterson, Jason J. Saleem,
Shilo Anders, and Steven Asch

Introduction

Medical decision making is complex in the best of circumstances. In the context of a busy outpatient clinic such as the ambulatory care unit of a Veterans Administration (VA) hospital, judgments about which treatments to recommend must consider the patient's lifestyle, the patient's ability to understand and comply with the treatment regimen, as well as interactions among an often long list of conditions for which the patient is being treated. Providing quality care requires that the physician thoroughly review the patient history, discuss the current condition, and assess the patient's understanding level and ability to comply with recommended treatment regimens. Under optimal conditions, physicians are able to integrate information from recent and long-standing problems, assess the patient's current condition, conduct a conversation that includes both elicitation of information from the patient as well as communication of sometimes complex health information, and document the interaction—all in the space of a 30-minute exam.

Recognizing that optimal conditions rarely exist, the VA has been a leader in introducing innovative technologies intended to streamline medical informatics, aid healthcare providers[1] in following best practices, and increase patient safety. This chapter focuses on one particular innovation—clinical reminders intended to act as a decision aid to primary care providers. Computerized clinical reminders are intended to highlight aspects of preventative care and disease management that might be overlooked in the context of a busy exam. Each time the electronic patient record is accessed, a list of clinical reminders relevant to the patient appears. Normal workflow in an ambulatory care clinic includes frequent interruptions, coordination with intake, consultations with specialists, and in the case of residents in training, discussion with a senior physician, all of which compete for the physicians' time with and attention to the patient. Clinical reminders are designed to reduce the likelihood that something will "fall through the cracks" during a busy exam.

The clinical reminders are available from two different screens in the electronic patient record. A list of clinical reminders relevant to the patient at the time of the visit

---

1 Throughout this chapter, the term "provider" will be used to denote healthcare providers who conduct patient exams including physicians, physician assistants, and nurse practitioners.

appears on the cover sheet of the patient record to serve as a prompt while the physician plans what will be covered during the exam. The reminders are also available from the screen used to document activity in a progress note. As the provider interacts with the patient and makes notes about the current visit, she or he is encouraged to click on individual reminders, which brings up a dialog box. The dialog box includes questions to ask the patient, as well as recommended treatment options. The dialog box also provides edit fields to facilitate accurate documentation in the note.

While the potential benefits of such a system are significant, and in fact, some VA hospitals are showing an increase in compliance with some best practices (Demakis, Beauchamp, and Cull, 2000), it is generally understood that some healthcare providers within the VA do not use the clinical reminders. The objective of this research was to better understand barriers to use and generate design recommendations to increase the usability and effectiveness of the clinical reminder software. This work builds on previous research exploring barriers to the use of clinical reminders in the context of HIV clinics (Patterson, Doebbeling, et al., 2004; Patterson, Nguyen, et al., 2004).

The "Method" section of this chapter describes three methods, including a taxonomy development exercise, a survey, and ethnographic observations. Data from these three sources were used to triangulate on a set of findings framed around barriers to the use of computerized clinical reminders. The "Design Seeds and Best Practices" section of this chapter details a set of recommendations intended to ameliorate these barriers. The final section of this chapter provides a discussion of this project in the larger context of the VA system.

Method

To better understand the use of clinical reminders, the research team conducted parallel analyses. One portion of the effort focused on examining the reminders themselves, categorizing them into a functional taxonomy intended to clarify the intended objectives and areas of responsibility of the actions specified by individual reminders within the healthcare delivery system. In the second portion of the effort, the research team administered a survey and conducted ethnographic observations to better understand the barriers and facilitators to clinical reminder use. Based on our findings, we have developed a set of design ideas intended to transform the reminders into a more helpful tool for the healthcare team.

Functional Analysis

The research team obtained printouts describing each of the 65 reminders currently in use at one site. Each printout consisted of the clinical reasoning and logic for when the reminder is applicable to a patient, as well as other information related to the reminder, such as the author(s) of the reminder, and editorial history. The reminders were sorted into categories, based on the objective of each reminder as inferred by the research team. Our goal in developing this taxonomy was to infer the function behind the various reminders, in order to compare the logic of its design with its actual use.

Categories were then discussed with a VA physician who was able to articulate the rationale behind many of the reminders, resulting in reformation of some categories and reclassification of some reminders. This activity continued in an iterative fashion. In cases where reminders related to more than one category, a reminder was classified by its strongest fit to a specific category. As the research team visited other sites for observation, additional reminders were considered in the context of the taxonomy. Furthermore, clinical reminders from a non-VA hospital were added to the sample of reminders to be categorized.

The taxonomy development process served to familiarize the research team with the clinical reminders, providing important background knowledge that informed the barrier identification portion of the study.

Barrier Identification

Survey Data A survey was administered at a week-long computer-training meeting for VA employees (Militello et al., 2004). Both clinicians and clinical computer professionals attended the meeting. Volunteers were recruited via a general announcement. A total of 261 respondents completed the survey, representing 104 of 142 VA medical facilities across the United States. The questionnaire was comprised of four open-ended questions appended to a quantitative survey (See Fung et al., 2004, for a discussion of the quantitative portion of the survey). The four open-ended questions focused on clinical reminder training, difficulties, benefits, and surprises:

Q1: What has helped you learn and incorporate automated clinical reminders in patient care?

Q2: What makes use of automated reminders difficult in patient care?

Q3: Has an automated clinical reminder ever helped you deliver care more effectively? If yes, please give an example.

Q4. Have you ever been surprised by the actions of any computerized clinical reminders in CPRS? If yes, please give an example.

Ethnographic Observations A team of three researchers visited a total of four VA hospitals in different geographic regions of the country. Researchers acted as unobtrusive observers of both the intake and patient exam. Nurses were observed interacting with clinical reminders as they conducted the intake portion of the patient visit. Physicians, physician assistants, and nurse practitioners were observed as they prepared for, conducted, and documented patient exams. Interviews were conducted opportunistically during breaks in the day. Every effort was made to ensure that the clinical environment was not disrupted by the presence of the researchers.

A total of 32 nurses and 55 providers (including physicians, physician assistants, and nurse practitioners) were observed. Participants were selected opportunistically at each site. Observers recorded extensive time-stamped field notes, which were later typed to facilitate qualitative data analysis.

Results from the taxonomy development, survey, and ethnographic observations are discussed below in the "Findings" section.

Findings

Functional Taxonomy

Analysis of reminders for placement in a functional taxonomy revealed the complexity of the reminders system. Many reminders could easily fit into more than one category. Further, many classification schemes could be used to categorize the reminders.

With our goal in mind of inferring the function behind the various reminders, we found that nearly all reminders could be grouped into two general categories of "Standardize Care" and "Reduce Error." Eleven subcategories were articulated and defined within these broad two categories (see Figure 11.1):

Category	Perceived Benefactor
Standardize Care	
1. Clarify responsibility	Provider
2. Assign and enforce responsibility	Administration
3. Educate regarding national or local guidelines	Provider
4. Enforce guidelines	Administration
Reduce Error	
5. Simplify documentation	Provider
6. Force documentation	Administration
7. Prevent forgetting well-known actions	Provider
8. Keep track of when specific procedures/tests are due	Provider
9. Identify patient is in a group	Provider
Other	
10. Collect information	Research Team
11. Broadcast patient at risk	Provider

Figure 11.1   Clinical reminder functional taxonomy

1. *Clarify responsibility* refers to reminders that highlight a generally routine aspect of care to be carried out by a specific role within the healthcare team. For example, at one site a depression screening reminder was assigned to the intake nursing staff.
2. *Assign and enforce responsibility* refers to reminders that are used to emphasize a less routine, and sometimes controversial aspect of care, to be carried out by a specific role within the healthcare team. For example, some sites use a reminder that requires the provider to remind the patient to use sunscreen. While many providers report that they routinely discuss the importance of sunscreen with patients independent of the reminder software, the software implies a greater level of accountability and allows the facility to track how often the sunscreen reminder is satisfied.
3. *Education regarding national or local guidelines* describes reminders that are designed to disseminate recently developed or previously unsystematically implemented practices based on guidelines. For example, a post-traumatic stress disorder screening was implemented based on national guidelines regarding treatment of Gulf War veterans.

4. *Enforce guidelines* refers to reminders that are designed not to just disseminate guidelines, but also to ensure that they are followed. The reminder includes an implied tracking or enforcement function. For example, a reminder might suggest that every patient, regardless of age should be counseled to have a flexible sigmoidoscopy in which a scope is used to conduct an examination of the rectum and colon. In this case, some physicians believe that for patients with a short life-expectancy, the flexible sigmoidoscopy adds unnecessary stress, particularly if the patient is unlikely to be able or willing to withstand treatment for colon cancer if it were detected. Others believe all patients should be counseled to have this colon cancer screening procedure done. The reminder, as implemented at one site, falls clearly on the side of the latter. The reminder also suggests that compliance with the reminder on the part of the healthcare provider will be tracked, implying therefore that non-compliance will be noted as well.

5. *Simplify documentation* refers to reminders that are designed to streamline record keeping. Often the reminder includes a dialog box with radio buttons and checkboxes that reduce the amount of free-text the provider or intake nurse must enter.

6. *Force documentation* describes reminders that require entry of documentation at a level of detail not valued by all providers, or even available for all patients. For example, an influenza vaccine reminder implemented at one site requires the provider to enter the exact date of the patient's last flu shot. Often the patient cannot remember the exact date of the previous vaccination. For some providers, noting that the vaccine was received in the fall of the previous year is an adequate level of detail. The reminder software, however, requires that a specific date is entered in order for the reminder to be satisfied.

7. *Prevent forgetting well-known actions* refers to reminders that address routine aspects of care. This includes a wide range of activities including reminding patients to wear a seatbelt and to get hepatitis C screening.

8. *Keep track of when procedures/tests are due* refers to reminders that aid the healthcare team in keeping track of events that occur at a specific interval. For example, one reminder encourages the provider to conduct a diabetic foot exam. This reminder appears only when the diabetic foot exam is due for the individual patient.

9. *Identify patient is in a group* refers to reminders that highlight important characteristics about the patient that make the patient eligible for certain treatment options. For example, one reminder is designed to identify women and remind the provider to offer a yearly Pap test to screen for cervical cancer. Another identifies patients who have had active-duty military experience and encourages the provider to conduct military sexual trauma screening.

10. *Collect information* refers to reminders that are designed to gather data, often for clinical research. For example, one site implemented a reminder directing providers to ask all patients diagnosed with hepatitis a series of questions about behaviors that put a person at risk for hepatitis. The data, while not directly relevant to the treatment of hepatitis, were used in a study to better understand risky behaviors with the intent of designing educational programs.

11. *Broadcast patient at risk* describes only one reminder identified in this study, and therefore may be considered an exception. At one site, a reminder was created to send a broadcast message to all providers that a residential patient suffering from dementia was missing, and likely wandering somewhere in the VA facility. This reminder is quite different from the others in that it is not relevant to the patient that the provider is treating at the moment, but instead an urgent issue for the larger facility.

Additional sorting of these functions focused on the perceived benefit for each function, as indicated in Figure 11.1. The left column of the figure depicts the functional taxonomy, and the right column shows the primary benefactor for each of these functions or objectives. While some functions are clearly intended to directly aid the healthcare team in providing care to the patient, other functions are more likely to be perceived as benefit to the administration or other concerns (that is, a research team). Previous research in the other industries indicates that users of any technology are more likely to adopt the technology if they perceive a direct benefit to themselves in fulfilling their jobs. On the other hand, if the people who suffer the deficits of the technology are not the same people who realize its benefits, problems in user acceptance are likely to occur. Norman refers to this as Grudin's Law: "When those who benefit are not those who do the work, then the technology is likely to fail, or at least, be subverted." (Grudin, 1987, quoted in Norman, 1993, p. 113). Thus, our observations suggest that those reminders that are designed with the individual provider in mind will be more successfully implemented and adopted than those designed to aid the organization in tracking and enforcement tasks, or those designed for an external research team.

The taxonomy development exercise was useful in examining current reminders to better understand potential barriers to use. Those that appear to benefit someone other than the provider are likely to raise questions about the value of this tool from the perspective of a busy healthcare provider. Further, the taxonomy provides a context for developers of future clinical reminders to consider what types of functions can be effectively accomplished via computerized clinical reminders.

Survey Findings

Responses to the four open-ended survey questions provided insight into the clinical reminders from the provider perspectives. Questions addressed issues of training available for learning to use the clinical reminder software, barriers to use, examples of benefits of the clinical reminders, and automation surprises. Responses to Question 1 indicate that support for learning to use clinical reminders software was highly variable across sites. Some respondents reported access to formal training as well as access to a clinical and/or computer support personnel with expertise in using clinical reminders who served as an SME to the clinic. Others report access to neither. Some felt forced to use the software without adequate training. Thirty percent of respondents attributed their own learning to self-teach strategies such as taking time to practice, taking advantage of templates made available, using web help and the documentation associated with the clinical reminders.

Regarding factors that make clinical reminders difficult to use in patient care, 50 percent of the respondents cited workload as the biggest barrier. Other factors include perceived lack of relevance, hard-to-use software, reminders that were not applicable to the patient, and insufficient training.

When asked for examples of situations in which clinical reminders aided physicians in providing care more effectively, a broad range of situations were described. These were sorted into ten categories. Of these categories, six suggested benefit directly to the provider in providing care. One category included examples in which providers were reminded to help patients with secondary issues such as sunscreen and seatbelt use. Two categories suggested benefit to the organization in administrative tracking tasks.

Table 11.1   Summary findings from survey data

Q1: CR Training		Q2: Barriers		Q3: CR Benefits		Q4: Automation surprises	
Category	%	Category	%	Category	%	Category	%
Internal support	32	Workload	50	Remind about infrequent actions	30	Inaccurate diagnostic code	43
Self-motivation	30	Relevance	20	Comply with recommended actions	25	Unexpected effect from resolution option	14
Formal training	22	Hard to use	12	Support information scanning	10	Reminder did not apply	11
Negative reinforcement	11	Not applicable	9	Discover unexpected information	8	Unexpected effects by following advice	7
Other	3	Training	9	Increase consistency in actions	5	Reminder did not recognize data	7
				Reduce the time to take an action	4	Other	18
				Focus on secondary issues	7		
				Help track performance	4		
				Encourage more accurate documentation	3		
				Other	3		

Findings from the survey are summarized in Table 11.1. Within each column, response categories are listed along with the percentage of responses that appeared in each category.

Ethnographic Observation Findings

Using traditional qualitative data analysis methods, data from the ethnographic observations were analyzed for emerging themes. Observation notes were sorted into emergent categories, which were then discussed by the research team. As a result of the discussion, some categories were merged and others split into logical subcategories. This process continued in an iterative fashion. Within each category, themes were extracted relating to common practices in the use of clinical reminders, barriers to use, and facilitators to use.

Of these themes, we chose barriers to reminder use as a framework for considering potential changes to the software. Barriers to reminder use identified via ethnographic observation were sorted into five categories including workflow, coordination, workload, lack of flexibility, and usability. (See Saleem et al., 2005, for a detailed discussion of the barriers to and facilitators of reminder use.)

Workflow   Data indicate that, in many cases, clinical reminders have not been smoothly integrated into workflow. They continue to be viewed as an extra task. Many healthcare providers are confident that the items in the reminder list are simply standard care and unlikely to be forgotten or overlooked. Working through the reminder dialog boxes then becomes an additional task that is sometimes out of sync with the rest of the exam.

Coordination   Observation data suggest that the clinical reminders were not designed adequately to support coordination between intake and the patient exam in all circumstances. For example, one area of confusion in some sites is whether a specific reminder is to be completed during intake, or by the provider during the patient exam.

Workload   Although there are many contributors to workload beyond the design of the reminder interface, our focus is on the reminders themselves. The number of reminders in use varies widely from site to site. Observation data indicate that a larger number of reminders was associated with a higher perceived workload. As the list of reminders in the system becomes longer, entering new patients becomes increasingly unwieldy as most reminders are due at the initial or second visit. If there is not time to work through the entire list with the new patient, the list of reminders carries over and can compound over time.

Lack of flexibility   The clinical reminder software is not able to take into account important variables such as the patient's life expectancy, lifestyle, and support system. Because of this, the provider's clinical judgment differs at times from the recommendations of the clinical reminders (which are generally based on national guidelines). In addition to these conflicts between context-based judgment and generic national guidelines, clinical reminder users reported instances in which inapplicable reminders appeared; and situations in which the logic behind the reminder seemed to be faulty, resulting in unpredictable or counterintuitive responses. These disconnects can take on inflated importance as some hospitals maintain statistics on each provider's compliance with clinical reminders. Some providers perceive that they are being "dinged" every time they fail to follow the recommended actions in the reminder. In many cases, users find that they have no obvious reporting mechanism to communicate problems and disagreements with the clinical reminders, or a clear source to consult in order to better understand the logic behind the reminder in question. This lack of flexibility in the clinical reminder system increases frustration and reduces trust in the software. Some users perceive the clinical reminders to represent a challenge to their own clinical judgment.

Usability   Usability is a constantly evolving issue as a system is fielded, refined, and improved over time. Observations and interviews highlighted specific usability issues that had become barriers to some clinical reminder users. These usability issues included difficulty accessing the reminders, as well as the design of the reminder dialog box itself.

Although we did not find 100 percent overlap in the barriers identified in the survey data and those identified via ethnographic observation, the findings were consistent across both methods. It is not unusual for observations to reveal additional information to that obtained via self-report.

Design Seeds and Best Practices

After thorough exploration of the qualitative data and discussion of these emergent themes, design sessions were held in which the research team met to generate ideas or *design seeds* (Patterson et al., 2001; Woods, 1998) intended to minimize the barriers to reminder use. We use the term "design seed" to indicate that these are not fully developed design concepts, but rather seeds that will likely need to develop and perhaps adapt to fit within the existing software architecture. The design seeds are linked to specific barriers in order to retain important context needed to communicate the intent behind each design seed.

Workflow Integration

Integrating the clinical reminders into core workflow is key to increasing use of the clinical reminders. As long as they are perceived to be an additional task, reminders are likely to be shed when workload increases (Patterson et al., 2001). This is particularly true from a provider's point of view. The intake conducted by the nurse is relatively well scripted, but the provider must budget time and weigh priorities for each patient exam. A typical workflow can be broken down into five phases as depicted in Figure 11.2. The provider generally reviews previous notes including consults, and begins to plan for what will be covered during the exam. This can be in the form of a mental plan, a handwritten list, or even entries in a new progress note. Interactions with the

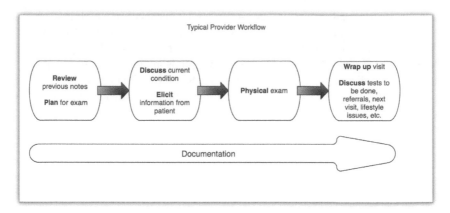

Figure 11.2   Typical provider workflow

patient often begin with a discussion of their current condition and any complaints. The provider asks a series of questions to elicit information from the patient. At some point, a physical exam tailored to the patient's condition is conducted. Often, before the patient leaves, the provider wraps up the visit, checking to make sure that all the patient's questions have been answered and going over any new medications, tests to be administered, consults ordered, and appointment to be made. At any point in this process, including after the exam is complete and the patient has left, the provider may document aspects of the exam in the new progress note.

Of course, there is considerable variability with regard to how much time is spent in the different phases and even where they occur. Some providers prefer to document the exam at their own workstation, separate from the exam room. Others make notes in the exam room, while the patient is present. The variability depends on a range of factors including time available, personal preference of the provider, as well as the workspace and scheduling conventions at a particular hospital. In some cases, reviewing notes and planning for the visit may occur before the patient arrives. In other cases, the provider is reviewing previous notes and formulating a plan for the visit as they interact with the patient.

Some providers indicated that they prefer to begin a new note before the patient arrives. Others routinely create and complete the documentation during the exam. Some prefer to wait until the patient has left before filling in the details of the notes. Many were observed to document aspects of the exam at many different stages, depending on the patient and time pressure.

In spite of this variability, recognizing the different stages of the exam has important implications for design. Reminders relevant to a particular phase may benefit from different display and interface features. For example, the current reminder design requires the provider to open a new note before interacting with the reminders, limiting the provider's ability to manipulate and organize the reminders during the review and planning phase. Further, some of the reminders require that the provider obtain information from the patient, so they must be completed during the exam. This is awkward for providers who prefer to maintain as much direct contact as possible with the patient during the exam and complete documentation afterward. Interface features to streamline this part of documentation would reduce the intrusiveness of the reminder software. Other reminders include educational information the provider should supply to the patient, and are therefore most useful during the wrap-up phase.

While redesigning the entire reminder system was beyond the scope of this project, we recommended two interface elements intended to increase the likelihood that providers would be able to fold the reminder activities into existing workflow rather than tacking them on as an extra task. The first of these was to provide a more flexible display of reminders. Particularly during the review/planning phase and the wrap-up phase, a checklist format is more compatible with workflow than the current interface. The new interface allows providers who wish to integrate reminder activities with other aspects of care into a single list to be used during the wrap-up to check that all planned activities have been completed. This checklist interface is provided in addition to the existing reminder interface. Figure 11.3 depicts this flexible display of information. The checklist contains the clinical reminders, as well as other key information about the patient such

Visit Checklist

Due Clinical Reminders: 9
   N - Advance Directive
   N - Colorectal Cancer Screei
   N - Depression Screen
   N - Hypertension Screen/BF
   N - Influenza Vaccine 65
   N - Pain Screen
   N - Tobacco Use Screen
   P - Hypertension BP>140/9C
   P - Pneumococcal
Allergy: Povidone Iodine,
      Penicillin
BP 160/95 11/22/04
Expired meds: 2
Advance Directive
Last Lab: Theophylline 11/22/04
Last PC Appt. 09/08/04
Last consult: PTSD 11/16/04

Additional Items

Patient Actions

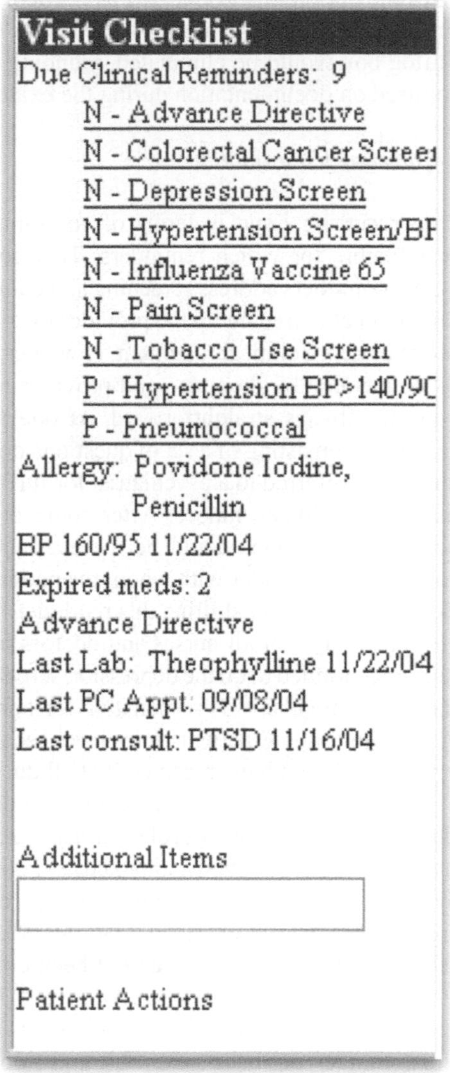

Figure 11.3   A flexible checklist reminder interface

as allergies, most recent vitals, and so on. In addition, there is a field in which the provider can add any additional items to the checklist that he or she wants to make sure to address.

A second interface recommendation was to integrate those reminders that require the provider to elicit information from the patient into the notes templates commonly used by some providers. For the providers that choose to use the template, screening questions would appear as part of the template. This would allow the provider to document answers to screening questions primarily by clicking on radio buttons with the option to add free text as needed. By integrating these screening questions into the template, the need to move from the template to the note to the reminders dialog box would be eliminated, reducing the amount of time the provider must be focused on documentation during the exam.

Supporting Coordination

One recurring issue in terms of coordination was confusion regarding who was responsible for which reminders. This confusion was observed at multiple sites. Some reminders address screening and education interventions which can be handled during intake, freeing up the provider to spend time on aspects of care which require more in-depth clinical judgment. However, differentiating between intake reminders to be handled by the nurse and provider reminders to be handled during the exam was not always straightforward. At one site for example, a depression-screening reminder consisting of a set of questions designed to identify depressed patients who should be referred to a psychiatrist for further evaluation and treatment, was initially assigned to intake nurses. After some time, it was decided that the depression-screening questions could be more effectively administered in the private exam room by the provider with whom the patient already had an established relationship. Several weeks after shifting this reminder from a nursing responsibility to a provider responsibility, not all nurses and doctors in the facility had made the switch. Some nurses continued to do the depression screening. Some providers continued to believe that someone else was responsible for completing the depression-screen reminder. As a result, some patients may have received the depression screening twice and others might not have received it at all during a particular visit.

In addition to the immediate problem of efficiently and effectively administering depression screening, this role confusion also had negative consequences for the busy healthcare team. To many, it looked as if others were not completing all of their tasks. Whether this was perceived as shirking responsibility, or in a more compassionate light as an issue of overload, team members were faced with a decision as to how to handle a reminder that had not been completed, but was not their responsibility. Potential actions in this situation include 1) completing the depression screen even though the individual does not believe it is his or her responsibility, and 2) leaving the reminder until the next visit and hoping it will be completed then. Both are legitimate responses as the provider may assume the intake nurse was busy and feel that they are helping out by completing the reminder (or vice versa). The intake nurse may feel that they are overstepping to complete a reminder assigned to the provider, and that it is up to the provider to prioritize the needs of the patient. This ambiguity regarding who is responsible for which reminder led to confusion, as well as resentment for some, both of which are strong detractors from smooth team performance.

To reduce role confusion, we recommended a simple change to the interface, clarifying whether each reminder should be handled by a nurse during intake or by the provider during the exam. This interface element was observed in use at one observation site and was implemented with success. Figure 11.4 illustrates this recommendation.

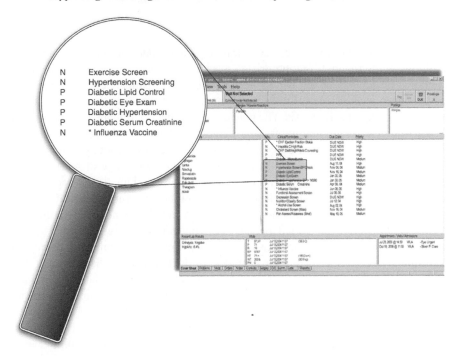

Figrue 11.4   Interface clarifying reminder responsibility

Notes: N indicates that a reminder is intended to be completed by the nurse during intake. P indicates that a reminder is to be completed by the provider during the patient exam.

Reducing the Workload Burden Associated with Reminders

Every hospital site we visited reported that reminders add to the workload of the healthcare team. For some, the value in terms of improved patient care seems limited and does not seem to warrant the additional workload. This perception of extra work with little value must be changed if reminders are to be used as intended. Our primary recommendation associated with this barrier to use was to reduce the number of reminders. One hospital site, in particular, reported to us that they deliberately maintained a short list of potential reminders. They have articulated specific criteria for inclusion. For example, the reminder must address an aspect of care that is recognized nationally in evidence-based standards and it must be an aspect of care the hospital has targeted for improvement. Selective strategies for choosing reminders such as this one are likely to help in equalizing the perceived workload-value equation.

It seems important to acknowledge that there are very important aspects of care that can be better addressed via means other than the reminders. For example, some hospital sites include reminders addressing issues such as seat belt and sunscreen usage. While it is important that the provider discuss these lifestyle issues with patients, prominently displayed posters might serve the purpose better than computer-based reminders.

Reducing Rigidity

As with any software tool, there are limited types of data the reminders monitor, resulting in occasional inappropriate recommendations. In addition to these functional limitations of the software, there are also circumstances in which error occurs, either because of a software bug or because of an incorrect data input such as erroneous coding of a diagnosis. The goal of the reminders software is to support and aid the healthcare team, leaving issues of clinical judgment to the provider. Current implementation, however, is perceived by some to be much more peremptory. At some hospital sites, statistics are kept regarding what percentage of reminders are completed by each provider, with the implied goal of reaching 100 percent. Some providers feel penalized for exercising their own clinical judgment when it conflicts with the reminder recommendation. It should be noted that none of the hospital sites visited reported an actual penalty for noncompliance with reminders, but the perception remains in some cases. One step toward reducing this perception would be to introduce a means for providers to document their disagreement with the reminder recommendation. This explicitly acknowledges that the reminder system will be wrong at times and that those who maintain the reminders software are interested in understanding where conflicts occur and how they are handled. We recommend adopting the feedback system developed at one hospital site visited. Users are able to easily document problems as they occur. Further, users receive an email acknowledging their input. This type of feedback system provides at least two benefits: 1) it explicitly acknowledges that the reminders software will be wrong at times, reducing the perception that users are expected to follow reminder recommendations 100 percent of the time; and 2) it provides feedback both to software professionals about bugs in the system and to clinical staff about controversial standards and common exceptions. A sample feedback tool interface is illustrated in Figure 11.5. Here the provider simply enters their contact information and enters whatever issues has arisen via free text. An email goes directly to the computer-support team (which may include a clinical adviser as well).

A second recommendation intended to reduce the perception of the clinical reminders software as a rigid tool that is difficult to apply in a realistic setting was to provide users information about the logic behind the reminder recommendation. Users reported confusion about how different responses might impact the reminder software. By clarifying the logic behind the reminder, as well as how the software will respond to different actions, users will be better able to understand the strengths and limitations of the software. We recommend making this information available on request via a simple button click. Users would then be able to examine a specific reminder when an issue arises—the only time this information is likely to be important to them. Figure 11.6 depicts a sample interface making the reminder logic visible. In this case, by clicking on clinical maintenance button (available on the existing interface), a box appears with information including the conditions for which the reminder is relevant, the rationale for the reminder, and the outcome associated with each of the options provided in the reminder dialog box.

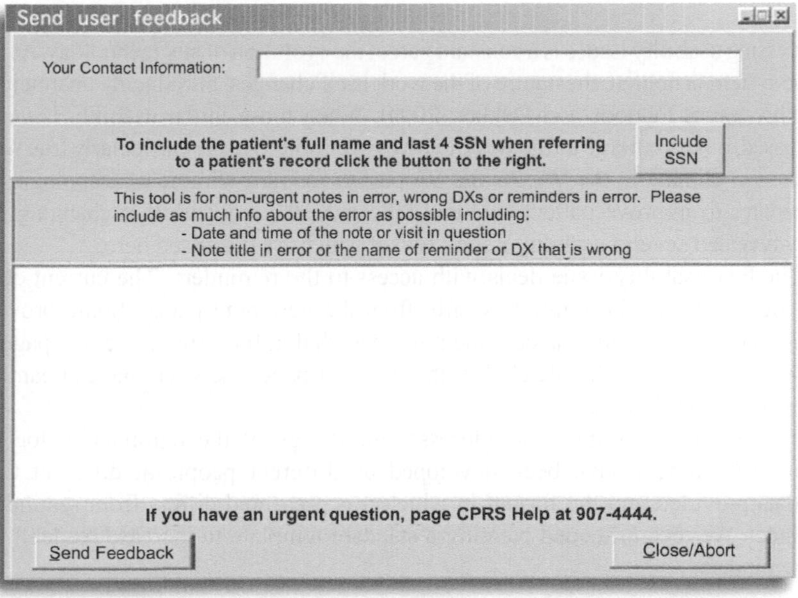

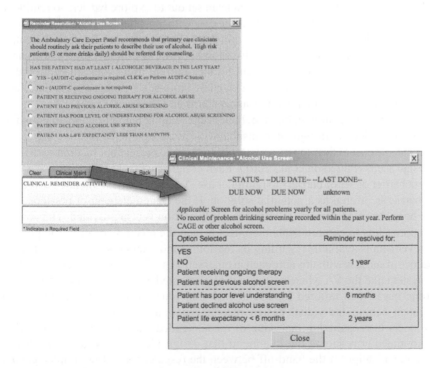

Send user feedback

Your Contact Information:

To include the patient's full name and last 4 SSN when referring to a patient's record click the button to the right.   Include SSN

This tool is for non-urgent notes in error, wrong DXs or reminders in error. Please include as much info about the error as possible including:
- Date and time of the note or visit in question
- Note title in error or the name of reminder or DX that is wrong

If you have an urgent question, page CPRS Help at 907-4444.

Send Feedback   Close/Abort

Figure 11.5   **Sample feedback tool interface**

Figure 11.6   **Making reminder logic visible to users**

Usability

Addressing usability issues is a constant part of the evolution of any technology. As soon as the system is fielded, the nature of the work itself changes, introducing unanticipated usability issues (Woods and Dekker, 2000). When those initial usability issues are addressed, others emerge as the work evolves over time. This is particularly true within the current climate in the VA. As the VA pushes forward with new technologies and procedures to improve patient care and efficiency, the work itself is changing. This study revealed several usability issues, two of which are discussed here.

The first usability issue deals with access to the reminders. The current design provides access to the reminders only from the new note pages. Some providers found that they wanted to access the reminders dialog box from the cover page. By allowing access via a double click from the cover page, users will have streamlined access to the reminders.

The second usability issue addresses the design of the reminder dialog box. Because reminders have been developed by different people at different times, the data provided and the format in which it is presented differs from reminder to reminder. We recommended creating a standard template to increase readability of the reminders.

Discussion

At the onset of this project, the research team set out to explore barriers to reminder use and recommend interventions. As is the case with most fielded systems, however, a complete redesign of the clinical reminders software was not feasible. Potential changes are limited by the existing software architecture as well as resources available to rework an existing software system. Further, current users who are comfortable with the existing interface will be penalized by large-scale changes to the interface as these users will need to unlearn usage patterns and create new ones.

Faced with this common, but nonetheless challenging human-factors dilemma of creating and selecting design elements that will have a significant positive impact, at an acceptable cost to developers and users, we chose to present our recommendations as design seeds (Patterson et al., 2001). Each design seed includes a discussion of a current barrier or set of barriers to reminder use. A design element intended to reduce that barrier is offered as a potential solution. If the design element or portions of it have been observed in use elsewhere, that information is also provided. The goal of this approach is to communicate the intent behind the recommendation and present whatever data is available regarding its feasibility and likelihood of success (that is, evidence of success with this approach in other settings). This is particularly important in a system like the VA where software is implemented across the entire VA hospital system, but is then tailored for use at each individual hospital site. By providing the intent behind individual design recommendations, we hope to aid software developers in recognizing the value of recommended changes. Intent information is also aimed at reducing the likelihood that design ideas will be distorted due to miscommunication in the hand-off between the researchers and the implementers. The goal is to facilitate implementation that reduces actual usage barriers.

Finally, it should be noted that clinical reminders hold great promise for increasing compliance with evidence-based practice and improving patient care. This software-based tool addresses support of key macrocognitive functions; supporting planning activities for an individual exam, as well as real-time decision making and replanning that occur during the patient exam. Many of those providers who have already integrated the reminders into their workflow report that they find the reminders helpful and that the reminders increase the likelihood that targeted aspects of care will be addressed in a timely manner. Changes to the software and the way it is administered can potentially increase the usability and usefulness of the clinical reminders for those providers who have not yet adopted them or use them only occasionally.

Acknowledgements

This research was supported by the Department of Veterans Affairs, Veterans Health Administration, Health Services Research and Development Service (TRX 02-216, CPI99-126 and CPI01-141, and Management Consultation Project). The views expressed in this article are those of the authors and do not necessarily represent the view of the Department of Veterans Affairs. A VA HSRandD Advanced Career Development Award supported Dr. Asch and a VA HSRandD Merit Review Entry Program Award supported Dr. Patterson.

References

Demakis, J.G., Beauchamp, C., and Cull, W.L. (2000). "Improving residents' compliance with standards of ambulatory care: Results from the VA cooperative study on computerized reminders." *Journal of the American Medical Association*, 284 (11): 1411–16.

Fung, C.H., Woods, J.N., Asch, S.M., Glassman, P., and Doebbeling, B.N. (2004). "Variation in implementation and use of computerized clinical reminders in an integrated healthcare system." *American Journal of Managed Care*, 10, Part 2: 878–85.

Gruden, J. (1987). "Social evaluation of the user interface: Who does the work and who gets the benefit." *Proceedings of the INTERACT 87: IFIF Conference on Human–Computer Interaction* (Stuttgart, Germany).

Militello, L.G., Patterson, E.S., Tripp-Reimer, T., Asch, S.A., Fung, C.A., Glassman, P., Anders, S., and Doebbeling, B. (2004). "Clinical reminders: Why don't they use them?" *Proceedings of the 48th Human Factors and Ergonomics Society Meeting* (Santa Monica, CA: HFES).

Norman, D. (1993). *Things that Make us Smart: Defending Human Attributes in the Age of the Machine* (Reading, MA: Addison-Wesley).

Patterson, E.S., Doebbeling, B.N., Fung, C.H., Militello, L.G., Anders, S., and Asch, S.M. (2004). "Identifying barriers to the effective use of clinical reminders: Bootstrapping multiple methods." *Journal of Biomedical Informatics*, <www.elsevier.com/locate/yjbin> accessed January 2005.

Patterson, E.S., Nguyen, A.D., Halloran, J.M., amd Asch, S.M. (2004). "Human factors barriers to the effective use of ten HIV clinical reminders." *Journal of the American Medical Informatics Association*, 11 (1): 50–59.

Patterson, E.S., Woods, D.D., Tinapple, D., and Roth, E.M. (2001). "Using cognitive task analysis (CTA) to seed design concepts for intelligence analysts under data overload." *Proceedings of the Human Factors and Ergonomics Society 45th Annual Meeting* (Santa Monica, CA: HFES).

Saleem, J.J., Patterson, E.S., Militello, L.G., Render, M.L., Orshansky, G., and Asch, S.A. (2005). "Exploring barriers and facilitators to the use of computerized clinical reminders." *Journal of the American Medical Informatics Association*, 12 (4): 438–47.

Woods, D.D. (1998). "Designs are hypotheses about how artifacts shape cognition and collaboration." *Ergonomics*, 41: 168–73.

Woods, D.D. and Dekker, S.W.A. (2000). "Anticipating the Effects of Technological Change: A new era of dynamics for human factors." *Theoretical Issues in Ergonomic Science*, 1 (3): 272–82.

Chapter 12

Macrocognition and Experiential Learning in Surgical Teams

Simon Henderson

Introduction

Human beings, who are almost unique in having the ability to learn from the experience of others, are also remarkable for their apparent disinclination to do so.

Douglas Adams, 1990, p. 114

Background

This chapter discusses work conducted by QinetiQ to develop a system for enabling surgical teams to learn effectively from their own experience. Its aim is to contrast theories of organizational learning with the practicalities involved in applying a consultative model to develop enhanced learning processes for teams that work in complex settings. While the performance of surgical teams has generally received a great deal of research attention (for example, Fletcher et al., 2002; Carthey, de Leval et al., 2003; Plasters, Seagull, and Xiao, 2003; Nemeth, 2003; Musson and Helmreich, 2004; Healey, Undre, and Vincent, 2006; Undre et al., 2006), this chapter focuses specifically on the mechanisms by which such teams learn from their own experience. In so doing, the chapter will describe the development and application of a process of "Team Self-Review" for surgical theater teams, and will examine how the naturalistic approach adopted throughout the study has uncovered empirical support for the concept of macrocognition.

The chapter begins with an overview of theories of individual, team, and organizational learning, that address in particular the role of experience. Key theoretical considerations are drawn out, and the chapter then discusses the development of processes and tools to support experiential learning within surgical teams and departments. Particular attention is paid to the practicalities involved in the implementation of such initiatives. Finally, the role of macrocognition in understanding and enhancing organizational learning in general will be considered, and a potential area for development within the metacognitive framework identified.

Origins of Study

In 2000, a UK Government expert group focusing on the topic of learning from adverse events in the National Health Service (NHS), and chaired by the Chief

Medical Officer, produced a report entitled "An organisation with a memory" (Chief Medical Officer, 2000). The report focused on failures to learn from experience within the NHS, and suggested that a number of major incidents and accidents could have been avoided had previous lessons been identified and acted upon. The report suggested that the etiology underlying many incidents often involves common, recurring patterns of failure, relating largely to breakdowns in teamworking and poor decision making by individuals and teams. To quote from the Foreword to the report, written by the (then) Secretary of State for Health: "Too often in the past we have witnessed tragedies which could have been avoided had the lessons of past experience been properly learned…most distressing of all, such failures often have a familiar ring, displaying strong similarities to incidents which have occurred before and in some cases almost exactly replicating them" (ibid., p. v).

In 2003, QinetiQ (a UK defense and security company) was contracted by the UK National Patient Safety Agency (NPSA) to develop and evaluate approaches for enabling surgical teams to learn more effectively from their own experience. To paraphrase the title of a paper by Burke et al. (2004, p. i96), we were asked to develop a method by which a team of surgical experts could turn themselves into an expert surgical team.

The NPSA is charged with addressing many of the issues raised in "An organisation with a memory." As part of their response to this challenge, the agency has sponsored three initiatives at the Royal Cornwall Hospital in Truro, each geared towards a different aspect of patient safety. These have partly been inspired by similar initiatives within the aviation industry (for example, Brannick et al., 1995; Helmreich, Merritt, and Wilhelm, 1999; Prince and Jentsch, 2001), and consist of the introduction of Team Resource Management education and training (analogous to the aviation Crew Resource Management programs); Close-Call Reporting (analogous to the aviation near-miss reporting systems) and Team Debriefing (analogous to military After Action Review processes). QinetiQ was contracted to facilitate the development of a system to support the third strand, team debriefing, as a result of previous work on enhancing After Action Review processes in the UK Army and the development of team debriefing approaches for Kent County Constabulary firearms teams.

The chapter begins by outlining theories of individual, team, and organizational learning, and drawing out key principles that underpin the empirical study described later.

Learning from Experience: Theories of Individual, Team, and Organizational Learning

Theories abound as to how learning occurs at individual, team, and organizational levels; and with regard to how such learning may be supported most effectively. Four theories will be addressed here, that focus in particular on the role of experience—as both a product of learning, and as a cognitive factor that shapes the learning process itself. These theories consist of Kolb, Boyatzis, and Mainemelis' Experiential Learning Theory (2000); Ausubel's theory of meaningful learning (1963); shared mental models (Kraiger and Wenzel, 1997), and networked expertise (Hakkarainen et al., 2004) that itself draws from a range of other fields of study.

Kolb et al.'s experiential learning theory provides a holistic model of the learning process and a multilinear model of adult development, both of which are consistent with what is generally known about how people learn, grow, and develop. The theory is called "experiential learning" to emphasize the central role that experience plays in the learning process, an emphasis that, according to the authors, distinguishes this theory from other learning theories. The term "experiential" is used therefore to differentiate experiential learning theory both from cognitive learning theories, which tend to emphasize cognition over affect, and behavioral learning theories that deny any role for subjective experience in the learning process.

Kolb and Kolb (2005) purport that experiential learning theory draws on the work of prominent twentieth-century scholars who gave experience a central role in their theories of human learning and development—notably John Dewey, Kurt Lewin, Jean Piaget, William James, Carl Jung, Paulo Freire, Carl Rogers, and others—to develop a holistic model of the experiential learning process and a multilinear model of adult development. The theory is built on six propositions that are shared by these scholars:

1. Learning is best conceived as a process, not in terms of outcomes.
2. All learning is relearning. Learning is best facilitated by a process that draws out the students' beliefs and ideas about a topic so that they can be examined, tested, and integrated with new, more refined ideas.
3. Learning requires the resolution of conflicts between dialectically opposed modes of adaptation to the world, that is reflection and action—and feeling and thinking.
4. Learning is a holistic process of adaptation to the world, not just cognition but also feeling, perceiving, and behaving.
5. Learning results from synergetic transactions between the person and the environment.
6. Learning is the process of creating knowledge.

Kolb defines learning as "the process whereby knowledge is created through the transformation of experience. Knowledge results from the combination of grasping and transforming experience" (1984, p. 38). The experiential learning theory model portrays two dialectically related modes of grasping experience—Concrete Experience and Abstract Conceptualization—and two dialectically related modes of transforming experience—Reflective Observation and Active Experimentation. According to the four-stage learning cycle depicted in Figure 12.1, immediate or concrete experiences are the basis for observations and reflections. These reflections are assimilated and distilled into abstract concepts from which new implications for action can be drawn. These implications can be actively tested and serve as guides in creating new experiences.

An alternative view of experiential learning that focuses on the cognitive processes supporting the development of experience, is Ausubel's theory of meaningful learning (1963)—subsequently developed by others including Novak,

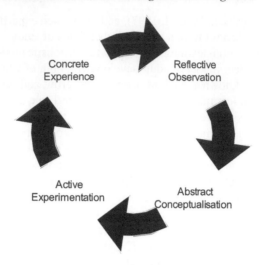

Figure 12.1 The experiential learning cycle
Source: Kolb, 1984.

Hanesian, and Gowin (Ausubel, Novak, and Hanesian, 1978; Novak and Gowin, 1984; Novak, 1998). Meaningful learning theory focuses on how individuals learn new concepts (defined as perceived regularities of objects or events, or of records of objects or events, that are designated a label) by assimilating them with relevant existing concepts and propositions, and subsequently integrating the concepts into their cognitive structure. Cognitive structure comprises an individual's organization of concepts and propositions. Consequently, if an individual has no recognized relevant concepts against which to integrate new concepts, learning by rote (or indeed, no learning at all) may occur. Learning is thus viewed as an idiosyncratic process, dependent upon an individual's prior experience and resultant knowledge.

Ausubel et al. (1978) suggest that three conditions must be met in order for meaningful learning to occur:

1. The subject matter to be learnt must be meaningful.
2. The learner must have a preexisting conception that is relatable to the new information to be learnt.
3. The learner must choose to learn meaningfully.

The process of assimilating new concepts and integrating them into cognitive structure is referred to as "subsumption," with existing concepts referred to as "subsumers." A key point of Ausubel's theory is the proposition that any learning task can be made meaningful if the learner is supported in bridging the gap between subsumers, and the new concepts they are seeking to understand. "Advance organizers" have been suggested by Ausubel as devices that may be used to support the introduction of a learning episode, such that learners may orient themselves to establish how new concepts fit into "the bigger picture"; and how these concepts link to concepts the learner already understands.

The field of shared mental models has extended the concept of individual learning to the level of teams and organizations. Mental models were first postulated Craik (1943), who proposed that the mind constructs "small-scale models" of reality that it uses to anticipate events, to reason, and to provide explanation. This concept was elaborated by Johnson-Laird (1983); Gentner and Stevens (1983), and Norman (1983, p. 7) who offered the following description: "In interacting with the environment, with others, and with the artefacts of technology, people form internal, mental models of themselves and of the things with which they are interacting. These models provide predictive and explanatory power for understanding the interaction." Mental models thus comprise cognitive structures used to encapsulate and represent events in an individual's environment, make sense of them and predict future events.

More recently, the notion of shared mental models (SMMs) in teams has received attention. Fleming et al. (2003) suggest that the majority of research on this topic has explored the relationship between SMMs and team performance and processes (for example, Kraiger and Wenzel, 1997; Stout et al., 1999; Mathieu et al., 2000). Such studies have generally found that teams whose members share similar knowledge structures regarding the task, the environment, equipment, member capabilities, and/or member interactions communicate more effectively and perform better than teams whose members do not share such knowledge (see Mohammed, Klimoski, and Rentsch, 2000, for a review of this literature). A more recent meta-analysis provides further empirical support for this relationship (Fleming and Griepentrog, 2003).

Both Tarnoff (1999, p. 8) and Fleming et al. (2003, p. 3) cite a paucity of research examining the origins and development of shared mental models. However, several studies that have investigated the training of mental models have found that training can indeed increase the quality and similarity of the mental models of team members (Smith-Jentsch et al., 2001; Minionis, 1994). When team members are said to share knowledge, coherence is assumed to occur at a collective level (thus mental model "sharedness" constitutes the similarity or overlap of knowledge held by members of the team). Fleming et al. (2003) suggest that the logic behind this assumption is that knowledge structures (in other words, experience) may be developed by way of the communication, interaction, and shared experiences of the individuals that make up the team (Stout et al., 1999). Other research (Pascual, Henderson, and Mills, 1998; Pascual et al., 2000) proposes that specific mental model constructs should only be shared when team-wide consistency or overlap on these constructs supports collective performance (for example, every team member must share a common view about the team's goals, if it is to achieve unity of direction). Having consistent understanding about other constructs (for example, what is happening in the environment) may potentially give rise to groupthink[1] and thus inhibit collective situational exploration and understanding (a diversity of views is thus especially important in the early stages of creative team tasks). Pascual et al. have thus suggested that team members should have *compatible* mental models.

---

1   "A mode of thinking that people engage in when they are deeply involved in a cohesive in-group, when the members' strivings for unanimity override their motivation to realistically appraise alternative courses of action" (Janis, 1979, p. 9).

In relating the experiential learning of individuals and teams to experiential learning in organizations, Argyris and Schön (1978) propose that organizational learning takes place within a defined framework that is created by the collective "theories-in-use" of the individual members of an organization. On this basis they have identified three levels of organizational learning: "single-loop learning," "double-loop learning," and "deutero-learning" (from the Greek, *deuteros*, meaning "secondary").

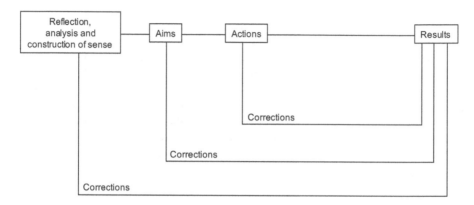

Figure 12.2   Single-, double-, and deutero-learning loops

Source: Fischer and Roben, 2004, after Argyris and Schon, 1978.

In Figure 12.2, the concept of deutero-learning is interpreted as organizational-level reflection, analysis and sensemaking about the individual learning processes that happened within the organization. In supporting this concept, Fischer and Roben (2004) suggest that this means that an individual within the organization firstly receives feedback (often by chance) concerning his or her performance so that he or she might draw conclusions from this feedback to avoid future mistakes and maximize future performance (single-loop learning). Secondly, organizational learning in the sense of double-loop and deutero-learning implies that the organization has created a *structure* (that is, configured the component parts, connections, and relationships that make up the organization) through which individual learning is permanently stimulated, documented, and evaluated. It is this organizational structure by which organizational learning is differentiated from pure individual learning—thus, organizational learning changes an organization's structures, not just its people.

Hakkarainen et al. (2004) have focused on the *process* of organizational learning, and suggest that it is founded on the evolution of networked expertise. They surmise that networked expertise refers to competencies that arise from social interaction, knowledge sharing, and collective problem solving. These are embedded in communities and organized groups of experts and professionals. Cognition and intelligent activity are not only individual and mental processes but ones which rely on socioculturally developed cognitive tools. These include physical and conceptual artifacts as well as socially distributed and shared processes of intelligent activity embedded in complex social and cultural environments. Networked expertise is constituted in interaction among individuals, communities, and larger networks supported by cognitive artifacts, and it co-evolves with continuously transforming innovative knowledge communities. Hakkarainen et al.'s concepts reflect a range of ethnographic and sociological fields of study that have come to prominence over the past two decades, including situated cognition (Suchman, 1987); distributed cognition (for example, Agnew, Ford, and Hayes, 1997; Hollan, Hutchins, and Kirsh, 2000); and communities of practice (Lave and Wenger, 1991; Wenger, 1998).

A number of salient points emerge from reviewing these theories:

- Learning and experience are tightly coupled. Learning generates and adds to experience, and experience helps shape the learning process itself.
- Learning occurs when new concepts are assimilated with existing concepts. This process provides a basis for supporting and enhancing learning.
- People construct mental models of the world to represent, understand, and predict it.
- Teams perform more effectively if the parts of their mental models requiring coherence are shared across team members.
- Learning organizations adapt their structure (that is, they reconfigure their component parts, connections, and relationships that make up the organization) to enable learning to occur across the organization.
- Organizational learning also occurs via the use and flow of sociocultural cognitive tools, and physical and conceptual artefacts across the organization.

The remainder of this chapter will employ these concepts to consider the practicality of developing organizational learning and expertise within surgical teams and theater departments; evaluation criteria identified will later be applied in order to gain insight into the impact of the learning processes developed.

Challenges, Barriers, and Enablers to Organizational Learning and the Development of Networked Expertise in Surgical Teams

Thus far, consideration has been given to a range of theories and factors that assist with understanding how experiential learning occurs at individual, team, and organizational levels. Attention will now turn to consideration of the practicalities involved in developing and implementing experiential learning processes at these levels with surgical teams and a theater department; beginning with an overview of our initial investigations.

Current Practice

Our initial investigations consisted of spending time gaining familiarity with the surgical teams working within one theater block in the Royal Cornwall, in order to understand how they (at the time) went about developing expertise and learning from experience. This was achieved through the use of observational studies of surgical procedures to study surgical teamworking; focus groups with theater staff to discuss learning mechanisms and associated barriers, and a range of one-on-one interviews to elicit specific individuals' opinions, experience, and expertise (these interviews employed some of the questioning techniques central to the Critical Decision Method (Klein, Calderwood, and MacGregor, 1989; Hoffman, Crandall, and Shadbolt, 1998). In addition, QinetiQ created and hosted a website that informed hospital staff as to the status of the project. We made available materials as they were developed, and we sought honest, open, and anonymous feedback on the issues discussed and trials conducted. The site served as a valuable conduit for gaining detailed and direct insight into learning processes occurring "on the shop floor."

From our initial investigations, it emerged that there existed little in the way of explicit structures and processes for enabling surgical teams to construct meaning from their own experience. We found that:

- Opportunities to learn from experience were limited due to the pressurized nature of the working environment (for example, time pressure and the stress caused by the drive to meet organizational targets gave rise to limited opportunity for teams and individuals to reflect on the day's events).
- Approaches to learning from experience were predominantly individualistic, and thus inconsistent across team members. Theater staff indicated that they would often reflect upon key events from the day during the drive home, or perhaps would talk through the day with their partner at home. At best, some staff would occasionally socialize after work, which would provide them with a cathartic opportunity to dissipate the day's stresses.
- Learning tended to be ad hoc and unstructured. Any discussions about how things went at work were entirely impromptu, and no formal procedures were employed. Consequently, such discussions would often cover a wide variety of topics, both related and unrelated to work. While the conversations could, under some circumstances, serve to build trust and establish common ground, they also potentially squandered valuable learning opportunities.
- Staff felt that the amount of personal effort required to participate in supporting organizational learning would not result in commensurate personal benefit. This view was compounded by a degree of cynicism from some staff, who felt that they would have to invest their own time in activities from which only "management" would derive benefit.

These factors were further compounded by a range of other challenges, as will now be discussed.

Challenges to Facilitating the Development of Team Expertise

Our primary challenge was to develop an approach by which surgical teams could examine, critique, and refine their constructive meaning as they sought to generate understanding from their own shared experience. Such a process would, in effect, support the team in moving from individual mental models to a shared mental model. Given the pressures inherent in the working environment, we felt that such processes had to be tightly integrated with, and minimally intrusive to their daily working practices; and that any process that increased workload would rapidly be rejected. Further, we identified a requirement to employ a standardized model, knowledge-base, process and common language through which such constructions could be captured, referenced, shared (between individuals and teams), reflected upon, and evaluated, and acted upon on a basis more rigorous and efficacious than the (entirely ad hoc) approach employed previously. In addition, the method had to cope with the challenges assessed below.

Time Pressure and Daily Schedule

Theater staff work in a highly pressurized environment with significant demands placed on their time in order to meet key organizational targets. In particular, there is a strong emphasis on completing the list of cases allocated to a particular team (usually for a morning, afternoon, or all-day list). Often lists overrun, staff end up missing lunch, and it is a regular occurrence for lists to extend into the evening. Given this situation, there seems little scope for introducing additional, deliberative processes through which organizational learning might be encouraged and enhanced. This issue pointed to the need for a method that could be tailored to meet the time available (from five minutes to several hours); that comprised a set of core processes which, should time be available, could be expanded to fill the gap. Such an approach contrasts with a fully featured, functionally bloated method that would regularly need to be compressed into the short period of time available.

Location for Holding Team Discussions

Space is at a premium in operating theaters, and the supporting infrastructure tends not to have many spare or unused rooms. In addition, as staff start to move out of a theater at the end of a list (for example, the anesthetist may accompany the patient to Recovery), other staff may move in (for example, to clean the theater down, or to set up for the next list). The only other potential locations were in the coffee room (which is intended as a place to get a break from work, and is also where staff from other theaters meet their colleagues socially), or the changing rooms (which are gender divided). As both of these locations were considered inappropriate, it was decided that review sessions should be held in the theaters, which also had the advantage that they could be held before theater staff started to drift apart; in addition, this location enables staff to refer to their artifacts and workspace.

The Impact of Ad Hoc Teams

Surgical teams often meet for the first time over the patient. While staff may know each other as colleagues, they do not necessarily work together on a regular basis. Consequently, they do not necessarily possess shared experience and a common mental model about taskwork and teamwork. Our approach to managing this problem was to place initial emphasis on facilitating individual learning geared around a common model, language, and structure. This enables all team members to employ a common reference set and structure to guide discussions when talking to their colleagues, and with which to frame and process knowledge gained relating to task and team performance.

Staff Migrations (People Leaving at the End of a List)

As indicated above, staff tend to move through the theater both during and after a case or list. By holding a debrief at the end of a list, immediately after closing, nearly all key staff would be regularly available to participate. In addition, we were keen to develop a process which could be led by anybody in the team (from the most senior surgeon to the newest student) and that could involve anyone who was available at the time. While there are undoubtedly some compromises with this approach (for example, in terms of the representativeness of views expressed) it seemed to maximize the opportunity available.

Obtaining Participation

The Impact of Role, Hierarchy, and Personality   There are a range of social and organizational factors that may inhibit individual participation in a group process. These include more junior staff feeling that their views are unworthy in relation to more senior staff in the room, people with particular roles believing that their ability to contribute to the greater collective learning is insignificant, or individuals who are, by nature, quiet, shy, or introverted, who do not wish to voice their opinion in front of the group. This highlighted the requirement to develop an approach that depersonalized the impact of these inhibiting factors, and that facilitated the difficult interpersonal parts of the process on behalf of the staff.

Facilitating Equal Participation

Where several people had opinions, insights, or knowledge to impart to the group, the potential existed for those with the loudest voice to end up shouting the most. Consequently, we were challenged to develop a method that could facilitate equal participation between the team members, to provide a voice for all team members, and to establish a democratic process for establishing how the session should be run.

Maintaining Focus

The key challenge here was to keep a team's review process focused on those things that are important, without allowing the conversation to wander onto unrelated yet time-consuming topics. While it is recognized that socialization can be critical to the development of trust and mutual respect between team members, such processes were not considered central to the review processes, and could be pursued via alternative means (for example, during coffee breaks). Focus was supported by the construction of a structured checklist and process for guiding the topics discussed (drawn from a model of effective team behaviors, as discussed later in this chapter) and a method for how the topics should be addressed.

Avoiding Monotony

If a team were to be posed the same set of five questions at the end of every list, the questions would soon become monotonous and repetitive, and the team members would become desensitized to them. This problem pointed to the requirement to develop a method that was flexible and reconfigurable throughout many permutations, potentially drawing upon novel and random permutations if possible.

Managing Time

When time is limited within the theater day, there is the possibility that a debrief held mid-day might have an impact on the time available for cases and lists held after it. Consequently, there is a need for close monitoring of time in order to keep a review on track, and our approach was to assign a role of "timekeeper" to one of the debrief participants. While curtailing a useful discussion may potentially impact learning opportunities (and may also lead to team-member frustration if an issue is not "bottomed-out"), the method will have at least served to initiate open dialogue on the topic that can be picked up again later, should time and opportunity permit. Thus, even if the method is compromised by time pressure, it offers distinct advantage over the previous "do nothing" alternative.

Management of Interpersonal Conflict and Personal Culpability

Clearly, any process for learning from experience must be sensitive to the management of interpersonal conflict and/or personal failure or embarrassment, while at the same time ensuring that critical issues can be raised and discussed constructively. We were thus challenged to develop procedures and tools that depersonalized the review process, and that assisted in defusing any interpersonal conflict. The procedures also required a means for facilitating clarification of participation rules at the start of a review session, and the allocation of specific debrief roles to control session flow.

Capturing Issues and Actions

In order that both positive and negative lessons be captured and shared, we developed a simple paper-based recording log that captured issue-action plans arising from the debrief. The log is produced by a person assigned the role of "scribe," and is handed to the Theater Manager at the end of the list. The Theater Manager then acts upon those actions as appropriate, and passes any that are relevant up the organizational chain. The top issues (for example, recurring themes, or any significant points) are then raised at the monthly Clinical Governance meeting, which is attended by all theater staff. In this way, all theater staff are exposed to the lessons identified in other teams, both positive and negative. Individuals who have contributed ideas are also able to see that their issues have been recognized and potentially acted upon, and that their contribution has been fed back anonymously to all theater staff.

Effecting Change

We envisaged that change resulting from learning could take place at two levels (these levels reflect the learning-loop modes described by Argyris and Schön (1978)). The first is at the local level, where individuals or teams engage in change in order to bring about personal, collective, or organizational benefit. Such change would include simple modifications to individual behaviors (for example, a runner or scrub nurse remembering to check a particular surgeon's personal instrument preferences, together with checks on the standard instrument set required for a forthcoming procedure), or changes to collective behavior (such as the team engaging in a morning pre-brief to plan the day), and would tend to be localized, fairly low effort, and of minimal or no cost (although it may include changes to local goals). The second form of change is that which requires managerial approval (for example, a formal change to policy—such as establishing a procedure for the use and placement of warning signage for patients with MRSA[2]) or the commitment of funds. Such change requires that the issues captured at the local level are provided in an appropriate form to senior hospital management. As part of the process for converting issues identified into lessons learned, we proposed a formal review procedure for assessing the issues and actions arising from the local debrief sessions. This procedure was to be owned by the Theater Manager, together with other staff involved in the governance process.

Managing Confidentiality and Legal Issues

The management of confidentiality and legal issues is critical when an adverse event occurs which is significant or catastrophic. While the debriefing method produced is able to link into formal investigative methods, we agreed with the hospital and the NPSA that the development of methods for managing these issues was outside the scope of our work.

The development of procedures and tools for enabling surgical teams and a theater department to learn from their own experience proved to be considerably challenging. Our initial discussions with theater staff uncovered a wide range of practical barriers to implementing procedures and tools for supporting learning, and we thus had to deliver and implement a simple, usable, yet efficacious approach that avoided, overcame, or managed the barriers identified. The approach we delivered is now described.

Team Self-Review (TSR) as a Process for Supporting the Development of Team Expertise

The approach we developed to enable surgical teams and theater departments to learn effectively from their experience draws together and builds upon knowledge and approaches developed under a range of previous QinetiQ research, including:

---

2   Methicillin-resistant staphylococcus aureus—a specific strain of the *Staphylococcus aureus* bacterium that has developed antibiotic resistance to all penicillins.

- Understanding the drivers of collocated, distributed, and ad hoc team performance (for example, Pascual, Henderson, and Mills, 1995; Mills et al., 1999a, 1999b). This work provided a framework for linking critical team behaviors to team performance.
- Studies to investigate the role of mental models in team performance (for example, Pascual et al., 1998; Pascual and Henderson, 1998; Pascual et al., 2000) and the development of methods for supporting and enhancing the development of shared mental models in teams (for example, Pascual, 1999; Pascual et al., 1999). These studies provided an outline set of approaches for enabling team members to externalize their individual mental models (through a process of briefing and debriefing), thereby making them available for assimilation with other team member's models.
- Work to understand how and where teams go wrong (for example, Walkinshaw et al., 2002). This work provided a useful set of indicators of the potential presence of problems within a team; together with an understanding about the causes of poor performance and team failure.

The resultant approach, known as Team Self-Review (the name reflecting the fact that the team reviews it s own performance—as opposed to having the process facilitated by a person external to the team) comprises two core processes:

- The TSR Pre-brief
- The TSR Debrief

The pre-brief is held at the beginning of a team "session" (that is an operating list, case, training exercise, or other event) and the debrief is held at the end of the session. The pre-brief aims to review and clarify the plan for the session ("What do we need to know? What are we going to do? What should we do if…?"), and the debrief aims to assess and review team performance in the session that has just occurred ("How well did we do? What should we do differently in future?"). Both techniques are characterized by four defining features:

- They are structured team discussions.
- They provide an opportunity for every team member to have a voice in the team.
- They provide an opportunity to project and reflect as a team.
- They provide an opportunity to learn as a team.

The TSR Pre-brief

Many teams experience problems, confusion, and communication breakdowns because team members do not have a clear understanding from the outset of the plan and what is expected of them (Walkinshaw et al., 2002). To help reduce the likelihood of these difficulties occurring, the pre-brief is designed to:

Royal Cornwall Hospitals **NHS**
NHS Trust

Theatre Team Briefing Guide

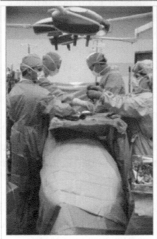

The importance of briefing
Many teams experience problems, confusion and communication breakdowns because team members do not have a clear understanding of the plan and what is expected of them.

To help reduce the likelihood of these difficulties occurring, all team members should be clear from the outset on:
- the plan (tasks and goals),
- how they are expected to contribute to and fit into the plan (their roles and responsibilities), and,
- where and how they need to co-ordinate their activities with others (how to co-operate and communicate effectively with fellow team members).

A briefing consists of a rapid 'get-together' to ensure that there is shared understanding in the team concerning the above details and that 'everybody is singing from the same hymn sheet' before they begin their tasks.

A briefing serves to:
- Enable clarity of direction;
- Facilitate better co-ordination between team members;
- Reduce the risk of problems and breakdowns occurring;
- Develop contingency plans for if things go wrong;
- Clear misunderstandings;
- Create a culture of open communication;
- Make everybody feel part of the team.

Briefing guidelines
- The member of the team who knows most about the patients to be treated in a list should conduct the briefing. Typically, in most cases this will be the Surgeon or the Anaesthetist.
- Typically a briefing will take between 5 and 10 minutes to conduct.
- A briefing should be a 2-step process:
  o Step 1 – the team is provided with key information concerning the plan for the list and cases.
  o Step 2 – the team is provided with the opportunity to ask any questions or raise any concerns they have.
- Use the format presented overleaf ('ILCS-QC') to structure a Theatre Team Brief.

Ad-hoc teams – Briefing to foster professional familiarity
An ad-hoc team is one where the majority of team members have not worked with each other previously. Consequently there is a lack of familiarity in the team. Ad-hoc teams can experience difficulties due to the fact that the people in them make unwarranted assumptions about each other. To counter this situation when in an ad-hoc team, there are some basic questions that all team members should ask themselves:

- Am I clear on who is going to lead and direct me?
- Am I clear on who I should be leading and directing?
- Am I clear on my role and responsibilities in this team?
- Am I clear on other team member's roles and responsibilities?
- Am I clear on who I should be supporting / working with most closely and how?   *Ad-hoc team questions.*
- Do I know how I fit into the plan for the list?
- Do people in this team use the same professional language as me?
- Am I familiar with the procedures they follow and the equipment they use?
- Am I clear on what will be expected of me in an emergency or if things go wrong?

If the answer to any of the questions is 'no', or if team members don't know the answer to a question, there is the potential for misunderstanding and problems. Consequently, these issues should be raised and discussed in the team before starting a list to ensure there is clear and shared understanding across team members.

QinetiQ

Figure 12.3a TSR Pre-briefing guide (front)

Briefing format: ILCS-QC

STEP 1 - Brief the team on important information (ILCS)

INTRODUCTION	Introductions • Introduce new members to the team (Name, role, any other necessary details). • Check if the team is 'ad-hoc'. An ad-hoc team is one where the majority of team members have not worked with each other previously. If this is the case, run through the ad-hoc team questions list presented overleaf. Debrief 'Issues and Actions' reminders • If necessary, provide reminders on relevant issues or actions that arose in previous debrief sessions (e.g. things to improve, change or do differently as highlighted and discussed in the debrief previously).
LIST	List details • Duration, i.e. AM, PM or Full Day. • Number of cases. • Any changes / modifications to original list. • Any current uncertainties on list (e.g. whether or not cases will occur) and means for ensuring that updates are communicated and shared across the whole team. • Estimated time of list completion and any potential delays to be aware of.

CASES
 	
Cycle through these details for each case.
	Case details
	• Describe the cases on the list in case order. For each case provide details (*as necessary*) on:
	
	   o Patient, Op type, duration. Name of patient, operation type and estimated duration of operation.
	
	   o Patient's physical and psychological details. If necessary, additional details concerning the patient's physical and psychological make-up (e.g. size and shape, psychological state etc.).
	
	   o Clinical information. Relevant clinical information regarding patient from surgeon/s and anaesthetist/s (blood, test results of importance etc.).
	
	   o Surgical approach. Key technical points to note and details on specific requirements for the approach (e.g. table or patient positioning etc.).
	
	   o Equipment requirements. Any specific equipment requirements above and beyond what was originally requested.
	
	   o Anticipated difficulties. Forewarning on potential difficulties, problems, risks, 'things to look out for' and possible changes to the planned surgical approach should certain circumstances arise. If relevant, contingency plans for dealing with potential problems should also be discussed.

STAFFING	Who, what, where, when • Clarify which staff will be working on each case and in what capacity (on Surgical, Anaesthetics and Nursing sides – e.g. who will be scrubbing for each case etc.) • Check that everyone is clear on his or her role and responsibilities.

STEP 2 - Let the team ask questions or raise concerns (QC)

QUESTIONS / CONCERNS	Any questions? Any concerns? Ask team if they have any questions or concerns they would like to raise (if possible, go round the team checking with each team member).

Figure 12.3b TSR Pre-briefing guide (back)

- Enable clarity of direction from the outset;
- Facilitate and strengthen coordination within the team and with other dependents outside of the team;
- Increase team awareness of risks and hazards and improve problem spotting and error trapping;
- Consider and develop contingency and mitigation plans and actions for problem areas (that is, where things could go wrong, what to do if they do);
- Allow team members to raise queries/concerns and clear any misunderstandings;
- Encourage a team culture of open and honest communication and assertive questioning;
- Increase sense of team identity (that is make everybody feel part of the team with a valid role, perspective, and opinion).

The TSR pre-brief takes five to ten minutes to conduct (a timeframe dictated primarily by staff availability at the beginning of a list) and the structure is presented in Figures 12.3a and 12.3b.

The TSR Debrief

The TSR debrief provides an opportunity and structure for the team to talk about teamwork and performance during the previous session. It provides an opportunity for every team member, irrespective of rank or status, to address their concerns or questions. Further, it enables the team to identify, implement, and monitor behavioral changes to enhance its performance.

A variety of TSR debriefing techniques were developed (fifteen in total) that map to different environmental constraints and pressures (for example, time available, familiarity of staff, critique depth required, and so on). In this way, the surgical team is effectively provided a "tool bag of techniques" from which to choose a TSR debrief appropriate for a particular context. Additionally, having this range of debriefing approaches available meant that sessions would be less monotonous, as structures and methods would be different between reviews, even though the same core topics would be probed.

The teamwork model shown in Figure 12.4 was developed specifically for surgical teams to provide the topics for review. The model uses an analogy of "team health." There are five key "areas of team health," each comprising three "health dimensions." The model has its origins in the Team Dimensional Training (TDT) work conducted at the Naval Air Warfare Centre Training Systems Division (Smith-Jentsch et al., 1998) and has been further developed to reflect the understanding derived from a wide range of previous QinetiQ team research (including the research activities cited earlier).

Each debriefing technique (techniques are referred to as "controls," as they provide the framework for controlling the flow of a debrief) presents a different way of structuring a review session based upon all, or a chosen subset, of these health dimensions. Figure 12.5 shows a sample of TSR cards that have been developed for facilitating some of the debriefs (one card has been produced to represent each

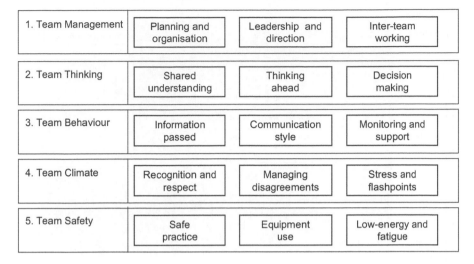

1. Team Management	Planning and organisation	Leadership and direction	Inter-team working
2. Team Thinking	Shared understanding	Thinking ahead	Decision making
3. Team Behaviour	Information passed	Communication style	Monitoring and support
4. Team Climate	Recognition and respect	Managing disagreements	Stress and flashpoints
5. Team Safety	Safe practice	Equipment use	Low-energy and fatigue

Figure 12.4   Dimensions of team health

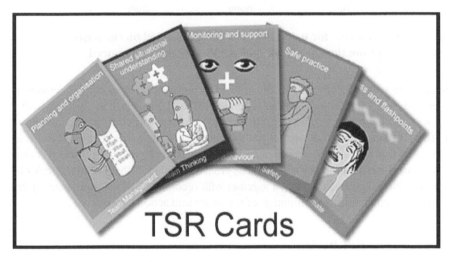

Figure 12.5   Example TSR cards

of the 15 team health dimensions); the back of each card contains a pertinent set of probes designed to get the team talking and thinking. The different controls comprise alternative ways in which the cards can be used (and their questions asked) and include approaches that feature an element of chance in order to stimulate creative thought (for example, team members might pick a card at random from which to ask questions to the rest of the team); card ranking, sorting, and grouping approaches, and various sampling and comparative approaches for selecting questions. Figure 12.6 shows the spread of the controls against the time available in which to conduct the review, and the depth of learning supported.

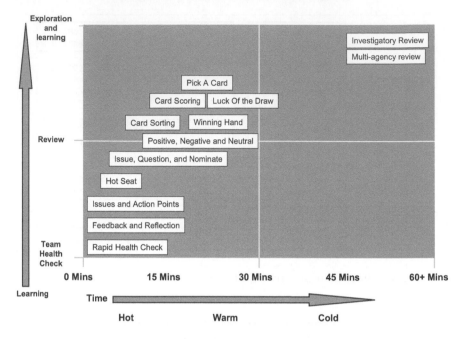

Figure 12.6   Controls for using the TSR cards, selected on the basis of the time available and depth of learning required

The TSR debrief is designed to facilitate two key processes; (a) *identifying* lessons to be learned, and (b) generating actions to implement (and thus *learn*) the lessons. In this way, *lessons identified* are converted into *lessons learned* via modifications to individual and team behavior. The TSR debrief is also concerned with packaging these outputs in such a way that they can be shared outside the team, thereby enabling teams working in the same department to learn vicariously from each other. Consequently, a series of recording sheets has been designed together with recommendations for managing the collection, feedback, and exploitation of the issues and actions generated.

Impact of Team Self-Review

Following several pilot studies, the TSR tools were tested with theater staff over twenty-one lists (covering approximately 80 operations) during the course of one week. In this week alone, over 80 relevant teamwork, performance and safety-related lessons, issues and actions were raised and acted upon. A number of positive changes to process and procedures resulted, and staff morale was self-reported as increasing dramatically. While a Hawthorne effect[3] cannot be entirely ruled out, due to the initial

---

3   An increase in worker productivity resulting from the psychological stimulus of being singled out and made to feel important. Named after studies conducted in the 1920s by Vannevar Bush (reported by Roethlisberger and Dickson, 1939, pp. 14–28) on factory workers at the Western Electric Company Hawthorne Works, Cicero, Illinois.

novelty of the method during this time, an independent review of the impact of TSR conducted 12 months after it was introduced showed that it was continuing to have a positive impact on behavior and attitudes (National Audit Office, 2005).

The Royal Cornwall Hospital has continued to employ TSR on a regular basis since its initial assessment, and further learning and process change has resulted. Figure 12.7 shows a subsequent coding of the issues raised through TSR conducted by theater staff during the period September–November 2003 (that is, when theater staff were running the TSR system themselves) with positive issues generating actions for reinforcement and repetition, and negative issues generating actions for corrective behavior. These metrics were adopted as part of the clinical governance process for tracking the local impact of TSR.

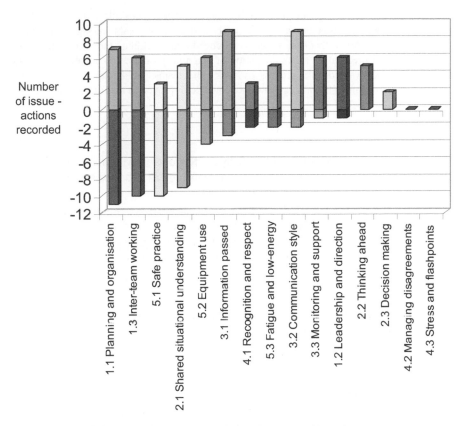

Figure 12.7   Coding of issues raised through team self-review

The TSR process has also been rolled out to a second theater block, and the hospital is considering its introduction to other departments. The project has gathered a great deal of press and publicity, and its impact has been cited by the Chief Executive of the National Health Service, as an example of successful organizational learning in the NHS (Crisp, 2003).

In addition, in 2005 the UK National Audit Office (NAO) conducted a review of the project (National Audit Office, 2005) as part of their assessment of improvements to organizational learning within the National Health Service since their original 2001 report. The NAO concluded:

> ...staff in the theatre complex exposed to team self-review showed statistically significant improvement in teamwork climate and some improvement in the safety climate (using the Safety Attitude Questionnaire, a reliable and formally validated research tool used in over 350 hospitals worldwide) than staff who were not offered debriefs or did not attend. [ibid., p. 16]

Further:

> ...briefing and debriefing had a positive impact on nontechnical skills and patient safety; pre-session briefing was important for safety and effective team management (those interviewed reported that the process improved teambuilding and communication, and enhanced preparation and anticipation of potential problems for theatre lists); and debriefing was valued as a process by which the teams could learn from problems encountered during lists and plan how care could be improved in the future. [ibid., p. 16]

Evaluation

Table 12.1 considers how the use of TSR by the Royal Cornwall matches Fischer's (2001) criteria for evaluating organizational learning (as described earlier in this chapter).

In short, the adoption of TSR by the Royal Cornwall appears to meet most of the criteria stipulated by Fischer for the achievement of organizational learning. The system has been taken-up, is being maintained, and is having a positive impact.

In the next section, the relationship between macrocognition and organizational learning will be considered.

Macrocognition and Naturalistic Approaches to the Study of Organizational Learning

This chapter has described an approach to enhancing the development of organizational learning in surgical teams via a process of Team Self-Review. In conducting the fieldwork underpinning this work, much has been learned about the individual and shared cognition that is essential to effective working within the complex cognitive system of an operating theater. Many of the processes identified as key to effective theater practice reflect the high-level functions and supporting processes considered central to macrocognition (Ross et al., 2002; Klein et al., 2003)—these are discussed shortly. It is suggested that this results primarily from the consideration of macrocognition as high-level cognitive processes that are usually situated within the context of team collaboration (Klein et al., 2003)—a situation and context fundamental to operating theaters. Further, such macrocognitive functions and processes impact directly on team performance in

Table 12.1   Measures of organizational learning

No.	Criteria	Evidence
1	Organizational work routines are being evaluated and improved.	Staff and management allocate time to evaluating their work routines at the end of lists. Emergent actions are recorded, followed up by the theater manager via the corporate governance process, and closed.
2	Formal and informal learning processes are being evaluated and improved.	The learning processes are modified and improved through experience. For example, the metrics employed for categorizing issue-action pairs was revised after a few months in order to provide additional insight.
3	Transformations are occurring in the culture of the organization.	Attitudinal improvements in theaters have been assessed independently using standardized culture assessment tools.
4	Knowledge is being created within the organization, at different levels (not only by the managers and senior staff), and it is being shared within the organization.	Lessons are recorded using standardized recording sheets; communicated to others via the corporate governance process; and acted upon locally and across the organization. In addition, the introduction of TSR has also been linked to an increase in levels of close-call reporting (reported by Crisp, 2003).
5	Learning from the environment is encouraged and systematically evaluated. The results are assimilated and accommodated to the organization's objectives and local constraints and opportunities.	Senior management are seen to drive, support, and act upon the outputs of the learning process. Metrics are being employed, and the learning process leads to improvements against higher-level organizational measures (such as the hospital's to improve safety culture).

Source: adapted from Fischer, 2001.

Table 12.2   Relevance of macrocognitive functions to surgical team performance

Macrocognitive function	Relevance to surgical team effectiveness
Decision making	Recognitional decision strategies are usually employed throughout the course of a surgical procedure; for example, the anesthetist employs naturalistic decision strategies to diagnose, monitor, and evaluate the patient's health throughout the surgical procedure; the scrub nurse will watch continuously the surgeon's activities in order to pre-empt and prepare the equipment he will need next. When unusual or unexpected circumstances arise during surgical procedures, strategies are employed to diagnose the situation, and to generate an appropriate course of action based upon the recognition of salient cues, the activation of past experience, and the use of mental simulation for course-of-action evaluation.
Sensemaking/ Situation assessment	Sensemaking, defined as "the process of fitting data into a frame and fitting a frame around the data" (Sieck et al., 2004) occurs throughout the surgical process as the team continuously monitors and makes sense of the situation and events surrounding them, and talks about it. We found that different frames were sometimes brought to the same problem by different people; while this was usually useful for bringing different perspectives to bear, it could on occasion also lead to problems (such as staff making assumptions about other team members' levels of awareness).
Planning	Planning is a core activity essential to preparing the theater for smooth running throughout the day; and to ensuring that each case is conducted successfully. We found that planning occurs across the team; and that this process was significantly enhanced when pre-briefing took place. In particular, pre-briefing enabled pre-defined plans for dealing with different types of operation to be customized to the situation at hand (thus pre-briefing supported the process of replanning). The activity was seen to support risk identification and management, and to ensure that all team members have a shared understanding as to what is expected of them.
Adaptation/ replanning	Adaptation and replanning occur within a team when things, literally, are not going to plan. In our study, adaptation was most often seen to occur when a team encountered an unexpected situation, or when it was clear that the current plan was not working (in other words, plans enabled people to recognize when to be surprised). Adaptation and replanning occurred both at a team level (for example, the team changed the processes it was employing to conduct the operation) and at an individual level (for example, one team member realized that they needed to do things differently in light of current circumstances).

Table 12.2 continued   Relevance of macrocognitive functions to surgical team performance

Macrocognitive function	Relevance to surgical team effectiveness
Problem detection	Problem detection was seen to occur across the entire surgical team, often via the detection of anomalies or deviations from expectancies. It was apparent that problems were detected using a wide range of sensory cues, including visual, auditory, tactile, and olfactory.
Coordination	Coordination is critical to performance within a surgical team composed of different specialists. Coordination will be required within the team (across, for example, the surgeon(s), anesthetist, scrub nurse, runner, and so on) and with other teams (for example, with porters and recovery teams). These individuals will be required to synchronize their activities across time and space in order to conduct a successful surgical procedure.

the operating theater, and thus should be included as targets for organizational learning. Table 12.2 describes the relevance of key macrocognitive functions to surgical team effectiveness.

Six key processes have been identified that support the macrocognitive functions. These consist of maintaining common ground, developing mental models, uncertainty management, turning leverage points into courses of action, attention management, and mental simulation and storyboarding. Table 12.3 considers each function in terms of its relevance to surgical teams.

Reflections on Macrocognition and Organizational Learning

A key issue that has emerged from this study is that there exists something of a schism between the theory and the practice of facilitating organizational learning in teams that work within highly politicized, organizationally complex, and cognitively challenging settings. The practical challenges encountered in implementing "at the coalface" and sustaining a learning system within this context (as described in the concerns and challenges sections of this paper) are significant, and yet they do not appear to be addressed in any detail in the organizational learning literature reviewed. Nonetheless, it is these factors that impact most directly on the success of any intervention of this nature.

The study of macrocognitive phenomena offers the potential for guiding future study and design of organizational learning systems within such settings. Indeed, the naturalistic fieldwork conducted under our studies has uncovered strong evidence for, and examples of, macrocognition within the context of team performance; the macrocognitive framework has further helped us to cluster and make sense of our findings. The study of macrocognition identifies and bounds together a broad range of high-level cognitive functions and processes that are essential to team and organizational performance—thus macrocognition provides useful pointers to those areas where experiential learning is most critical.

Table 12.3   Relevance of macrocognitive process to surgical team performance

Macrocognitive process	Relevance to surgical team effectiveness
Maintaining common ground	Maintenance of common ground facilitates mutual understanding, coordination, and effective relationships between team members. In a surgical team, common ground enables team members to coordinate their activities through the use of common technical and non-technical language. We also found that Team Self-Review supported the establishment and maintenance of common ground by promoting common structures for conceptualizing team performance, and a language for discussing it.
Developing mental models	Mental models comprise both in-situ dynamic models of the situation and team status; and experiential models of accumulated experience and meaning. They are based on organized shared knowledge, that helps collaborators form accurate explanations and expectations about the task and each other, thus helping them coordinate explicitly. We found that Team Self-Review enabled teams to derive both meaning from their experiences, and a common model and language for discussing it. We also found that a range of tools were used to assist the formulation and maintenance of shared understanding and mental models across a team, including the use of patient records, aides-mémoire listing surgeons' instrument preferences, and the utilization of a state-board to record various aspects of an operation as it progressed (such as the marking of swab counts, as discussed previously).
Uncertainty management	It became clear during our study that uncertainty could occur at any point throughout the patient journey, from the initial medical diagnosis, during the surgical and subsequent treatment process, right through to post-operative monitoring and the aftercare process. Key uncertainty management activities within the operating theater included establishing processes for reducing the opportunity for uncertainty to arise (for example, physically marking surgical sites on the patient and confirming this indication on multiple occasions (including with the patient) prior to the operation commencing); processes for recognizing when uncertainty is present (for example, identifying inconsistencies on the status whiteboard); and processes for reducing or coping with uncertainty when it had been identified (for example, performing additional in-situ checks, such as taking an x-ray, to gather more information).

Table 12.3 continued   Relevance of macrocognitive process to surgical team performance

Macrocognitive process	Relevance to surgical team effectiveness
Turning leverage points into courses of action	Course-of-action generation in the operating theater is a fundamental component of surgical decision making. Capitalizing on opportunities when they arise enables a surgical team to enhance its performance, thereby delivering maximum value to the patient, and increasing the likelihood of a successful outcome. We also found that pre-operation assessment of risk (for example, via a case pre-brief and discussion) could support the identification of other leverage points that might be utilized in course of action generation.
Attention management	Each member of the surgical team will have a broad range of cues that they are actively monitoring in the environment, and they are also attentive to a range of other intrusive (and potentially distracting) indicators. A core skill is the ability to recognize cue patterns to support sensemaking, thus providing a basis for recognitional decision making and course-of-action generation. We found that a key to successful surgical team performance is the active management and balancing of attention among the patient, other members of the team, supporting equipment, and also other people from outside of the team. Indeed, situational awareness can potentially become compromised if an appropriate balance is not maintained.
Mental simulation and storyboarding	Mental simulation and storyboarding techniques support diagnostic processes, sensemaking, and evaluation. By constructing narrative structures, information concerning past, present, and future can be linked together, and incompleteness, ambiguities, conflict, and any anomalies can be detected. Mental simulation can be employed to evaluate the story, and to see how well it holds up under different conditions. It can also be used to play through a proposed course of action in order to assess its strengths and weaknesses, and to generate expectancies about future events that can subsequently direct attention and monitoring. We found that mental simulation was regularly used to assist with playing-through forthcoming events, and identifying potential problems before they arose.

While learning is inherent in a number of the functions and processes that comprise macrocognition (such as the development of mental models and sensemaking), it is not currently identified as a macrocognitive process in its own right. As research at the macrocognitive level of description advances, the relationship between experiential learning and macrocognitive functions is likely to warrant further attention.

Acknowledgements

The author would like to acknowledge the support of Matthew Mills and Carol Blendell, whose significant efforts contributed directly to the success of the TSR project. In addition, thanks is also extended to all the theater staff at the Royal Cornwall Hospital Trust (especially to Dr James Boyden, Dr Adrian Hobbs, Dr Alan Bleakley, and Ms Linda Walsh) for their support and enthusiasm throughout the project; and to the NPSA for their vision in sponsoring and supporting such innovative safety-oriented research.

References

Adams, D. and Carwardine, M. (1990). *Last Chance to See*. London: William Heinemann.

Agnew, N.M., Ford, K.M., and Hayes, P.J. (1997). "Expertise in context: Personally constructed, socially selected and reality relevant?" In P.J. Feltovich, K.M. Ford, and R.R. Hoffman (eds), *Expertise in Context*. Cambridge, MA: AAAI Press/The MIT Press (pp. 219–44).

Argyris, C. and Schön, D.A. (1978). *Organizational Learning: A Theory of Action Perspective*. Reading, MA: Addison-Wesley.

Ausubel, D.P. (1963). *The Psychology of Meaningful Verbal Learning*. New York: Grune and Stratton.

—— (1968). *Educational Psychology: A Cognitive View*. New York: Holt, Rinehart and Winston.

Ausubel, D.P., Novak, J.D., and Hanesian, H. (1978). *Educational Psychology: A Cognitive View*. New York: Werbel and Peck.

Brannick, M.T., Prince, C., Salas, E., and Stout, R.J. (April 1995). "Assessing aircrew coordination skills in TH-57 pilots." In C. Bowers and F. Jentsch (chairs), *Empirical Research using PC-based Flight Simulations*. Symposium conducted at the Eighth International Symposium on Aviation Psychology, Columbus, OH.

Burke, C.S., Salas E., Wilson-Donnelly, K., and Priest, H. (2004). "How to turn a team of experts into an expert medical team: Guidance from the aviation and military communities." *Quality and Safety in Health Care*, 13: i96–i104.

Carthey, J., de Leval, M.R., Wright, D.J., Farewell, V.J., and Reason, J.T. (2003). "Behavioral markers of surgical excellence." *Safety Science*, 41: 409–25.

Chief Medical Officer (2000). *An Organisation with a Memory. Report of an Expert Group on Learning from Adverse Events in the NHS, chaired by the Chief Medical Officer*. London: The Stationery Office.

Craik, K. (1943). *The Nature of Explanation*. Cambridge: Cambridge University Press.

—— (2003). *Introduction to National Patient Safety Agency*, Special Edition, *Health Service Journal*, November.

Fischer, M. (2001). "How to conceptualise and investigate learning in the learning company?" In M. Fischer and P. Roben (eds), *Ways of Organizational Learning in the European Chemical Industry and the Implications for Vocational Education and Training. A Literature Review*. Bremen: Institute for Technology and Education. University of Bremen (pp. 5–14).

Fischer, M. and Roben, P. (eds) (2004). *Organizational Learning and Vocational Education and Training—An Empirical Investigation in the European Chemical Industry.* Brermen: Institute for Technology and Education. University of Bremen.

Fleming, P.J., Wood, G.M., Ferro, G., Bader, P.K., and Zaccaro, S.J. (2003). "The locus of shared mental models in teams: Whence does the sharedness come?," Paper presented at the Eighteenth Annual Conference of the Society of Industrial and Organizational Psychology, 11–13 April 2003, Orlando, Florida.

Fletcher, G.C.L., McGeorge, P., Flin, R.H., Glavin, R.J., and Maran, N.J. (2002). "The role of nontechnical skills in anaesthesia: A review of current literature." *British Journal of Anaesthesia*, 88: 418–29.

Gentner, D.A. and Stevens, A.L. (eds) (1983). *Mental Models.* Hillsdale, NJ: Lawrence Erlbaum Associates.

Hakkarainen, K.P.J., Palonen, T., Paavola, S., and Lehtinen, E. (2004). *Communities of Networked Expertise: Professional and Educational Perspectives* (Advances in Learning and Instruction Series). Amsterdam: Elsevier Science.

Healey, A.N., Undre, S., and Vincent, C.A. (2006). "Defining the technical skills of teamwork in surgery." *Quality and Safety in Health Care*, 15 (4): 231–4.

Helmreich, R.L., Merritt, A.C., and Wilhelm, J.A. (1999). "The evolution of crew resource management training in commercial aviation." *The International Journal of Aviation Psychology*, 9 (1): 19–32.

Hoffman, R.R., Crandall, B., and Shadbolt, N. (1998). "Use of the critical decision method to elicit expert knowledge: A case study in cognitive task analysis methodology." *Human Factors*, 40 (2): 254–76.

Hollan, J., Hutchins, E., and Kirsh, D. (2000). "Distributed cognition: Toward a new foundation for human–computer interaction research." *ACM Transactions on Computer–Human Interaction (TOCHI), Special Issue on Human–Computer Interaction in the New Millennium, Part 2*, 7 (2): 174–96.

Hutchins, E. (1995). *Cognition in the Wild.* Cambridge, MA: The MIT Press.

Janis, I.L. (1972). *Victims of Groupthink.* Boston, MA: Houghton Mifflin Company.

Johnson-Laird, P.N. (1983). *Mental Models.* Cambridge: Cambridge University Press.

Klein, G. (1992). "Using knowledge engineering to preserve corporate memory." In R.R. Hoffman (ed.), *The Psychology of Expertise: Cognitive Research and Empirical AI.* Mahwah, NJ: Lawrence Erlbaum Associates (pp. 170–90).

Klein, G., Phillips, J.K., Rall, E., and Peluso, D.A. (2006). "A data/frame theory of sensemaking." In R.R. Hoffman (ed.), *Expertise Out of Context: Proceedings of the 6th International Conference on Naturalistic Decision Making.* Mahwah, NJ: Lawrence Erlbaum Associates.

Klein, G., Ross, K.G., Moon, B.M., Klein, D.E., Hoffman, R.R., and Hollnagel, E. (2003). "Macrocognition." *IEEE Intelligent Systems*, 18 (3): 81–5.

Klein, G.A., Calderwood, R., and MacGregor, D. (1989). "Critical decision method for eliciting knowledge." *IEEE Transactions on Systems, Man, and Cybernetics*, 19 (3): 462–72.

Kolb, A.Y. and Kolb, D.A. (2005). "Learning Styles and Learning Spaces: Enhancing Experiential Learning in Higher Education." *Academy of Management Learning and Education*, 4 (2): 193–212.

Kolb, D.A. (1984). *Experiential Learning: Experience as the Source of Learning and Development.* Englewood Cliffs, NJ: Prentice-Hall.

Kolb, D.A., Boyatzis, R.E., and Mainemelis, C. (2000). *Experiential Learning Theory: Previous Research and New Directions.* Cleveland, OH: Case Western Reserve University.

Kraiger, K. and Wenzel, L.H. (1997). "Conceptual development and empirical evaluation of measures of shared mental models." In M.T. Brannick, E. Salas, and C. Prince (eds), *Team Performance Assessment and Measurement: Theory, Methods, and Application.* Mahwah, NJ: Lawrence Erlbaum Associates (pp. 63–84).

Lave, J. and Wenger, E. (1991). *Situated Learning: Legitimate Peripheral Participation.* New York: Cambridge University Press.

Mathieu, J.E., Heffner, T.S., Goodwin, G.F., Salas, E., and Cannon-Bowers, J.A. (2000). "The influence of shared mental models on team process and performance." *Journal of Applied Psychology*, 85: 273–83.

Mills, M.C., Pascual, R.G., Blendell, C., Molloy, J.J., and Verrall, N.G. (1999a). "Leading distributed and ad-hoc teams: A teamwork guide for team leaders." DERA, DERA/CHS/MID/CR980262/1.0. Unpublished.

—— (1999b). "Supporting distributed team interaction." DERA, DERA/CHS/MID/CR980262/1.0. Unpublished.

Mohammed, S., Klimoski, R., and Rentsch, J.R. (2000). "The measurement of team mental models: We have no shared schema." *Organizational Research Methods*, 3: 123–65.

Musson, D.M. and Helmreich, R.L. (2004). "Team training and resource management in healthcare: Current issues and future directions." *Harvard Health Policy Review*, 5 (1): 25–35.

National Audit Office (2005). "A safer place for patients: Learning to improve patient safety." Report by the Comptroller and auditor general, HC 456 Session 2005–2006, 3 November 2005. London: The Stationery Office.

Nemeth, C. (2003). "How cognitive artefacts support distributed cognition in acute care." Symposium on Insights from Technical Work Studies in Healthcare. Human Factors and Ergonomics Society, National Conference, Denver, CO. 13–17 October 2003.

Norman, D.A. (1983). "Some observations on mental models." In D.A. Gentner and A.L. Stevens (eds), *Mental Models.* Hillsdale, NJ: Lawrence Erlbaum Associates (pp. 7–16).

Novak, J. and Gowin, D.B. (1984). *Learning how to Learn.* New York and Cambridge: Cambridge University Press.

Novak, J.D. (1998). *Learning, Creating and Using Knowledge.* Mawhah, NJ: Lawrence Erlbaum Associates.

Pascual, R.G. (1999). "Tools for capturing and training shared understanding in teams." In *People in Control: an International Conference on Human Interfaces in Control Rooms, Cockpits and Command Centres*, Bath, 21–23 June 1999. ISSN 0537–9989.

Pascual, R.G. and Henderson, S.M. (1998). "An investigation of the relationship between experience, shared mental models and team performance in Police Armed Response Units." Poster session at the Fourth Naturalistic Decision Making Conference, Warrenton, VA.

Pascual, R.G., Henderson, S.M., and Mills, M. (1995). "Foundations for Enhancing Military Team Interaction." DERA, DRA/LS (LSC1)/89T4.002/95/1/3. Unpublished.

—— (1998). "Understanding and supporting team cognition." DERA/CHS/MID/ CR980122/1.0. Unpublished.

Pascual, R.G., Mills, M.C., Blendell, C., and Molloy, J.J. (1999). "Understanding and supporting team cognition." DERA, DERA/CHS/MID/CR990127/1.0. Unpublished.

—— (2000). "The role of mental models in team effectiveness." DERA/CHS/MID/ CR000213/1.0. Unpublished.

Plasters, C.L., Seagull, F.J., and Xiao, Y. (2003). "Coordination challenges in operating-room management: An in-depth field study." *AMIA Annual Symposium Proceedings*; pp. 524–8 <http://www.pubmedcentral.nih.gov/picrender.fcgi?tool =pmcentrez&blobtype=pdf&artid=1480348> accessed 3 March 2007.

Prince, C. and Jentsch, F. (2001). "Aviation crew resource management training with low fidelity devices." In E. Salas, C. Bowers, and E. Edens (eds), *Improving Teamwork in Organizations: Applications of Crew Resource Management Training*. Mahwah, NJ: Lawrence Erlbaum Associates (pp. 147–64).

Roethlisberger, F.J. and Dickson, W.J. (1939). *Management and the Worker*. Cambridge, MA: Harvard University Press.

Ross, K.G., McHugh, A., Moon, B., Klein, G., Armstrong, A., and Rall, E. (2002). *Year One Final Report: High-Level Cognitive Processes In Field Research*. #02TA1-SP1-RT1. Advanced Decision Architectures Collaborative Technology Alliance.

Sieck, W.R., Klein, G., Peluso, D.A., Smith, J.L., and Harris-Thompson, D. (2004). *FOCUS: A Model of Sensemaking*. Fairborn, OH: Klein Associates, Inc.

Smith-Jentsch, K.A., Campbell, G.E., Milanovich, D.M., and Reynolds, A.M. (2001). "Measuring teamwork mental models to support training needs assessment, development, and evaluation: Two empirical studies." *Journal of Organizational Behavior*, 22: 179–94.

Smith-Jentsch, K.A., Zeisig, R.L., Acton, B., and McPherson, J.A. (1998). "Team dimensional training: A strategy for guided team self-correction." In J.A. Cannon-Bowers and E. Salas (eds), *Making Decisions under Stress: Implications for Individual and Team Training*. Washington, DC: APA Press (pp. 271–97).

Stout, R.J., Cannon-Bowers, J.A., Salas, E., and Milanovich, D.M. (1999). "Planning, shared mental models, and coordinated performance: An empirical link is established." *Human Factors*, 41: 61–71.

Suchman, L. (1987). *Plans and Situated Actions: The Problem of Human–Machine Communication*. Cambridge: Cambridge University Press.

Tarnoff, K.A. (1999). "An Exploratory Study of the Determinants and Outcomes of Shared Mental Models of Skill Use in Autonomous Work Teams." Dissertation submitted to the Faculty of Virginia Polytechnic Institute and State University in partial fulfilment of the requirements for the degree of PhD in Management. Blacksburg, VA.

Undre, S., Sevdalis, N., Healey, A.N., Darzi, A., and Vincent, C.A. (2006). "Teamwork in the operating theatre: Cohesion or confusion?" *Journal of Evaluation in Clinical Practice*, 12 (2): 182–9.

Walkinshaw, O., Outteridge, C., McCorquodale, B., and Henderson, S. (2002). "Team pitfalls: A guide to how and where teams can go wrong." QinetiQ, October. Unpublished.

Chapter 13

Macrocognition in Systems Engineering: Supporting Changes in the Air Traffic Control Tower

Peter Moertl, Craig Bonaceto, Steven Estes, and Kevin Burns

Introduction

In an effort to alleviate congestion, many airports are seeking means to increase capacity, that is, the number of aircraft that can move through the system in a given period of time. Such means typically include procedural revisions, the construction or reconfiguration of taxiways, the construction or reconfiguration of runways, or changes in tower equipage. Reconfigurations that entail the construction of new runways and taxiways will almost always require procedural revisions, and they may also entail additional changes, such as the construction of satellite control towers or the reconfiguration of an existing control tower. While such improvements may increase capacity, they may also significantly change air traffic controller decision making and communication. So before such procedural changes are made, it is necessary to explain and predict the effects they will have on air traffic controller performance. Moreover, changes in procedures and equipment should be informed by an understanding of the human performance limitations in the current environment.

To adequately assess human performance aspects in air traffic control, we are looking for methodologies and a level of description that allows us to adequately analyze the tasks of air traffic tower (ATCT) controllers in their work environment. Such methods need to be administered in the confined environment of a control tower over sometimes extended periods of time while causing minimal disruption. Whereas we have found that complete task analysis methods (see, for example, Kirwan and Ainsworth, 1992) are desirable, cost and accessibility constraints together with the fact that such findings can frequently not be ported to other towers directed us toward other alternatives. Specifically, we have found the macrocognitive approach useful in this domain and have started adjusting task analytical methods to assess the macrocognitive functions in the ATCT domain. Specifically, our selection of methods is guided by the fact that air traffic control procedures occur not only under routine, but also under non-routine operations where unusual conditions frequently facilitate the occurrence of errors. Macrocognitive functions can be assessed as part of a complete task analysis but as we will demonstrate, can be assessed in isolation.

Thereby, controller workload and task completion time are important and quantifiable aspects of controller performance, and can be the results of cognitive modeling. We therefore adopted cognitive modeling as one approach to predict the impact of change on controller workload and task completion time. But while cognitive modeling allows the prediction of change for routine events, we also found the need to assess the effects of change in controller performance under non-routine operating conditions. This led us to select the critical incident analysis method to shed light on factors underlying non-routine events and errors, allowing us to determine which factors may persist or worsen with a new set of procedures. Yet another important aspect of performance is how controllers coordinate their activities and communicate. For this purpose, we developed a controller coordination analysis framework that leverages and combines previous research in that area. The coordination analysis framework is applied to the specific task analytical constraints in ATC.

We will provide examples for these methodologies in the following section. One question, however, that arises is how do we know if our methods were successfully employed or not? We do not address this question here in detail and acknowledge that other methods may also serve the requirements of systems engineering. However, we found the presented methods useful to capture macrocognitive functions in ATCT and communicate human behavioral constraints effectively to facilitate change. Therefore we judge methods as successful not only if they support the research tasks but furthermore help port the research findings into a different community to ultimately facilitate the change. We found that macrocognition is an approach that can be adjusted to this requirement.

Macrocognition and Systems Engineering

The air traffic control tower (ATCT) work domain involves a full spectrum of macrocognitive functions, including naturalistic decision making, sensemaking and situation awareness, planning and replanning, problem detection, and coordination. These macrocognitive functions require the execution of processes that include the formation and use of mental models, the management of uncertainty, and, perhaps of foremost importance in the ATCT domain, the management of attention.

The process of attention management is crucial in domains such as air traffic control, where controllers must track and manage multitudes of objects that interact in confined spaces and whose status and position is constantly changing. This steady stream of information requires controllers to actively seek relevant information to maintain a sufficient level of situation awareness. Moreover, there are demands on controllers to make decisions quickly and efficiently to maintain adequate system performance, while also balancing the sometimes competing need to maintain safe operations.

Thus, to engineer safe and effective procedures and systems in the ATCT domain, it is necessary to model and understand the domain, which is comprised of both the controllers themselves and the work environment, at multiple levels of analysis. The methods that comprise our multifaceted approach are designed to address the bulk of the relevant macrocognitive functions and processes, but we do not ignore the microcognitive building-blocks that underlie these functions (cognitive modeling may be characterized as ecological or applied microcognition). Rather, our methods are designed to first qualitatively describe and model controllers and the environment, therefore providing an "explanatory" function. The methods then secondly also facilitate quantitative evaluations that can be used to make predictions about how well a proposed set of systems and procedures will work, therefore providing a "measurement" function.

Beginning with the explanatory function, current system bottlenecks are first identified by generating a qualitative understanding of the system, with special consideration of human roles and dependencies. User and observer narratives are used to describe this function at the macrocognitive level. The formulation of cognitive and behavioral attributes in macrocognitive terminology is understandable not only by the research community but also by stakeholders (for example, management) and users (for example, other air traffic controllers). We found that explanatory descriptors that stay just within the research or engineering community (such as some microcognitive approaches) are less likely to lead to successful procedural or system redesigns that meet commonly agreed-upon objectives.

The measurement function is essential for performance prediction and therefore highly relevant for system change acceptance. While the explanatory function lays the groundwork and establishes shared terminology, it does not by itself provide the arguments for organizational decision making concerning investment decisions. In our experience, the predicted benefits of a system or procedural redesign are the strongest arguments for implementation decisions. Estimation of these benefits is an inherently quantitative task and therefore relies on the measurement of system or procedural enhancements with special consideration of the associated cognitive and performance improvements. It is necessary to show that safety, efficiency, and capacity gains will result from system or procedural redesigns. Systems-engineering design questions that can be addressed by such quantitative methods include:

- What different information is needed in a redesigned work place and how does this relate to bottlenecks in controlling traffic?
- How much time do controllers spend scanning their environment to extract information and how much time could be used for capacity enhancing tasks?
- How many operator errors have been associated with uncommon runway usage patterns and what is the expected reduction of error rates by usage of memory aids?

In this chapter, we show the application of the methods which address both explanatory (qualitative) and measurement (quantitative) levels of analysis, in the context of supporting change in the ATCT domain. First, we provide a brief overview of air traffic control tower operations and some of the macrocognitive functions that controllers perform, and we then discuss the cognitive analysis and modeling, critical incident analysis, and coordination analysis in further detail.

Overview of US ATCT Operations

There are five types of ATCT positions in the United States: flight data, clearance delivery (or a combined flight data/clearance delivery position), ground control, local control, and tower supervision. The flight data controller's primary responsibility is to manage flight plan information for departures and coordinate this information with other controllers. The clearance delivery controller is primarily responsible for the pre-departure coordination of flight plan information with pilots. Ground controllers control and coordinate the movement of aircraft on the airport surface, whereas local controllers are responsible for controlling aircraft landings and departures. A tower supervisor is responsible for managing the overall control operations in the tower.

Control towers at larger airports may have additional controller positions, such as a ground metering controller whose main responsibility is to sequence departure aircraft on the ground. An air traffic manager may also be present to relay traffic constraints of the surrounding airspace to the other controllers. Communication between controllers and pilots occurs via radio, with different controller positions communicating over different radio frequencies.

The specific configuration of control towers varies widely, and is dependent on characteristics of the airport layout, traffic, geographic location, winds, and other factors such as noise restrictions. So operations at one tower may vastly differ from those at another.

Coordination and communication between different positions typically occurs either face-to-face or via flight strips. Flight strips are small pieces of paper that contain the most relevant information about a flight, including its departure fix, and departure and estimated arrival time, aircraft equipage and flight plan information. Depending on the tower's configuration, additional systems may be used for coordination and dissemination of air traffic management information.

Prior to departure, an aircraft will file a flight plan indicating its intended flight path, the location and time of departure, and the arrival airport. As a departing aircraft readies for pushback from the gate, the clearance delivery controller communicates the flight plan to the pilot (although this function is sometimes performed via electronic datalink).

After a departing aircraft pushes back from the gate and reaches the active movement area outside the ramp area, a ground controller issues it taxi instructions. These instructions include the sequence of taxiways and runways to use to reach the departure runway, such as "American 123, runway 35L, taxi via Bravo, Alpha, hold-short of runway 29." Bravo and Alpha refer to taxiways, and the "hold-short" instruction indicates that the pilot is required to stop and ask for clearance prior to crossing runway 29 and continuing to taxi to runway 35L.

As the taxiing aircraft stops at runway 29 and waits for a crossing clearance, the ground controller will coordinate with a local controller who controls the traffic on runway 29 and ask if it is safe for the aircraft to cross the runway. Alternatively, the ground controller may instruct the pilot of the taxiing aircraft to switch to the local controller's radio frequency and receive a crossing clearance directly from the local controller. In this case, the required coordination with the local controller is moved from the ground controller to the pilot.

After the aircraft crosses runway 29 and reaches the departure runway 35L, a local controller issues a departure clearance, such as "American 123, cleared for take-off on 35L." The aircraft then rolls onto the runway and initiates the take-off. After take-off, the pilot switches radio frequencies to initiate contact with the departure controller who issues clearances to the now-airborne flight. The departure controller is located in the Terminal Area Approach Control Center (TRACON) that services the airport.

Changes in ATCT procedures

A specific example of a proposed change in the air traffic control operations is provided by Domino et al. (2006). They investigated the implications of changing departure procedures when using new wind-detection and wake vortex-prediction technologies. Wake vortices are tornado-like flows of air that are produced by the wings of aircraft especially during slow flight maneuvers and under high-lift configurations. These wake vortices are produced by any aircraft and are directly proportional to the weight of the aircraft. Wake vortices can, under specific conditions, have detrimental effects on the flight characteristics of a following aircraft. For example, in 2001, an Airbus A300 crashed after take-off behind a Boeing B747 due to the overcorrecting actions of the copilot to counteract the effects of the wake vortices from the large aircraft ahead. To control the impact of wake vortices on successive aircraft, controllers apply wake-vortex separation between aircraft during departure and arrival situations. Essentially, controllers ensure sufficient spacing between successive aircraft to allow wake vortices to dissipate. For example, Figure 13.1 shows the separation criteria for aircraft departing from parallel runways.

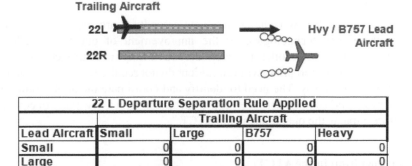

22 L Departure Separation Rule Applied			
Trailing Aircraft			
Lead Aircraft **Small**	**Large**	**B757**	**Heavy**
Small	0	0	0	0
Large	0	0	0	0
B757	2min or 5nmi	2min or 4nmi	2min or 4nmi	2min or 4nmi
Heavy	2min or 5nmi	2min or 5nmi	2min or 5nmi	2min or 4nmi

Figure 13.1   Wake vortex separation requirements for departures on parallel runways

These separation criteria between aircraft directly impact the arrival and departure capacity of airports. Therefore, efforts are under way to determine if "smarter" wake vortex-separation rules can be identified. For example, under certain conditions such as depicted in Figure 13.2, winds are such that the wake vortices of a departing aircraft do not travel to an aircraft on a parallel runway. Therefore, the wake vortex-separation criteria for departing aircraft could be relaxed or even removed under such conditions.

The recognition of those conditions requires specific technologies that are currently being developed (see, for example, MIT Lincoln Labs, 2006).

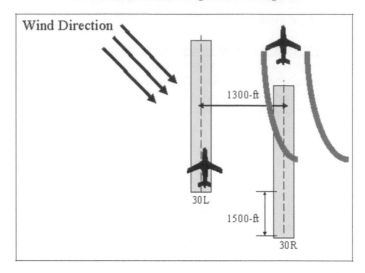

Figure 13.2   Specific wind conditions may eliminate impact of wake vortices on departure aircraft on parallel runway

In addition to such new technologies, new procedures are required for ATCT to utilize these new technologies for the improvement of operational efficiency. Specifically, the two controllers that work the two runways need to exchange different information when appropriate wind conditions do not require wake vortex separation on the parallel runway. The need to identify and coordinate the applicability of the appropriate wind conditions as well as the interaction with a new piece of equipment, therefore impacts the operations in the ATCT.

Macrocognition in the ATCT

The provided example indicates the required coordination between controller positions as function of wind conditions and the application of new control procedures. A description of the interdependencies between these processes in macrocognitive terms is necessary to maintain an understanding of the context in which the task is performed. Lower-level, microcognitive descriptions may be able to only depict subcomponents of the whole and therefore not provide sufficient understanding of the situation. The overview of ATCT operations and the provided example give opportunities for macrocognitive descriptions. An example of naturalistic decision making in the ATCT are decisions by local controllers to issue landing clearances on closely spaced parallel arrival runways. ATCT situation assessments include the measurement of the expected brake effectiveness of aircraft after landing on a runway in poor weather conditions. ATCT planning adapts to events such as runway closures, wind changes, or changes of traffic flows due to inclement weather. ATCT problem detection includes the identification of unexpected aircraft positions or misunderstood clearances and is an ever-present aspect of ATCT responsibilities for safe operations. Multiple operators in the ATCT coordinate their responsibilities when aircraft transition from the ramp area to the active movement area on taxiways and then onto runways.

The examples illustrate the inherently macrocognitive approach that is required to describe the ATCT work environment. The remainder of this paper discusses the three methods that we used for the assessment of macrocognitive functions and processes. As mentioned above, the first method, cognitive modeling, originates from a microcognitive research tradition but can be adapted to macrocognition. We found this method effective not only for explanation but also for the prediction of cognitive and behavioral performance under routine ATCT operations. The second method, critical incident analysis provides us with an understanding for non-routine operations, shedding light on shortcomings in the macrocognitive function distribution and allocation within the ATCT. The third method, coordination analysis, describes the collaborative nature in the ATCT.

Cognitive Analysis and Modeling

Cognitive modeling can be used for making predictions about human performance across a wide range of routine tasks. In the case of air traffic control procedures, cognitive models can be useful both as a predictive tool and as a unique measurement instrument because they enable objective, predictive measurement of mental workload and work time (that is, the time required to complete a task) before a procedural revision is actually enacted. This allows the analyst to identify potential pitfalls with a set of procedures and to address specific issues either early in the design process or in preparation for more rigorous evaluations in a simulation environment.

Below, we discuss our cognitive analysis and modeling methodology, which entails cognitive task analysis, field observation, cognitive model construction, and generation and analysis of model-based performance predictions. Our methodology begins with the development of a basic understanding of the controller's task through cognitive task analysis and ultimately provides outputs that include objective measures of cognitive workload and work times.

Initial Cognitive Task Analysis (CTA)

The initial CTA provides the starting point upon which we base the construction of more detailed cognitive models. The CTA acts as a decision inventory, enumerating all the decisions controllers must make over the course of each of their identified tasks. Each decision is further annotated with environmental conditions, including when the decision is made, the tools used to support the decision (for example, a radar system), and a high-level classification of that decision (for example, a planning decision). Since extensive work has been done on CTA in the aviation domain (for example, Alexander et al., 1989), we adapted that work for use as a starting point in our analysis.

Our aim for the CTA model is that it is readily understandable by domain practitioners, because once we have developed a baseline CTA model we enlist their

These (observable) actions are taken	if the AC is located at...	based on this information...	which is used for...
Clear for Takeoff (headset) → Mark departure heading → FS angled in bay	Hold Line Position and Hold On txwy approaching rwy	Location of previous arrival. Status of other AC (Flight Strips) Location of previous departure (BRITE) Previous departure AC type (Flight Strips)	Monitoring conflict/sequencing Sequencing/conflict Departure time/wake vortex/conflict Departure time/wake vortex

Figure 13.3
Excerpt from CTA

help to actually complete it. Since access to controllers is often limited, we have found it much easier to teach a controller how to complete a CTA than it is for a controller to convey all of the salient points about controlling traffic within a limited amount of time. Moreover, allowing controllers to complete the CTA allows them to review and rethink answers, possibly adding details that were previously missed. The following figure provides an example of the analysis result.

In this example, we see one of many tasks a local controller—the controller responsible for giving take-off and landing clearances in the tower—must complete. In the far left column, we note actions that are observable. This is important as our ability to recognize and classify these actions will be crucial in the observation stage. Aircraft (AC) location, seen in the second column, is needed for scenario development as will be described later. The third column provides information on which data sources are needed to make complete the take-off clearance task and the final column simply provides a high-level description of what each data source is used for. The initial CTA is important because it provides a primer to the domain which aids in preparation and execution of the next phase in our methodology: tower observation.

Tower Observation

While the initial CTA introduces the domain, actual observation contextualizes it. Observation of the controllers allows us to strengthen a macrocognitive technique like CTA into a tool in which the general observations on cognition are situated with a real environment. With this in mind, we move to structured observations in the existing air traffic control tower (before any procedural revisions are enacted). We use observation as a tool to capture metrics about key controller decision-making events highlighted in the initial CTA under a variety of airport operating conditions (for example, day, night, high-traffic load periods, low-traffic load periods). Specifically, we are interested in determining when such events occur, how often they occur, how they occur in sequences with other events, how long they take, and what tools are used to facilitate them. We are also interested in assessing how factors such as traffic load impact these metrics. Understanding these factors is important because it will aid in the development of realistic traffic scenarios for the models to interact with. As a secondary goal, we use observation to refine our initial CTA model. The ultimate use of our observational results is to inform our cognitive modeling development and to help validate baseline model predictions of controller performance.

To support observational data collection in the tower, we developed a PDA-based software application that allowed us to keep up with the rapidity with which controllers perform their tasks. The application allows the analyst to record each type of controller task identified in the initial CTA (for example, using a tool such as the controller's radar display, issuing a command to an aircraft, or communicating with another controller) as the controller performs them. When a task is performed, the analyst clicks a button corresponding to the task, and the application records the task—along with the time it occurred—in an event log.

Figure 13.4 below shows the screen that appears when the software is first launched. It provides the analyst with several drop-down boxes to configure the software for the current data collection objective. The analyst is able to select the Application type (the Timer is selected to signify task events are accurately time stamped, or the Linker is selected to signify that it is only important to record task events in the order with which they occur). The analyst also selects the controller position that will be observed (for example, Local, Ground, or Ground Metering), as well as the current configuration of the airport (for example, Plan A). Figure 13.4 shows that tool.

Analysis of the data collected during the observation phase provides a general understanding of situated cognition—that is, an understanding of both the decision making and the environment in which that decision making takes place. If only the current environment were of interest, then a complete analysis of the observation data may be sufficient to provide an understanding of routine operation in the air traffic control tower. However, our goal is not only to understand and model the current environment, but to also compare the current environment to a future environment. The future environment may include not only modified procedures, but also modification of the physical geography of the environment through the addition of new runways and taxiways or the reorientation of existing ones. In the absence of a simulated environment to evaluate proposed changes with actual controllers, an environment with simulated controllers and predictive capabilities is required. This brings us to the next phase of our methodology: cognitive modeling.

Cognitive Modeling

Cognitive models are representations of humans that interact with and react to the environment in the same way a human would. Further, they are constrained in the same way human cognitive, motor, and perceptual resources are. The aim of the model is to replicate or predict how a human would interact with artifacts while completing a task. Because cognitive models are based in microcognitive theory but embedded in a naturalistic environment, cognitive modeling may be characterized as applied microcognition in that it shares characteristics of both macro- and microcognitive techniques.

Figure 13.4   The task event recorder screen

For our cognitive modeling efforts, we use a GOMS (Goal, Operators, Methods, and Selection Rules) technique that we have tailored to the air traffic control domain. More specifically, we use a GOMS variant known as Natural GOMS Language (NGOMSL) (Kieras, 1996) with a modification (as discussed in Estes and Masalonis, 2003; Estes, 2001) that provides a measure of working memory load.

To provide some insight into the contents of these models, it is helpful to review a specific task for which a model has been constructed. Consider the task of a controller transferring a pilot to a new frequency (known as Transfer of Communications, or

TOC). The TOC task occurs when the controller who currently has responsibility for an aircraft determines that the aircraft under consideration has arrived at a location in which it is necessary to transfer control of the aircraft to the next air traffic controller. When the pilot is given a TOC clearance, he is being given instruction to begin communicating with the next controller. When the controller notices the aircraft is in position to be transferred to the next controller in line, the controller first offers control of the aircraft to the next controller by typing a command into the radar workstation. Once transfer of control is accepted by the next controller, the current controller completes the process by contacting the pilot and verbally issuing a transfer of communications clearance. The pilot will repeat that clearance back to the controller so that the controller may verify that the pilot correctly understood the radio frequency on which to contact the next controller. This task description is translated into an NGOMSL cognitive model in Figure 13.5.

Method for Goal: Initiate Handoff
  Step 1. Scan for eligible aircraft
  Step 2. Store eligible aircraft in working memory
  Step 3. Decide: If no eligible aircraft THEN Return with Goal Accomplished
  Step 4. Accomplish Goal: Enter Handoff Command
  Method for Goal: Enter Handoff Command
        Step 1. Hands to Keyboard
        Step 2. Type <next_sector> <space> <aircraft id>
        Step 3. Verify Correct
        Step 4. Keystroke Enter
        Step 5. Return with goal accomplished
  Step 5. Accomplish Goal: Transfer of Communication
  Method for Goal: Transfer of Communication
        Step 1. Recall next controller frequency from Long Term Memory
        Step 2. Hands to radio button
        Step 3. Keystroke button
        Step 4. Verify Line Clear
        Step 5. Speak Clearance (Delta 123, contact Jacksonville Center on 123
              point 4, good day
        Step 6. Keystroke button
        Step 7. Return with goal Accomplished
  Step 6. Accomplish Goal: Verify Readback
  Method for Goal: Verify Readback
        Step 1. Listen (Roger, contact Jacksonville Center on 123 point 4, Delta
              123)
        Step 2. Decide in readback frequency does not equal issued frequency
                    Then Accomplish Goal Transfer of Communication
        Step 3. Verify Correct
        Step 4. Return with goal accomplished
Step 7. Return with goal accomplished

Figure 13.5   NGOSML model of transfer of communication task

Each step in the NGOMSL model has a 50-millisecond cycle time associated with it. Further, most steps have an execution time associated with it as well. For example, it takes about 400 ms to move the hands to the keyboard or 280 ms to press a button. By adding up these times, we can calculate a total execution time for the task. Likewise we can determine working memory load for the task by deriving the information that must be contained in working memory to complete each step in the task. Again, as an example, in order to speak the transfer of communications clearance, the controller must have, at a minimum, the following information in working memory: Aircraft ID, the controller to which the aircraft is being transferred, and the frequency of the controller to which the aircraft is being transferred.

In the case of modeling new procedures, we construct cognitive models for both the current procedure (for use as a baseline) and the future procedure. For example, in a data link environment, the controller would uplink the TOC clearance rather than speak it. Armed with our models, it would be possible to determine changes in work time (how long it takes to complete the TOC task) and mental workload both with and without a data link. Our observational data provide a means with which to validate the baseline model predictions.

Model-Derived Outputs

Human performance metrics generated by the models include mental workload and work time using a new set of procedures. As an illustration, a typical result from an analysis for a busy US airport is shown in Figure 13.6. These results are from cognitive models of tower air traffic controllers responsible for issuing take-off and landing clearances for a variety of different runway configurations. Note that the red line in the workload profiles indicates a level of seven working memory chunks. This is shown simply as a reference and is not meant to imply that anything above that line is "too high."

Using the output from the models, comparisons of workload and work time utilizing different sets of procedures and systems can be made quickly. With further analysis, models may also provide insight into questions of staffing. For example, the models may be used to answer questions about the number of controllers that would be needed if intersecting runways are used instead of mixed-use runways.

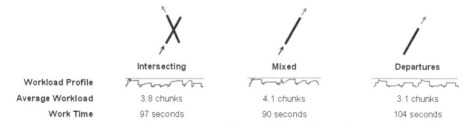

Figure 13.6   Cognitive model outputs for tower controller

In summary, the effort of creating an initial CTA and then performing field observation allows for both an understanding of key decisions and common task sequences as well as insight into what that decision making looks like when situated within a real environment. This data can then be used to create cognitive models that not only provide insight into the current environment, but make predictions about working with the future environment as well. The output of these models provide for a comparative analysis of changes in work time and workload under routine, error-free operations. This then leads to the logical question, "What happens in non-routine operations?" To answer this question, we employ the next method in our trio: critical incident analysis.

Critical Incident Analysis

The cognitive analysis and cognitive modeling provide us with an understanding of how controllers perform their jobs and allow us to make predictions about aspects of controller performance under fairly routine operating conditions. While much of control does indeed involve carrying out fairly routine procedures, such analyses and models are not particularly well-suited to predict things that are difficult to model and anticipate, including nuanced aspects of operations like misunderstood clearances, impacts from poor weather, work-arounds for dealing with equipment malfunctions, dealing with shift transitions, the use of memory aids, and a host of other factors (Wickens, Mavor, and McGee, 2003). Moreover, such factors and their interactions in complex and unexpected ways may have detrimental effects on performance and safety.

Thus, to extend the analysis, we aim to gain a greater understanding of non-routine conditions that have given rise to incidents such as runway incursions.[1] These events may arise from situational factors such as weather, runway configurations, and frequency congestion, as well as from human performance-related factors such as fatigue and degraded situation awareness. For example, when dealing with a particularly high volume of traffic, a controller may "lose the picture," fail to anticipate the future location of an aircraft, and issue a clearance that will place that aircraft too close to another aircraft. Armed with an understanding of the factors that lead to such events with a current set of procedures, we aim to then identify similar factors that may persist using the new set of procedures and cause unsafe operating conditions.

The approach we adopt is a three-pronged one. First, we review incident reports for airport surface movement events such as runway incursions that have occurred at the existing airport using the current set of procedures. Incident reports may be culled from the Aviation Safety Reporting System database, the National Transportation Safety Board Aviation Accident and Incident database, as well as from any incident report files that the airport maintains. In the course of our incident analysis, we develop an incident categorization scheme and assess how frequently certain types of incidents occur. Table 13.1 below shows a sample incident categorization scheme.

---

1   A runway incursion is "any occurrence in the airport runway environment involving an aircraft, vehicle, person, or object on the ground that creates a collision hazard or results in a loss of required separation with an aircraft taking off, intending to take off, landing, or intending to land" <http://www.awp.faa.gov/ops/awp600/runway/definitions.pdf>.

Table 13.1   Sample incident categorization scheme

Environmental factors	Pilot-controller communication factors	Controller-related factors
Intersecting runways in use	Non-standard or non-specific phraseology used	Issued clearance with obstruction present
Parallel runways in use	Instructions issued rapidly	Issued clearance to wrong aircraft
Aircraft landing and departing from the same runway	Responded to instructions issued to a different aircraft	Controller-controller mis-coordination
Poor weather conditions	Misinterpreted instructions	Workload excessive
Wind conditions	Misheard instructions	Failed to visually scan runway
	Read-back not given or not corrected	Anticipated separation error
	Frequency congestion	

If possible, we next perform a similar incident analysis on an airport that has procedures in common with the proposed set of procedures, or is at least configured in some way that is similar to the proposed airport configuration. The results of these initial incident analyses enable us to gain an understanding of the factors (or combination of factors) that typically underlie such incidents.

Based on the results of our initial incident analysis, we then conduct interviews with controllers that allow us to gain a greater understanding of the underlying contextual and human performance-related factors behind operational errors. This analysis goes beyond the understanding that can be gained from an analysis of incident reports alone. If possible, we conduct critical incident interviews with controllers at each airport using the Critical Decision Method (CDM) (Klein, Calderwood, and Macgregor, 1989).

The first step of the CDM is to assist the interviewee in selecting an incident that suits the purposes of our analysis. Interviewees are then asked to describe the incident in its entirety with little interruption from the interviewer. Next, the interviewee and interviewer step through the incident together. At each decision point, the interviewer uses a series of probe questions to identify the factors that have led to the current decision. Table 13.2 opposite outlines a possible set of probes that may be employed to suit the purposes of our analysis. The construction of incident probe questions and selection of appropriate incidents for analysis is informed by both the important factors identified in the initial incident analyses as well as by the potential pitfalls identified in the cognitive modeling analysis.

Table 13.2   Example critical decision probe questions

Probe type	Probe content
Cues	What information sources were you relying on to make your decision?
Goals	What were your specific goals and objectives at the time?
Basis of choice	How was this option selected/other options rejected?
Errors	What mistakes are likely at this point? Did you acknowledge if your situation assessment or option selection were incorrect?
Hypotheticals	If a key feature of the situation had been different, what difference would it have made in your decision?

Source: Klein, Calderwood, and Macgregor, 1989.

For example, one controller-related factor that we determined to be important was anticipated separation errors. Anticipated separation errors occur when controllers fail to anticipate the required separation or miscalculate the impending separation between aircraft (Cardosi and Yost, 2001). Anticipated separation is a traffic flow management strategy that controllers use where they issue clearances with some type of obstruction present, but, based on their judgment, they predict that the obstruction will be clear by the time the clearance is actually executed (for example, aircraft x will be off the runway and out of the way by the time aircraft y lands). This increases the controllers' workload, as they must not only maintain awareness of the current situation, but they must also use that awareness to make predictions about future situations. We determined that this strategy is likely to be relied upon when traffic loads are high, when the same runway is being used to both land and depart aircraft, or when two intersecting runways are both in use. Thus, in our critical decision interviews, we would be sure to elicit scenarios from controllers where anticipated separation was a factor, and we would also address situations that may arise utilizing a new set of procedures where controllers would be likely to make anticipated separation errors.

Another important set of factors that we identified in initial incident analyses were those related to pilot-controller communication, a subset of which are shown in Table 13.1. Frequency congestion was an oft-cited communication factor underlying a host of incidents. It often results in blocked transmissions, incomplete messages, repeated communications, and misunderstood instructions. Safety measures such as readbacks (that is, the pilot "reads back" the instruction from the controller so the controller is assured that the instruction was heard correctly) are often dropped so that more instructions can be transmitted during busy periods (Cardosi and Yost, 2001). However, our analysis determined that frequency congestion was not only a function of the business of the airport, but also of the complexity and number of instructions necessary to properly coordinate the movement of aircraft. Again, when performing critical decision interviews, we would be sure to ask about factors that contribute to frequency congestion, strategies for dealing with it, and the impact of those strategies on performance and safety. Moreover, we would also identify factors in the new configuration that may lead to an increased amount of communication to properly coordinate aircraft, and investigate potential mitigating technological or procedural solutions.

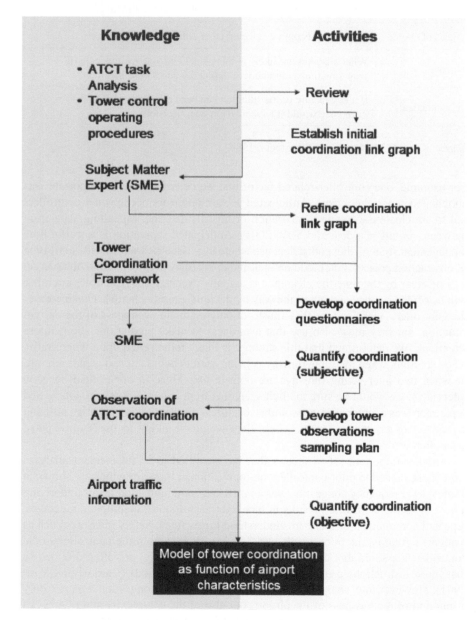

Figure 13.7   Methodology for the assessment of air traffic control tower coordination

Controller Coordination Analysis

Another important area identified for analysis and not particularly well addressed by either cognitive modeling or critical incident analysis is controller coordination. Koros et al. (2006) point to coordination between controllers as an oft-used strategy to alleviate high complexity. As in any team environment, the tasks of one team member depend on the tasks of other team members and workload may shift among team members. An analysis of coordination between and among team members should reveal the interdependencies in the overall tasks. To perform this analysis, we developed a coordination analysis method that leverages previous work. Our method, summarized in Figure 13.7 and described in some detail below, was developed in cooperation with air traffic control subject matter experts and postulates a procedure based on best practices in the field.

Coordination has been defined as management of dependencies between activities (Mallon and Crowston, 1993). In the aviation domain, Peterson, Bailey, and Willems (2002) developed a descriptive framework for en-route controller coordination. This framework categorizes coordination into three dimensions: coordination topics (for example, coordination concerning traffic), grammatical form (for example, questions, commands), and communication expression (for example, verbal, non-verbal).

The Peterson et al. (2002) framework was used as a starting point for the development of our ATCT coordination framework. Coordination topics can be extracted from Alley et al. (1987), who had performed a task analysis of ATCT operations, and from the Standard Operating Procedures (SOP) document of an ATCT. SOPs outline the responsibilities and duties of tower controllers. A subject matter expert from an ATCT should also review the initial list and added additional topics. We established a list of 55 coordination topics for the local control, ground control, ground metering, and clearance delivery positions. Coordination topics include such things as "respond to requests for transfer of control" and "plan and issue ground movement instructions."

A subset of these coordination topics (for example, forward flight plans, establish departure sequence) is represented using link charts in Figure 13.8 and Figure 13.9 that show tower coordination as defined in a SOP. Numbers on these charts refer to the SOP-defined responsibilities but are renumbered for illustrative purposes.

The links in Figure 13.9 describe the face-to-face communication between tower positions. The initial list of coordination topics can then be used to elicit operational knowledge about control tower coordination from air traffic controllers who serve as subject matter experts (SMEs). The purpose is to derive an understanding of the aspects of coordination in the context of changing tower procedures. First, narratives about the initially identified coordination topics are collected from the SMEs. The discussion can be structured around a card-sorting technique (for example, Cooke, 1999). SMEs first sort coordination topics into two piles of either high or low subjective importance. After the initial sorting, SMEs articulate the rationale for their importance ratings and describe the coordination topics in more detail. These narratives can reflect a wide variety of experiences of controllers in their control towers. Over time, however, the SMEs seem to converge onto similar dimensions in their descriptions.

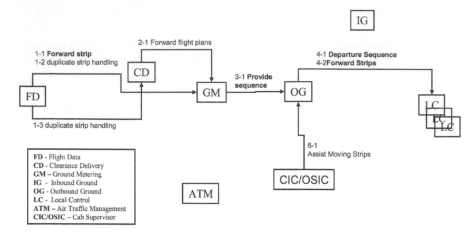

Figure 13.8   Example of flight strip coordination in an ATCT

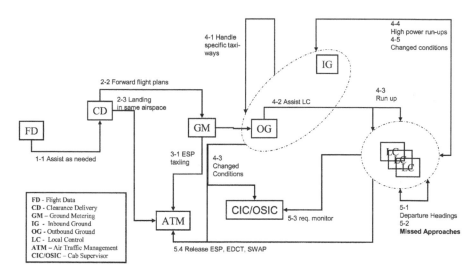

Figure 13.9   Example of face-to-face coordination between controller positions in an ATCT

As example, we distilled seven key dimensions for coordination that are summarized in Table 13.3 and described opposite.

The first two dimensions are simply the coordination topic, and the controller position (or positions) responsible for carrying out the coordination.

The third dimension indicates the coordination frequency, quantified on a five-point scale with values of very rarely, rarely, occasionally, frequently, and very frequently.

Table 13.3   Dimensions of coordination

	Coordination dimension	Example
1.	Topic	Active runway crossings
2.	Controller positions	Ground and local controller
3.	Estimated frequency of coordination	Very frequent
4.	Estimated time criticality of coordination	Immediate
5.	Coordination dependencies	Visibility Conditions
6.	Estimated workload of coordination	Low, no follow-up activity required
7.	Medium of coordination	Face-to-face

The fourth dimension describes coordination time criticality, reflecting the timeframe which SMEs determine a coordination request should be responded to. Example values on the time criticality scale are immediate (less than five seconds), high (five to ten seconds), medium (ten to thirty seconds), and low (thirty seconds or more).

The fifth dimension describes events and conditions that influence coordination. These include environmental conditions such as visibility, weather, winds, traffic loads, as well as other airport characteristics such as runway or taxiway closures.

The sixth dimension describes the level of coordination workload. Coordination workload is defined as the amount of follow-up activity required for a specific coordination topic. It is rated on a scale with values of easy, medium, and hard. Easy workload requires no or very little additional coordination with other controllers after the initial coordination, medium workload requires some follow-up coordination, and high workload requires substantial follow-up. Workload is also judged as high when more than two controller positions are involved.

The seventh dimension describes the coordination medium. SMEs reported using face-to-face communication, radio communication, telephone communication, and communication using flight progress strips.

After establishing the coordination dimensions, SMEs then quantify each coordination topic along each dimension (for example, active runway crossings are performed by ground and local controllers, occur very frequently, have low workload, and so on). After this quantification, SMEs select the coordination topics that they deem as the most important and most likely to be impacted by procedural changes. Examples for the top five coordination topics are:

1. Resolution of conflicts for taxiing aircraft (for example, pilot fails to hold-short of intersection)
2. Active runway crossings
3. Runway crossing under Land-and-Hold-Short Operations
4. Response to requests for runway condition data
5. Response to air traffic flow restrictions.

Uses of the Coordination Framework

Our coordination framework is intended to guide the identification of coordination activities that may be impacted by changes in airport systems and procedures. Related to the above described example about the impact of modified wake vortex-mitigation procedures on the operations in the ATCT, field observations and a simulation study were performed to fill in the coordination framework (Domino et al., 2006). In field observations, it was determined that to coordinate the applicability of wake vortex procedures, controllers, rather than relying solely on face-to-face communication, they relied mostly on handing colored paper strip holders between controller positions. Controllers use such paper strips generally as representations for the flights they are handling (Flight Progress Strips or FPS). Such paper strips were inserted by the ground controller into the sequence of paper strips that the local controller handled to indicate which aircraft needed to be delayed due to a preceding wake vortex-producing aircraft on the parallel runway. Table 13.4 shows the filled-in dimensions for this coordination.

Table 13.4   Example of coordination framework for a new wake vortex mitigation procedure for departures

	Coordination dimension	Example
1.	Topic	Determine applicability of wake vortex mitigation criteria for departure of an aircraft on a closely spaced parallel runway
2.	Controller positions	Ground and local controller
3.	Estimated frequency of coordination	Very frequent; depends on the type of aircraft to be departed and the planning on the ground controller to depart heavies (and B757s) on the downwind runway to facilitate applicability of the proposed procedure
4.	Estimated time criticality of coordination	Medium; the time horizon reaches between the fixation of the departure sequence on both runways to the issuance of the departure clearance
5.	Coordination dependencies	Traffic type on the departing runway (that is, aircraft of heavy type, B757), wind conditions as indicated by wake vortex mitigation detection technology; approval of the ATCT supervisor to perform operations
6.	Estimated workload of coordination	Low, no follow-up activity required
7.	Medium of coordination	Colored paper strip holder

Based on these initial observations in the field, a simulation study was then conducted to quantify the impact that the new procedure had on operational variables such as throughput, controller workload, and coordination among controllers. In that simulation study, controllers controlled simulated air traffic on closely spaced parallel runways either with or without use of the advanced wake vortex-mitigation procedure. In addition, coordination was measured and it was found that in the advanced procedure condition, the amount of communication was reduced by about 18 percent compared to a baseline condition without the new procedure. This reduction was confirmation for a somewhat simplified coordination process between controllers as part of the procedure.

Other examples for the applicability of the coordination framework include the insertion of new communication media (for example, from face-to-face to radio, or to system automation) that can significantly impact tower coordination. "Active runway crossings," for example, are rated by the SMEs as frequent and highly time-critical but with little requirement for follow-on coordination. This coordination topic poses different requirements on the communication medium than, for example, "responses to flow restrictions," a coordination topic the SMEs deem to occur less frequently and with less time-criticality but with potentially significant amounts of follow-on coordination. In this example, the coordination framework can help guide the design of communication media to best support controller coordination. It can also identify coordination topics that will be most impacted by proposed communication media changes and should be examined in a simulation.

Our framework can also help determine the frequency of various coordination activities in a specific tower. Such knowledge can be important in determining which coordination activities are most important and should be evaluated in simulation. To perform this quantification, the coordination activity frequency is first estimated by tower controllers. These estimates are then used to derive a tower observation sampling plan by identifying which activities should be sampled to obtain empirical coordination frequencies that are independent from the initial estimation. Thereby, low-frequency activities are sampled less frequently than high-frequency activities to direct resources to their best use. Once coordination frequency data have been collected, these data can be correlated with actual airport surface traffic characteristics, allowing us to build computational models that predict how coordination frequency changes with changes in airport traffic characteristics.

Discussion: Levels of Descriptive Granularity

The multifaceted set of methods that we have selected to address human performance issues in the ATCT domain address the domain at multiple levels of descriptive granularity—from the macrocognitive to the microcognitive. In this section, we propose three levels of descriptive granularity and discuss the analytical merits of models constructed at each level.

At the finest level of granularity, units of behavior consist of observations within short periods of time (for example, less than two seconds). Such behavioral units are often not accessible in self-reports by human participants because units tend to be encapsulated within higher behavioral sequences. Examples for such behavioral or cognitive units include single eye-movements during a visual scan, words within clearances, or pieces of information that are retrieved from memory. We refer to this level as the microcognitive level of descriptive granularity.

Models and observations at this low level of granularity include the performance of task analyses and instrumented approaches to assess human behavior such as eye- and head-tracking equipment, video analysis, or response-time analysis.

At the next level of descriptive granularity, we consider behavioral units between a few seconds up to a few minutes. For example, Koros et al. (2003) and Koros et al. (2006) assessed the factors that contribute to complexity in air traffic control tower operations. Specifically, Koros et al. (2006) interviewed air traffic controllers to determine the strategies that they used in alleviating or reducing the identified complexity factors. In these interviews, the duration of the identified behavioral units ranged generally between a few seconds to several minutes. Examples for such strategies include "expediting traffic across runways," or "strategies to cross aircraft at specific taxiway runway intersections." Note that controllers describe most of these tasks in relation to the requirements of the environment and traffic. This level of description includes several sequences of microcognitive behavioral units (such as scan sequences, clearances, coordination between positions) and interacts behavior specifically with the environment. Because of this inherent interaction, macrocognitive descriptions generally lose interpretability when severed from environmental descriptions. We refer to this level of description as the macrocognitive level of description.

A third, higher level of descriptive granularity can be used to describe ATCT operations holistically. Metrics include the numbers of arrivals and departures within a specific time period. At this level, the ATCT is described as a system that may interact with other operational entities in the National Air Space such as other ATCTs, terminal air traffic control, or the FAA Command Center. We refer to this descriptive level as System level.

The macrocognitive level of description allows users to describe their activities in their own words because it meaningfully relates activities with the environmental conditions such as traffic and weather events. Furthermore, it can be assessed via non-instrumented observational methods because of sufficient activity duration. The macrocognitive approach therefore facilitates terminology that allows communication not only within a research community but also facilitates mutual understanding among stakeholders and users. Using the macrocognitive approach, findings can be formulated as arguments for organizational decision making.

Conclusion

We find the macrocognitive approach useful to provide both the terminology and a framework to facilitate procedural or system changes in the ATCT domain. The macrocognitive approach permits both explanation and measurement of cognitive and human behavioral performance in a naturalistic work environment.

The three methods that we reviewed in this chapter—cognitive modeling, critical incident analysis, and coordination analysis—allow us to make predictions about the effects of procedural changes on air traffic controller performance, and it provides results that may, for example, be used to guide the construction of representative scenarios for use in interactive tower simulations. Our development of a suitable set of methods demonstrates two fundamental lessons that we have learned in the course of our analysis: methods require customization to meet the goals of the analysis and the important features of the domain, and a combination of complementary methods is typically needed. We continue to adapt and refine our methods as we apply them in our work.

Acknowledgements

This work was supported by the MITRE Corporation's Center for Advanced Aviation System Development and MITRE's Technology Program.

References

Alexander, J.R., Alley, V.L., Ammerman, H.L., Fairhurst, W.S., Hostetler, C.M., Jones, G.W., and Rainey, C.L. (1989). *FAA Air Traffic Control Operations Concepts, Vol. 7: ATCT Tower Controllers* (Report No. DOT/FAA/AP-87-01). Oklahoma City, OK: FAA Civil Aeromedical Institute.

Alley, V.L., Ammerman, H.L., Fairhurst, W.S., Hostetler, C.M., and Jones, G.W. (1987). *FAA Air Traffic Control Operations Concepts, Vol. 5: ATCT/TCCC Controllers (CHG 1)* (Report No. DOT/FAA/AP-87-01). Washington, DC: Department of Transportation.

Cardosi, K. and Yost, A. (2001). *Controller and Pilot Error in Airport Operations: A Review of Previous Research and Analysis of Safety Data* (Report No. DOT/FAA/AR-00/51, DOT-VNTSC-FAA-00-21). Washington, DC: Federal Aviation Administration, Office of Aviation Research.

Cooke, N.J. (1999). "Knowledge elicitation." In F.T. Durso (ed.), *Handbook of Applied Cognition*. Chichester: John Wiley and Sons Ltd (pp. 479–510).

Domino, D.A., Lunsford, C., Moertl, P., and Oswald, A. (2006). *Controller Assessment of Procedures for Wake Turbulence Mitigation for Departures (WTMD) from Closely Spaced Parallel Runways, Mitre Product MP06W0000102R1*. McLean, VA: The MITRE Corporation.

Estes, S.L. (2001). "Cognitive theory in application: Evaluation of the Gulfstream PlaneView FlightDeck." Paper presented to the Human Factors and Applied Cognition Department. Fairfax, VA: George Mason University.

Estes, S.L. and Masalonis, A.J. (2003). "I see what you're thinking: Using cognitive models to refine traffic flow management decision support prototypes." *Proceedings of the 47th Annual Meeting of Human Factors and Ergonomics Society*. Santa Monica, CA: Human Factors and Ergonomics Society (pp. 610–14).

Kirwan, B. and Ainsworth, L.K. (eds) (1992). *A Guide to Task Analysis*. London: Taylor and Francis.

Klein, G.A., Calderwood, R., and Macgregor, D. (1989). "Critical decision method for eliciting knowledge." *IEEE Transactions on Systems, Man, and Cybernetics*, 19 (3): 462–72. New York:IEEE.

Malon, T.W. and Crowston, K. (1993). "The interdisciplinary study of coordination." *ACM Computing Survey*, 26 (1): 87–119. New York: ACM.

MIT Lincoln Lab (2006). "Wake Turbulence Mitigation for Departures (WTMD) Wind Forecast Algorithm and Safety Net Description." Unpublished manuscript, Massachusetts Institute of Technology, Lincoln Laboratory.

Peterson, L.M., Bailey, L.L., and Willems, B. (2002). *Controller-to-controller Communication and Coordination Taxonomy (C4T)* (Report No. DOT/FAA/AM-01/19). Washington, DC: Office of Aerospace Medicine.

PART III
Micro-Macro Relationships

Chapter 14

Bridging Macrocognitive/Microcognitive Methods: ACT-R Under Review

Claire McAndrew, Adrian Banks, and Julie Gore

Introduction

This chapter begins to confront the difficulties of validation currently faced by Naturalistic Decision Making (NDM) researchers and provides a methodological agenda that renews interest in the theoretical representativeness of the Recognition-Primed Decision Making (RPD) model (Klein, Calderwood, and Clinton-Cirocco, 1985) and the use of microcognitive methods. Reviewing the findings from our study, we support the potential value in using microcognitive modeling architectures such as ACT-R as validation tools in NDM. Specifically, we describe how the microcognitive focus of the ACT-R model enables the specification of micro processes that exist both within and between a number of macrocognitive phenomena. We suggest that value exists in drawing across the macrocognitive/microcognitive divide and thus, make a call for the bridging of this distinction.

One of the hallmarks of NDM's success has been the development of the RPD model (Klein et al., 1985; Klein, 1995). This model sought to provide a descriptive account of the process by which courses of action could be identified and applied without extensive option generation within time-pressured and ill-structured environments. Under such conditions, the RPD model depicts the process by which decisions are guided by prior experience (Klein et al., 1985). There are three levels of RPD that represent the interaction of experience with the increasing complexity of decision making. The simplest form of RPD execution termed "Simple RPD match" depicts the process by which decision makers use their experience to identify situations as representative of a class of problem. This identification leads directly to an appropriate course of action. The second level of RPD, "Developing a course of action," describes the process of action generation and the use of mental simulation to evaluate its suitability. Finally, "Complex RPD strategy" is the most elaborate of the three models and deals with the possibility of non-immediate situation recognition (see Figure 14.1). Under these non-routine conditions, the RPD model suggests decision makers handle the situation by seeking additional information and focusing attention more deeply upon existing cues. Once a coherent picture of the situation has been formed, plausible goals and courses of action are identified. The decision maker then examines the potential of the generated courses of action, adopting a satisficing strategy to select the first one that appears to work. The selected course of action is then evaluated using a process of mental simulation (Klein, 1992).

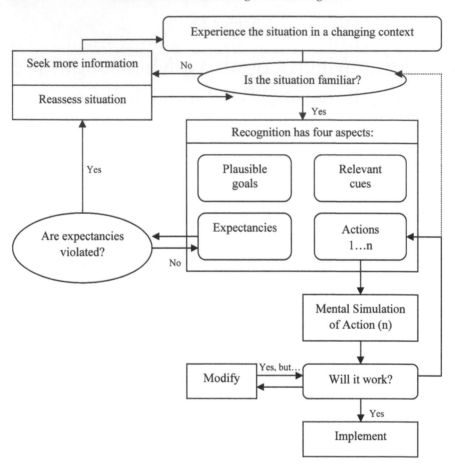

Figure 14.1 Complex RPD strategy

Source: Klein, 1995.

While the RPD model has been applied within naturalistic enquiry (see Drillings and Serfaty, 1997; Kaempf and Orasanu, 1997), theoretical validation is still under review. This chapter illustrates how ACT-R could be used to bridge microcognition and macrocognitive processes in RPD. In doing so, we first document the progress of RPD validation to date. We extend our examination of RPD validation by secondly considering the microcognitive basis of RPD. Third, we summarize a number of methodological barriers that exist within NDM that currently inhibit validation. Fourth, we outline ACT-R and our rationale for using it in the study of RPD. Building upon these arguments, we then present an illustrative study using ACT-R as a validation tool to bridge the microcognitive/macrocognitive divide. Finally, our results are discussed and a number of implications are drawn for macrocognition and for the future study of RPD.

Validation: Recognition-Primed Decision Making Model

Confirmation of recognition-primed processes has been largely rooted in frequency studies that document the prevalence of recognitional vs. analytical decisions. Although the existence of recognitional processing has been substantiated consistently across a variety of disciplines including: urban fire-ground commanders (Klein, 1995; Tissington, 2001), forest firefighters, tank platoon leaders, expert design engineers (Klein, 1995); electronic warfare technicians (Randel et al., 1994); pilots (Mosier, 1991) and Navy officers (Driskell, Salas, and Hall, 1994; Kaempf et al., 1992), "these findings are not a critical test of the model" (Klein, 1997. p 288). Studies of this type provide only limited understanding of the nature of recognitional processing or the theoretical representativeness of the RPD model. Indeed, Lipshitz and Pras (2005) have recently cautioned haste at frequency studies, urging understanding to fall upon the ways in which recognition-primed decision making is executed. On this front, evidence has started to emerge for the application of recognitional heuristics (Raynard and Williamson, 2005) and for RPD's proposition that option evaluation does not precede but follows choice (Lipshitz and Pras, 2005).

RPD Validation and Microcognition

One of the more significant discoveries for RPD validation is the finding that RPD is comprised of a consortium of decision heuristics that have already been examined within the microcognitive paradigm (Klein, 1995; Klein et al., 2003). Specifically, these include the availability and representativeness heuristics used to identify possible courses of action and the simulation heuristic to evaluate their potential. In response to this finding, Klein et al. (2003, p. 82) claimed that it was "...possible to trace [the] macrocognitive phenomena back to hypothetical microcognitive components." What is interesting in the context of current macrocognitive/microcognitive debates is the platform this creates for theoretical synthesis across these seemingly disparate levels of analysis. Before progressing to our arguments regarding how theoretical synchronization may be advanced methodologically, some of the methodological barriers facing NDM and validation of the RPD model are first outlined.

Methodological Barriers

While the NDM community has been witness to a number of interesting methodological developments (for example, see ACTA, Militello and Hutton, 1998), there still remain difficulties in executing critical tests of the RPD model. It is suggested that these difficulties are methodological in nature and are intimately tied to the limitations of qualitative approaches to the observation and analysis of experts' decision making. The methodological barriers are summarized as follows:

• Quantification and prediction
• Subjectivity
• Abstract representations and under-specification.

The first obstacle derives from the qualitative methods employed to observe and analyze experts' decision making processes. Although this is no barrier in the sense of the depth of description it affords, its downfall lies in its preclusion of quantification and thus prediction. This is marked most prominently by our second suggestion that decision protocols are open to a degree of subjective interpretation. The question of quantification and prediction within qualitative research more broadly, is fundamentally debatable and beyond the realms of this chapter. Our third point proposes that methods of this sort also have the effect of producing representations that are abstract in form, thus under-specifying the cognitive processes involved in decision making. Consequently, validation of the recognitional processes in the RPD model is an inherently subjective endeavour. In light of these methodological barriers, how may the RPD model be more successfully validated?

Microcognitive Methods: ACT-R

Building upon Klein et al.'s (2003) earlier suggestion of theoretical connections existing across macrocognitive and microcognitive levels, we suggest that the difficulties associated with RPD validation may be addressed by utilizing microcognitive modeling architectures. We propose ACT-R (Atomic Components of Thought-Rational) to be a constructive research tool that may inform RPD validation efforts with a higher degree of precision than methods currently available. ACT-R is a computational architecture that is designed to support the modeling of human cognition and performance (Anderson and Lebière, 1998). In short, it is a production system that operates via a series of production rules (condition-action pairs) that embody procedural knowledge and declarative structures. Our choice in ACT-R as a modeling environment resides in the findings of a recent study of offshore installation emergencies, which found that decision making was characterized by serially generated condition-action pairs (Skriver, 1998).

For the purpose of validating the theoretical prescriptions of the RPD model, it is anticipated that the ACT-R model may be able to contribute to our current state of knowledge. The requirement to specify more precisely the cognitive processes involved in a decision ensures that the relationships between goals, cues, expectancies and actions are more clearly understood. On this basis we envisage using ACT-R as a mechanism for bridging macro/micro levels of cognition.

Research using ACT-R applications to explore aspects of NDM has recently started to emerge. Modeling efforts include the representation of situation awareness in the military with battlefield commanders (Juarez and Gonzalez, 2003), within submarine environments (Ehret et al., 1997), and the modeling of "take the best heuristics" as opposed to the exhaustive comparison of options (Nellen, 2002). While these establish the validity of concepts used within NDM, they do not demonstrate the value in borrowing microcognitive methods to progress our understanding of macrocognitive phenomena such as RPD.

The purpose of this study was therefore to explore the potential use of microcognitive modeling architectures in the study of macrocognitive phenomena. Specifically, the focus of our investigation lay in the use of ACT-R as a validation tool of RPD processes. It was envisaged that the combination of naturalistic enquiry and microcognitive methods would allow both the quantification and validation of the RPD model while acknowledging the importance of context and realism (as defined by high risk, time pressure, ill-defined/multiple goals, and the use of domain practitioners). On this basis, we outline a macrocognitive model of decision making currently in development that uses ACT-R to demonstrate the potential value in bridging macrocognitive and microcognitive research (McAndrew, Banks, and Gore, 2004).

Method

Participants

Sixteen expert engineers and marketing personnel with an average of 16 years of experience within their respective fields participated in this study. The subject matter experts (SMEs) were recruited from within a multinational mobile communications organization, selected given the under-researched nature of NDM within organizations (Gore et al., 2006).

Design

This study used a triangulated design that incorporated qualitative analysis with the cognitive modeling architecture ACT-R. Two comparisons were permitted by this design: (i) quantitative comparison of the predictions of the ACT-R model to the SMEs' decision outputs as documented by their think-aloud protocols and (ii) qualitative comparison of ACT-R's trace outputs to the theoretical propositions of the RPD Model.

Materials

A paper and pencil scenario was developed in collaboration with a communications expert from the participating multinational mobile communications organization. During this development period, time was spent with the expert discussing potential scenarios in order to ensure both appropriateness and sufficient challenge. In brief, the final scenario focused upon the integration and release of a new feature into an existing wireless mobile product in a situation characterized by time pressure and high risks (see Figure 14.2).

The interdisciplinary nature of this scenario necessitated the use of engineering and marketing expertise as a reflection of the cooperative nature of decision making typically encountered with the release of new products within this industry. For example, the technical challenge of integrating the new feature could be achieved (at least hypothetically) if engineers were allocated infinite time resources. Without the restrictions induced by the time to market, the decision from an engineering perspective would have posed little challenge or risk. Likewise, the challenge of releasing a new feature into an emerging market is incomplete without the uncertainty associated with the development and integration of the market-specific feature.

Scenario

Your company is launching a major new ground breaking, highly important and prestigious wireless mobile product that is driven by one major global customer. As time to market is important in order to gain a competitive advantage in the marketplace, it is essential to meet the customer's timescales for product launch.

Recently, the customer has identified another previously untargeted market place (China) that could be captured by this product, but requires the inclusion of a major new feature in order to meet the needs of this market. Early penetration in the China market would both lead to a 25% increase in global unit sales and the generation of a longer-term market. If China is not included in the initial product launch then a competitor will capture your company's potential market share. The research teams have already identified this new feature and they have the technology in an advanced state of development. However, deployment of this feature within the product would still require effort for productionisation and field test.

You have been selected because of your specific expertise to form part of a multi-disciplinary team that has been formed to assess the feasibility of this feature in this product. Your goal is to determine:

- Whether the product should be launched as originally intended without this feature or,
- Whether this feature can be included albeit at some risk and allow product deployment in China.

Figure 14.2   Simulation scenario

Obviously, the following factors are important and should be considered when assessing the

risk and benefits: (i) time constraints for deployment, (ii) achievement of cost objectives and

(iii) meeting the customer's product launch time scales.

The proposal of this feasibility group will be presented to top-level management who will

ultimately approve this decision.

Task

Your task is to describe the processes you use for analysing and assessing this

situation, and to secondly, explain how you would ultimately reach a decision.

Figure 14.2 continued   Simulation scenario

Thus, the developed scenario attempted to re-create an environment whereby both product development and the time to market did not exist in isolation, but necessitated the interaction of engineering and marketing expertise.

In terms of the relationship between the design of the scenario and the elicitation of RPD processes, three points emerge. First, it is important to acknowledge that while an effort was made to create a realistic scenario and use SMEs with relevant experience; this scenario differs from the type typically used within NDM. RPD was developed to describe decisions made by experts working under time pressure and risk in situations for which they had a large experience base. The risk involved during this scenario was purely hypothetical and the time pressure artificially imposed. While it is unlikely that the SMEs' experiences of product releases parallel the depth of experiences used in previous NDM studies (for example, the highly concentrated knowledge-base of firefighters), it was anticipated that they would be of sufficient depth to elicit RPD processes. It was expected that the SMEs would utilize RPD processes preempted by their experience and knowledge of product releases, the technical integration of market-specific features, and their knowledge of the strategic importance of the emerging Chinese market.

Building upon this, our second point is that despite the scenario involving the integration of a *new* feature, we anticipated the elicitation of RPD processes to be driven by the SMEs' familiarity with the processes of integrating new features into existing products and the use of these modified products to access new markets.

Thus, we suggest that the exclusive focus of the scenario was not intended to lie with the intricate details of the new feature itself, but rather the recognition of this scenario existing as analogous to those previously encountered. As a reflection of this, the nature of the feature for integration was left underspecified.

As a final point, it was anticipated that the interdisciplinary design of the scenario would produce RPD processes within both disciplines. We maintain this stance on the basis that the generated scenario is representative of the nature of interdisciplinary projects conducted within this industry and the subsequent interaction and experience of engineering and marketing SMEs. It was therefore expected that any emergent RPD processes in response to the scenario (that is, cues attended to, goals identified, and expectancies and potential courses of action generated) would adopt a form united by their common experience and by a sequence reminiscent of the theoretical propositions of RPD.

Procedure

Think-aloud protocol methodology summarized SMEs' decision processing in response to the written scenario. SMEs were instructed to read the scenario and to verbalize any initial thoughts that came to mind during this period. Once the SMEs had read the scenario, they were asked to describe the processes they would use for assessing this situation and how they would reach an ultimate decision given the information provided. They were reminded that a decision had to be made within a timeframe of twenty minutes, which reflected the time allocated for each meeting between the interviewer and SMEs. If SMEs asked a question that required additional information, that is, for information regarding the feasibility of a particular situation assessment action, then the interviewer would respond in all cases by asking the SME to describe the process that would be followed hypothetically under both conditions. Each think-aloud protocol was tape-recorded and typically lasted between fifteen and twenty minutes.

Protocols were then transcribed and subjected to a content analysis. From this, key recognitional features and generated actions identified by each SME were then extracted from the protocols and documented according to the SMEs' familiarity with the scenario (see Table 14.1). An expert within the mobile communications field verified the existence of these codes. Using the categories documented in Table 14.1, each protocol was then recoded. Protocol recoding was also subject to cross-validation by an expert in order to ensure close adherence to the data. From this, 97 percent agreement was achieved between expert and researcher. The 3 percent of situation assessment actions that were ambiguously categorized were recategorized through reference to the original protocol. The final step of this process converted each recoded protocol into a production rule format. This formed the basis for the ACT-R model development described in further detail in the next section.

Table 14.1   **Recognitional features identified by SMEs**

	Plausible goals	Relevant cues	Expectancies	Actions 1...n
Situation Familiar (n=14)	• Make a decision whether to release product as originally intended or include feature and release in China • Meet customer's timescale for product launch in China • Gain short-term profit (extra 25% global unit sales) • Capitalize on potential long-term market share • Form cross-disciplinary team • Avoid high risks • Avoid long-term losses • Don't introduce an undeveloped product • Engineer product on time • Complete field test on time	• Feature maturity: Advanced state of development • Productionization required • Field test required • Presence of competition • Time constraints for deployment • Primacy of new feature for accessing Chinese market • Customer's timescale for launch • Importance of time to market for gaining competitive advantage • Threat of competitor gaining market share • Potential for long-term market • Strategic importance of Chinese market • Customer request for additional functionality • Release of original product in other markets	• Inclusion of feature will lead to the generation of long-term strategic market • Delay in launch will lead to lost sales • Inclusion of feature will lead to an increase in unit sales • Inclusion of feature in competitor's products • If China is not included in launch, will lose market share • Competitor will get market share if deadline is not met • Additional productionization feasible within launch date? • Immediate cost and gains of including feature • Required field test feasible within launch date? • Costs and long-term returns of accessing Chinese market	• Phase launch • Launch product in China with feature within deadline • Launch in China without feature within deadline • Negotiate timescale with customer • Shift all timescales • Delay decision

Table 14.1 continued   Recognitional features identified by SMEs

	Plausible goals	Relevant cues	Expectancies	Actions 1…n
Situation Familiar (n=14)		• Achievement of cost objectives: —Product development costs —Unit costs	• State of competitor's development • Early penetration of an immature product may be counter-productive • Extent of technology maturity within timeframe • Feature omitted from product—demands of market not met • Potential delay in launching product by including feature	
Situation Not Familiar (n=2)	• Form cross-disciplinary team • Gather more information: —Time constraints —Financial constraints —Engineering feasibility • Compare information to other Motorola projects • Conduct a risk assessment • Make a decision whether to release product with feature or not	• Customer requirements • Time constraints for deployment • Field test requirement • Engineering effort • Time to market • Available budget • Competition • Advanced state of feature development	• Product flaws will negatively impact on market share and reputation • Customer expectations of product flaws • Immediate benefits of including feature • Long-term benefits of including feature • Assessment of returns/forecast	• Compare conditions to other analogous situations • Seek more information • Launch product in China with feature within deadline (despite product flaws)

Results

ACT-R Model Development

It was envisaged that the development of the ACT-R model would serve as a representation of the SMEs' *actual* NDM processes. The construction of the ACT-R model was characterized by two stages. First, production rules generated from the SMEs' think-aloud protocols were converted into a generic representation of the decision processes used. In doing so, the production rules for each SME (that were anchored in specific statements from the scenario) were transformed into a language format predetermined by the components of "Complex RPD." For example an SME's production rule "IF customer's time-scale for launch and the advanced state of feature development has been identified as relevant THEN make an assessment of situation familiarity" would become "IF Relevant Cues (Initial Response) have been identified THEN make an assessment of situation familiarity." It is important to emphasize that while the mainstay of this process necessitated the use of "Complex RPD" categories, a number of additional categories were used. These were invoked as a means of representing the iterative nature of the decision process. For example, relevant cues were drawn upon at numerous points during the decision process and were thus differentially termed, for example, relevant-cues-initial, relevant-cues-feature, and strategic-relevant-cues. It is also necessary to state that the generic representation of SMEs' decision protocols permitted variability between SMEs' decision making processes. The production rule format could be viewed as analogous to a decision-tree approach, permitting successive deviations at each point of process. Thus, the generic representation was able to accommodate the decision processes even in the most extreme case whereby the scenario was identified as familiar and non-familiar. The conversion of the production rules into a language format in line with "Complex RPD" also assisted in the minimization of deviations created by engineering and marketing SMEs' differential attenuations to the scenario content. The second stage of developing the ACT-R model then transformed these production rules into ACT-R syntax.

RPD Model Validation

Validation was achieved by independently running the critical decision processes from SMEs' decision protocols through the ACT-R model. This permitted two comparisons: (i) examination of SMEs' actual courses of actions to ACT-R predictions (that is, whether the ACT-R model was able to replicate the courses of action SMEs generated), and (ii) consideration of the degree of similarity between ACT-R trace outputs and the theoretical propositions of the RPD model (that is, whether the decision processes underpinning SMEs' actual decision making, as represented by the ACT-R model, were reflective of those described by the RPD model).

Predicting Courses of Action: Protocol Data—ACT-R Model Fit

Using this process, the ACT-R model predicted appropriate courses of action for 81.3 percent of the SMEs' protocols (thirteen protocols). The remaining 18.7 percent (three protocols) that were incorrectly classified consisted of nonsensical actions. These percentages of agreement were derived from the comparison of courses of action from the think-aloud protocols to the decision output generated by ACT-R for each SME. It was anticipated that this approach would provide a statistic that represented the overall degree of data-model fit afforded by our ACT-R model.

Our assertion that the ACT-R model is representative of SMEs' *actual* decision making processes is proposed on the basis that the production rules underpinning the ACT-R model and the specification of parameters for each SME were drawn directly from SMEs' decision protocols. Thus, the ACT-R model was developed as a reflection of the *actual* decision making processes of SMEs. Figure 14.3 provides an example trace output from the ACT-R model for an SME's decision to "release the product in China with the new feature within the deadline." The linkage between the SMEs' think-aloud protocols, the generic representation of SMEs' decision processing and the trace output is outlined below.

First, Figure 14.3 provides evidence of the close coherence between the SMEs' protocols and the production rules underpinning the generic representation of SMEs' decision processing. For example, the think-aloud protocol begins by the SME identifying relevant cues from the scenario: "So, you are in the process of developing a product and you have a customer who comes in and says yes I want this product but I want this as well" and their correspondent assessment of this scenario as familiar "We...I guess that the risk that we are talking about in the scenario, we have been in this dilemma many, many times." Using the generic representation of SMEs' decision processing, the first section of the SME's protocol was coded under "Relevant Cues" as "Customer request for additional functionality." The coded protocol was then used to form Production rule 1, that is, "IF Relevant cues are identified from scenario (*Customer request for additional functionality*) THEN Make an assessment of situation familiarity." Note that both here and in Figure 14.3 the coded section of protocol is presented in parentheses for clarity. Extracts from the SME's original protocol are also documented.

Second, the linkage between the generic production rules and the trace output is demonstrated through our annotations. Completing our example, Production rule 1 is represented by the output "Relevant-Cues-Initial Fired and Assess-Situation-Familiarity Selected." It is important to emphasize that one of the benefits of this approach to modeling in ACT-R using a language format predetermined by Complex RPD is that it minimizes the effect of differential attenuations between SMEs to specific contents of the scenario, that is, the identification of different "relevant cues."

While the data discussed so far suggests the ACT-R model displays a high degree of predictive accuracy and data-model fit (that is, the model is representative of SMEs' decision processes), it does not address whether the decision processes are representative of RPD. At this point, it is necessary to outline our comparative analysis of ACT-R's trace outputs to the theoretical propositions of the RPD model.

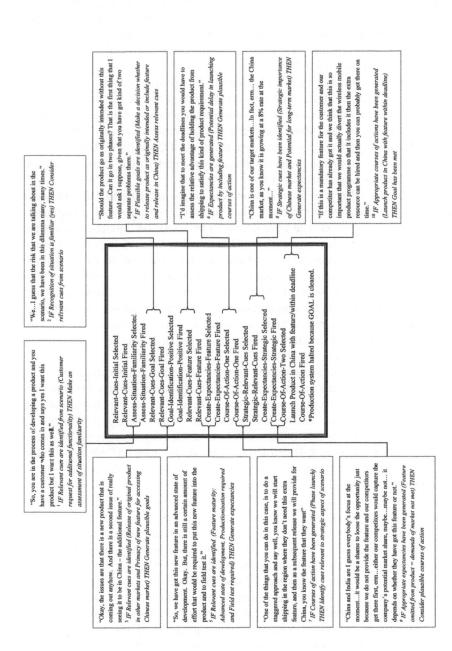

Figure 14.3   Example ACT-R trace output (Situation familiar)

Comparing ACT-R Model vs. Complex RPD

Qualitative comparison of the ACT-R trace outputs to the theoretical specifications of the RPD model allowed assessment of the representativeness of our microcognitive model of decision processing. Using the trace output derived from an SME in Figure 14.3 as an example of our comparison, a number of processes reflective of RPD were elicited. These are discussed in turn.

First, the trace output in Figure 14.3 shows the decision process to be marked by the early assessment of situation familiarity from cues that featured within the scenario. The trace output specifies the assessment of situation familiarity to follow the selection of initial cues from the scenario (Relevant-Cues-Initial and Assess-Situation-Familiarity). Conversely, the specifications of Complex RPD suggest that situation familiarity occurs in direct response to the experience of the situation and does not involve the use of relevant cues.

The ACT-R model as represented by the trace output in Figure 14.3 also provides insight into the specific interrelations that occur between the four components associated with recognition. The trace output from the ACT-R model organizes the four aspects of recognition as a series of linear pairs. What is important to note is that the trace outputs propose that not only do these components of recognition interact in a sequential manner, but that the extraction of relevant cues is not a one-off activity. For example, the productions that appear between Relevant-Cues-Goal and Course-Of-Action-One are summarized in RPD terms as follows: relevant cues-goal, goal-relevant cues, relevant cues-expectancies and expectancies-course of action. The Complex RPD model as it currently exists does not specify the manner in which these components are linked.

Our third finding elaborates upon the iterative processes that occurred within the trace output. It is interesting to note that the trace output reflects a separation of cues relevant to the development of the wireless feature and cues relevant to the strategic aspect of the scenario. This can be seen in the trace output through the emergence of two cues, expectancies and course-of-actions cycles (cycles starting Relevant-Cues-Feature and Strategic-Relevant-Cues). The specification of Complex RPD in its current form does not permit assessment of whether cycles of this nature occur.

A final point regarding the analysis of the trace output in Figure 14.3 was that the SME did not employ a discrete process of mental simulation. Rather, the SME's production rules coupled the generation of expectancies together with the mental simulation of courses of action. For example, Production rule 9 generates expectancies and uses those to mentally simulate its effect upon a course of action as a part of the selection process. This differs from the theoretical specifications of Complex RPD, which proposes that mental simulation follows the selection of a course of action.

Analysis of the trace outputs also enabled insights into the process of situation non-familiarity as Figure 14.4 demonstrates. As the Complex RPD model currently stands, this process of failed recognition is under-specified, concluding that the SME will identify the situation as non-familiar and then immediately seek more information. In this trace output, the SME follows a similar process as that of Figure 14.3 in making an initial judgment of situation familiarity based upon the assessment of relevant cues. Despite identification of the situation as unfamiliar, the

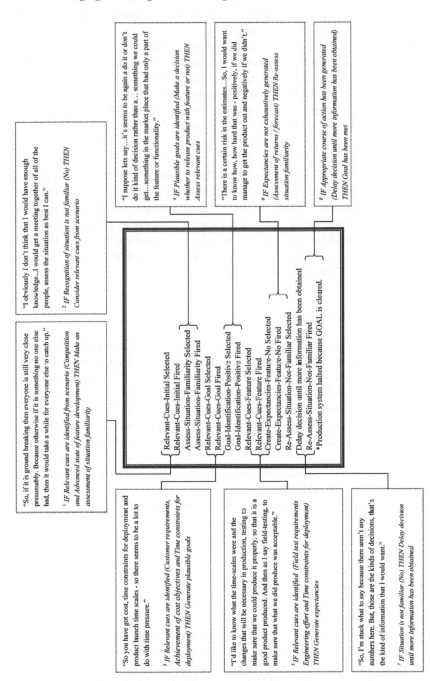

Figure 14.4 Example ACT-R trace output (Situation unfamiliar)

SME continues to attempt to draw relevant cues from the scenario to understand the decision goal. However, despite being able to extract relevant cues and the decisional goal, the SME was unable to generate appropriate expectancies and thus decided to reassess the situation (Create-Expectancies-Feature and Reassess-Situation-Not-Familiar). What the trace output in Figure 14.4 highlights is the importance of using these relevant cues and plausible goals in judgments of situation non-familiarity. It also provides converging evidence (in conjunction with Figure 14.3) to suggest that SMEs extract relevant cues prior to the assessment of situation familiarity.

In conjunction with the trace outputs obtained from the further fourteen SMEs who participated in this study, the trace outputs outlined above provide an evidence base for establishing the validity of a number of components specified within Complex RPD. Specifically this includes early assessments of situation familiarity, the generation of goals, the extraction of relevant cues, the development of expectancies, and the identification of appropriate courses of action. ACT-R has also afforded additional insights into the specific interrelations that exist between the components of Complex RPD and offered additional insights into the decision process in the event of situation nonfamiliarity.

Discussion and Conclusions

The underlying motive of this chapter was to explore how microcognitive methods such as ACT-R may be used to validate macrocognitive phenomena such as RPD. Our study extends the theoretical propositions of Klein (1995) and Klein et al. (2003) by empirically illustrating how macrocognitive processes may be traced back to their microcognitive components. The success of our ACT-R model as a tool for validation is most markedly characterized by the agreement between ACT-R's trace outputs to the RPD models' descriptions. By turning our attention to consider how specific microcognitive processes may link into higher-order macrocognitive phenomena, our case for validation and for bridging understanding across the macro/micro divide may be established.

Before doing so, a short divergence is necessary to consider the division of what we shall term generally for these purposes "cognitive components" into macrocognitive and microcognitive "camps." One aspect that struck us as authors was the inability to assign particular components of cognition such as "mental simulation" exclusively to one "camp." Based on Klein et al.'s (2003) list of "macrocognitive functions" (that is, NDM, sensemaking/situation assessment, planning, adaptation/replanning, problem detection, and coordination) and "supporting macrocognitive processes" (that is, mental simulation and storyboarding, maintaining common ground, developing mental models, uncertainty management, turning leverage points into courses of action and attention management) "mental simulation" would be termed a "macrocognitive supporting process." This conceptualization lies at odds with Klein's (1998) earlier suggestion that considered the availability, representativeness, and simulation heuristics to form the microcognitive components of the RPD model. While this theoretical incongruence is problematic for our discussion, it also suggests that the practicalities of distinguishing between macrocognitive and microcognitive phenomena are complex. We view this problem as a reaffirmation of our argument that the macrocognitive/microcognitive divide is artificial and thus unnecessary. For our current discussion, we therefore adopt the stance that particular "cognitive components" such as "mental simulation" transgress both levels of analysis and can both be considered as macrocognitive processes and deconstructed into their microcognitive counterparts.

Refocusing the discussion back to the examination of macrocognitive/ microcognitive relations that emerged from our ACT-R model four illustrative examples are outlined. Our first example concerns the macrocognitive function of situation assessment/sensemaking. The process delineated in the ACT-R model suggests the decision process to be marked by early assessments of situation familiarity. While this parallels the RPD model's description of recognitional processing, the ACT-R model differs in one key way, by suggesting the assessment of situation familiarity to be situated after the selection of the key relevant cues (marked by availability and representativeness heuristics). While the RPD model does not deny the role of relevant cues in assessing situation familiarity, it does so retrospectively, that is, the question of "Is the situation familiar?" precedes the identification of relevant cues. Our ACT-R model in conjunction with the SMEs' protocols has provided evidence to suggest that the extraction of relevant cues may occur at some point between "experiencing the situation in a changing context" and consideration of "situation familiarity." This makes intuitive sense on the principle that in order to conduct an assessment of situation familiarity, some element of cue identification must have previously occurred upon which to ground assessment. Thus, by using ACT-R to consider the sequence that microcognitive processes such as the extraction of relevant cues occur within higher-order macrocognitive phenomena such as situation assessment, increased specificity of RPD may be obtained.

Additional evidence for our case comes from instances whereby the situation was not viewed as familiar and the collection of more information was recommended. In these instances, our ACT-R model outlines a number of microcognitive processes that link into our understanding of the macrocognitive phenomena of situation assessment and sensemaking in two key ways. First, our model (in conjunction with the findings above) provides converging evidence to suggest decision makers extract relevant cues prior to the assessment of situation familiarity. Second, in the event that the situation is categorized as nonfamiliar, the ACT-R model outlines a process of sensemaking marked by an iterative cycle of microcognitive processing using relevant cues and goals. As the RPD model currently stands, this process of failed recognition is underspecified. It is interesting to note that the Recognition/Metacognition model (Cohen, Freeman, and Wolf, 1996) was developed with the objective of understanding the process of recognition. The Recognition/Metacognition model is built around two key propositions: (i) when the stakes are high, time is available, and the situation is unfamiliar or uncertain, a cycle of critical thinking occurs and (ii) if time is available, qualitatively different kinds of uncertainty are sought and dealt with in appropriate ways. Thus, it could be inferred that this iterative process was preempted by the existence of cues relevant to both the development of the wireless feature and the strategic aspect of the scenario. Thus, the complex and multifaceted (for example, engineering, marketing, and strategic) nature of the scenario may have necessitated the SMEs to deal with the decision in a cyclic manner. It is important to highlight that while the Recognition/Metacognition model echoes and consequently cross-validates the processes documented in our ACT-R model at a macrocognitive level, they remain underspecified and do not achieve the same level of specificity afforded by our microcognitive focus.

Our second example considers our contribution towards the understanding of supporting macrocognitive processes. Our ACT-R model exemplifies this claim by providing a number of macrocognitive outputs that informs our understanding of the processes and roles of decision making, uncertainty and mental simulation as defined by Klein et al. (2003). The microcognitive focus of the model enabled the specification of the micro processes that exist both within and between these macrocognitive components. For example, our model suggests that uncertainty (as characterized by the uncertainty surrounding penetration into the Chinese market and the feasibility of completing productionization and field test prior to the customer's deadline) is dealt with in an iterative way by utilizing relevant cues within successive rounds of expectancy generation and mental simulation. In the vein of Skriver's (1998) findings, the four aspects that denote the process of recognition and the generation of appropriate courses of action are organized in the decision process as a series of microcognitive linear pairs. This process is akin to the "progressive deepening" strategy outlined by De Groot (1965) and may have occurred as a function of the complexity, interdisciplinary and underspecified nature of the scenario. This practice of successively refining the problem as seen in our ACT-R model (that is, the integration of more detail, recognition of cues added themselves, and further specification in order to reach a satisfactory conclusion), is similar to what Goel and Pirolli have found with design problems in general (1989, 1992). The strategy is a result of working memory limitations and has also been found in chess in the early work of De Groot, in software design (Adelson, 1990; Kant and Newell, 1984), and in the design of vehicles (Greeno et al., 1990). It could thus be reasoned that the design nature of our scenario was not a classic RPD situation, and by implication may not have fully evoked RPD processes.

Our model also informs our understanding of the macrocognitive process by which leverage points are turned into courses of action. Complex RPD implies the use of relevant cues to be integral to the development of courses of action, but aside from the specification of goals and expectancies within the four components of recognition, it does not specify how precisely these components interact. The ACT-R model deconstructed this process into a series of microcognitive components: cues, goals, and expectancies. It proposed the identification of goals and expectancies to play an important mediating role in converting the relevant cues into leverage points and end courses of action. Our ACT-R model also used leverage points from the scenario as a means of structuring thought and was reflected in the iterative cycles of decision processing such as the consideration of cues relevant to the feature and the strategic element of the scenario. It is useful to consider here the parallels of this process to Montgomery's (1983, 1989) Search for a Dominance Structure (SDS) Theory. SDS stipulates that the definition of a dominance structure is achieved through the following stages: (a) pre-editing, (b) search for a promising alternative, (c) dominance testing, and (d) dominance structuring. In the event that the dominance structuring process is unable to find a satisfying solution, the SME starts again from the pre-editing stage or searches for a new promising alternative. It appears viable that SDS processes could provide a rationale for the occurrence of SMEs' iterative cycles of processing viewed in the ACT-R model and thus theoretically unite the four components of recognition in the RPD model.

A final example that demonstrates the value in modeling macrocognitive processes within the microcognitive framework of ACT-R considers the positioning of mental simulation within Complex RPD. Theoretical specification of Complex RPD proposes mental simulation to occur following the selection of a course of action and prior to its implementation. However, structuring the SMEs' production rules during the development of the ACT-R model revealed the concurrent coupling of expectancies with the mental simulation of potential courses of action for SMEs for whom the situation was familiar. Analysis of the specific contents of production rules used by SMEs supported this finding. The importance of expectancies and mental simulation for the generation of courses of action was also emphasized by SMEs for whom the situation was unfamiliar. In these instances, SMEs either failed to generate expectancies in an absolute sense, or generated expectancies which, although related to the scenario, did not set the stage for the development of appropriate courses of actions. This raises questions of the difference between expectancies and the mental simulation of action, especially when both precede the point of choice of action and appear during the cyclic iterations of our ACT-R model. Note that this also adds to the debate on the macrocognitive/microcognitive status of mental simulation presented earlier. Both of which serve as points worthy of future investigation.

Lessons Learned

While the ACT-R model has provided some interesting insights into the nature of the RPD process, there are a number of limitations associated with our approach. First is the element of artificiality and generalization that was imposed during the coding process. While this is a necessary precondition for the use of protocol data in ACT-R, it does mean that in some instances data may have been omitted from the model, which in studies of a more qualitative nature would otherwise have been included. Second, there was also an element of artificiality due to the specification of parameters at the start of the ACT-R model, namely by specifying the decision inputs and outputs for each production rule. Thus, by predetermining decision inputs and outputs, there is a consequential degree of incompleteness in our model. A final lesson learned stemmed from the difficulty in providing an accurate representation of relevant cues within ACT-R. The problem lay in the fact that these perceptual entities are difficult to articulate in a few descriptive words. Thus, a number of subjective inferences about the form of these perceptual entities were made, as adopted by Stanard et al. (2001).

Implications for Macrocognition

Notwithstanding the lessons learned from our study, this chapter has both defended theoretically and illustrated empirically a case for bridging the macrocognitive and microcognitive divide. Specifically, we have demonstrated the value in understanding the nature of macrocognitive functions and supporting macrocognitive processes by virtue of the microcognitive method ACT-R. Our study has also provided an evidence base for establishing the validity of a number of microcognitive components specified within the RPD model and the specific interrelations that exist between cognition at the macro and micro levels.

What are the implications of this study for the definition of macrocognition and the inherent divide between the macro and micro levels of analysis? While we have used the terms "macrocognition" and "microcognition" within this chapter, we suggest that it is perhaps an unnecessary distinction. Our inability to assign cognitive phenomena to either macrocognitive or microcognitive "camps" during our analysis exemplifies our case for viewing it as a somewhat artificial distinction. We suggest that the adoption of such terms is likely to stunt cross-fertilization and promote paradigmatic divides. Our case in favor of such a stance rests upon macro/micro debates that are echoed within other divisions of the Social Sciences and Management Studies (McAndrew and Gore, 2006). A recent symposium held at the British Academy of Management (2006) concerned with "connecting the micro and the macro" made a case for more holistic approaches as opposed to "paradigmatic territorialism" and "methodology wars" (Collinson, Jarzabkowski, and Thomas, 2006). To some degree, these fractures have already started to appear within our own research community. In a recent report entitled "High-level cognitive processes in field research," Ross et al. (2002) argue in favor of the use of the term "macrocognition" on the grounds that:

1. Microcognitive research dominates the field and presents little chance to progress the study of macrocognitive phenomena.
2. The interest in studying macrocognitive phenomena lies in linking the phenomena with the actual performance context. This is unlikely to be addressed within research of a microcognitive nature.
3. The study of macrocognitive phenomena will require different research methods which are marked by the dichotomy of completeness vs. precision.
4. Eventually the linkages between macrocognition and microcognition will become more transparent and we will gain an appreciation of how microcognitive processes are contributing to the larger macrocognitive picture and vice versa.

Indeed, traces of "paradigmatic territorialism" and "methodology wars" infiltrate this agenda. In response to the first claim, we argue that the occlusion of macrocognitive research by its microcognitive counterpart is an insufficient motive for the coining of the term "macrocognition." Cacciabue and Hollnagel's (1995) creation of the term "macrocognition" in response to their observation that the simulation of cognition was dominated by microcognitive models such as SOAR (Newell, 1990) and

ACT (Anderson, 1983) exemplifies this point. Moreover, our study demonstrates the value in using microcognitive research methods to progress the study of macrocognitive phenomena. Secondly, the claim that microcognitive phenomena are unlikely to be explored within performance contexts is one that is unlikely, but one that can be addressed as our earlier study demonstrates. Thirdly, and reminiscent of the methodological wars of Collinson et al. (2006), we suggest that while the study of macrocognitive phenomena do (on the whole) require different research methodologies, there is value in "borrowing" from the microcognitive research community, for example, our use of the ACT-R modeling environment. We agree with the final proposition of Ross et al. (2002), but suggest that this sits antithetically against propositions (i)–(iii) making the future rapprochement of macrocognition and microcognition increasingly unlikely. Rather, we propose like Klein et al. (2003 p. 81), that "the two types of description are complementary. Each serves its own purpose, and together they might provide a broader and more comprehensive view than either by itself."

Future Directions

In order to achieve further assessments of the representativeness of RPD theory as symbolic of recognitional decision processes, it may be fruitful to extend validation efforts twofold:

- *Modeling of error-free expert decision making*: Using microcognitive models, the modeling of other types of decisions is viewed as necessary to assess the explanatory value of RPD theory.
- *Goodness-of-fit of alternative theoretical representations*: Extending our understanding of the explanatory value of RPD, it would be valuable to compare alternative theoretical representations of NDM, for example, image theory to our generated ACT-R model.

Besides benefits for academia in affording NDM and RPD model respecification and development, this research also has a number of applied utilities such as the development of organizational decision support systems.

In terms of a macrocognitive agenda, it would be of value to explore in more detail how our understanding of naturalistic phenomena may be enriched by drawing upon other microcognitive methods. It may also be interesting to consider how, as Klein et al. (2003) have suggested, microcognitive research might benefit by being contextualized by macrocognitive theory and method.

This chapter demonstrates both theoretically and empirically the value of borrowing from microcognition for progressing our understanding of macrocognitive phenomena. Specifically, it fashions a theoretical argument in favor of using ACT-R to model RPD processes with the objective of bridging the macro and micro levels of analysis. The model provides a starting point for: (i) the modeling of more complex aspects of RPD, (ii) for validating what have typically to date been subjective judgments of data–model fit and finally, (iii) for increasing the specificity of RPD theory. Our arguments represent a call for the cessation of the macrocognitive/

microcognitive divide, proposing this to be an unnecessary and unfruitful distinction. A number of implications of our study are raised for the future progression of the field of macrocognition.

References

Adelson, B. (1990). "Modeling software design within a problem-space architecture." In R. Freedle (ed.), *Artificial Intelligence and the Future of Testing*. Hillsdale, NJ: Lawrence Erlbaum Associates (pp. 213–37).

Anderson, J. (1983). *The Architecture of Cognition*. Cambridge, MA: Harvard University Press.

Anderson, J.R. and Lebière, C. (1998). *The Atomic Components of Thought*. Mahwah, NJ: Lawrence Erlbaum Associates.

Cacciabue, P.C. and Hollnagel, E. (1995). "Simulation of cognition: Applications." In J.M. Hoc, P.C. Cacciabue, and E. Hollnagel (eds), *Expertise and Technology: Cognition and Human–Computer Interaction*. Hillsdale, NJ: Lawrence Erlbaum Associates (pp. 55–73).

Cohen, M.S., Freeman, J.T., and Wolf, S. (1996). "Metarecognition in time-stressed decision making: Recognizing, critiquing, and correcting: Decision making in complex environments." *Human Factors*, 38(2): 206–19.

Collinson, S., Jarzabkowski, P., and Thomas, R. (September 2006). *Connecting the Micro and the Macro*. Symposium held at the British Academy of Management Conference, Belfast, Ireland.

De Groot, A.D. (1965). *Thought and Choice in Chess*. The Hague: Mouton.

Drillings, M. and Serfaty, D. (1997). "Naturalistic decision making in command and control." In C.E. Zsambok and G. Klein (eds), *Naturalistic Decision Making*. Mahwah, NJ: Lawrence Erlbaum Associates (pp. 71–80).

Driskell, J.E., Salas, E., and Hall, J.K. (April 1994). "The effect of vigilant and hyper-vigilant decision training on performance." Paper presented at the Annual Meeting of the Society of Industrial and Organizational Psychology, Nashville, TN.

Ehret, B.D., Gray, W.D., Hutcheson, T.D., and Kirschenbaum, S.S. (1997). *ACT-R Models of Submariner Situation Assessment*. Retrieved 1 March 2006 <http://act.psy.cmu.edu/ftp/workshop/Workshop-97/Ehret/quick_index.html>.

Goel, V. and Pirolli, P. (1989). "Motivating the notion of generic design within information-processing theory: The design problem space." *Artificial Intelligence Magazine*, 10(1): 19–36.

Goel, V. and Pirolli, P. (1992). "The structure of design problem spaces." *Cognitive Science*, 16: 395–429.

Gore, J., Banks, A., Millward, L., and Kyriakidou, O. (2006). "Naturalistic decision making and organizations: Reviewing pragmatic science." *Organization Studies*, 27(7): 925–42.

Greeno, J.G., Korpi, M.K., Jackson, D.N., and Michalchik, V.S. (1990). *Processes and Knowledge in Designing Instruction*. (Tech. Report No. ED329558). Stanford, CA: Stanford University, School of Education.

Juarez, O. and Gonzalez, C. (2003). *A Cognitive Architecture for Situation Awareness (CASA)*. Retrieved 1 March 2006 <http://act-r.psy.cmu.edu/workshops/ workshop-2003/talks/3.pdf>.

Kaempf, J.L. and Orasanu, J. (1997). "Current and future applications of naturalistic decision making in aviation." In C.E. Zsambok and G. Klein (eds), *Naturalistic Decision Making*. Mahwah, NJ: Lawrence Erlbaum Associates (pp. 81–90).

Kaempf, G.L., Wolf, S., Thordsen, M.L., and Klein, G. (1992). *Decision Making in the AEGIS Combat Information Center*. Fairborn, OH: Klein Associates.

Kant, E. and Newell, A. (1984). "Problem solving techniques for the design of algorithms." *Information Processing and Management*, 28: 97–118.

Klein, G.A. (1992). *Decision Making in Complex Military Environments*. Fairborn, OH: Klein Associates.

Klein, G.A. (1995). "A recognition-primed decision (RPD) model of rapid decision making." In G. Klein, J. Orasanu, R. Calderwood, and C.E. Zsambok (eds), *Decision Making in Action: Models and Methods*. Norwood, NJ: Ablex (pp. 138–47).

Klein, G.A. (1997). "The current status of the naturalistic decision making framework." In R. Flin, E. Salas, M. Strub, and L. Martin (eds), *Decision Making Under Stress: Emerging Themes and Applications*. Aldershot: Ashgate (pp. 11–28).

Klein, G.A. (1998). *Sources of Power: How People Make Decisions*. Cambridge, MA: The MIT Press.

Klein, G.A., Calderwood, R., and Clinton-Cirocco, A. (1985). *Rapid Decision Making on the Fire Ground*. Fairborn, OH: Klein Associates.

Klein, G., Ross, K.G., Klein, D.E., Hoffman, R.R., and Hollnagel, E. (2003). "Macrocognition." *IEEE Intelligent Systems*, May/June: 81–85.

Lipshitz, R. and Pras, A.A. (2005). "Not only for experts: Recognition-primed decisions in the laboratory." In H. Montgomery, R. Lipshitz, and B. Brehmer (eds), *How Professionals Make Decisions*. Mahwah, NJ: Lawrence Erlbaum Associates (pp. 91–106).

McAndrew, C., Banks, A., and Gore, J. (2004). "Naturalistic decision making: Validation of the recognition-primed decision making model using ACT-R cognitive architecture." Unpublished MSc thesis, University of Surrey, UK.

McAndrew, C. and Gore, J. (August 2006). "Bridging paradoxes in managerial cognition: Contributions from naturalistic decision making." *Proceedings of the Annual Academy of Management Conference*, Atlanta, GA.

Militello, L.G. and Hutton, R.J.B. (1998). "Applied Cognitive Task Analysis (ACTA): A practitioner"s toolkit for understanding cognitive task demands." *Ergonomics*, 41(11): 1618–41.

Montgomery, H. (1983). "Decision rules and the search for dominance structure." In P.C. Humphreys, O. Svenson, and A. Vari (eds), *Analyzing and Aiding Decision Processes*. Amsterdam, The Netherlands: North Holland Press (pp. 343–69).

Montgomery, H. (1989). "From cognition to action: The search for dominance in decision making." In H. Montgomery and O. Svenson (eds), *Process and Structure in Human Decision Making*. Chichester: Wiley (pp. 23–49).

Mosier, K.L. (1991). "Expert decision making strategies." In R.S. Jensen (ed.), *Proceedings of the Sixth International Symposium on Aviation Psychology.* Columbus, OH: Department of Aviation, Ohio State University (pp. 266–71).

Nellen, S. (2002). *The "take the best heuristic" and ACT-R.* Retrieved 1 March 2006 <http://act-r.psy.cmu.edu/workshops/workshop-2002/talks/Stefani Nellen.pdf>.

Newell, A. (1990). *Unified Theories of Cognition.* Cambridge, MA: Harvard University Press.

Randel, J.M., Pugh, H.L., Reed, S.K., Schuler, J.W., and Wyman, B. (1994). *Methods for Analyzing Cognitive Skills for a Technical Task.* San Diego, CA: Navy Personnel Research and Development Centre.

Raynard, R. and Williamson, J. (2005). "Conversation-based process tracing methods for naturalistic decision making: Information search and verbal protocol analysis." In H. Montgomery, R. Lipshitz, and B. Brehmer (eds), *How Professionals Make Decisions.* Mahwah, NJ: Lawrence Erlbaum Associates (pp. 305–18).

Ross, K.G., McHugh, A., Moon, B., Klein, G., Armstrong, A., and Rall, E. (2002). Year one final report: "High-level cognitive processes in field research." Fairborn, OH: Klein Associates.

Skriver, J. (1998). "Emergency decision making on offshore installations." Unpublished PhD thesis, University of Aberdeen, UK.

Stanard, T., Hutton, R.J.B., Warwick, W., McIlwaine, S., and McDermott, P.L. (2001). *A Computational Model of Driver Decision Making at an Intersection Controlled by a Traffic Light.* Retrieved 11 November 2005, from <http://ppc.uiowa.edu/ drivindassessment/2001/Summaries/Driving%20Assessment%20Papers/65_stand.pdf>.

Tissington, P. (2001). "Emergency decision making by fire commanders." Unpublished PhD thesis, University of Westminster, UK.

Chapter 15

Reconceptualizing Expertise: Learning from an Expert's Error

Stephen Deutsch

Introduction

Errors made by the skilled aircrews of commercial aircraft are not infrequent (Klinect, Wilhelm, and Helmreich, 1999). While many aircrew errors are benign, some have consequential outcomes—an outcome in which an aircrew error is linked to an undesired aircraft state or to another error. More particularly, Klinect and colleagues (ibid.) found that while operational *decision* errors made up only 6 percent of the errors that they observed during their Line Operation Safety Audits, 43 percent of those errors had consequential outcomes.

Decision errors by skilled operators are a persistent problem that needs to be better understood. As new systems and procedures are designed or as current systems and procedures are updated, we are seeking the means to better prevent the incidence of error and to better support error detection and mitigation when errors do occur. An important goal within this research program has been to achieve a better understanding of the sources of human error in decision making on the flight deck. The real world problems being addressed are very much the subject matter of macrocognition (Cacciabue and Hollnagel, 1995; Klein et al., 2003). The research methods and tools described, while not elements of current cognitive engineering practice for user-centered system design (Roth, Patterson, and Mumaw, 2002), provide an in-depth approach to examining integrated human perceptual, cognitive, and motor capabilities and limitation, and their implications for systems and procedure design.

Simulation provides a safe and relatively inexpensive means to rigorously explore multiple dimensions of complex real-world scenarios involving multiple players. By including human performance models in a simulation, we can provide the means for the detailed examination of operator cue-processing and team and system interactions that are precursors to decision making and action selection. If the models are to be useful, they must produce human-like expertise while also exhibiting human-like limitations. Through multi-task behaviors much is accomplished, yet error can intrude on the performance of experts. To the extent that the models produce their behaviors as humans do, they can help us to gain better insight into how our perceptual, cognitive, and motor capabilities combine to produce these behaviors. Exploring alternate paths through a scenario adds to the list of situations and potential problem solutions that may be investigated to extend and refine previously acquired insights. The requirement to configure models that lead to realistic human-like behaviors demands focused attention on the sources for those behaviors.

In the present study, we reviewed and modeled the windshear accident of 2 July 1994 at Charlotte/Douglas International Airport as discussed in the National Transportation Safety Board Aircraft Accident Report (NTSB, 1995). In particular, our study focused on developing a better understanding of the factors contributing to the captain's planning and decision-making processes related to the threat posed by the weather cell during his approach to Runway 18 Right. With the first officer as pilot-flying, the captain briefed a plan to execute a missed approach that did not follow prescribed operating procedures. We will review the plan as briefed—"if we have to bail out…it looks like we bail out to the right"—and attempt to understand why the captain provided the brief that he did rather than plan for a microburst escape maneuver as advocated in his prior simulator training.

The modeling aspect of the current study has made contributions on two basic levels. First, the model fostered new insights into aircrew decision making that were sources of human error contributing to the accident. The modeling enabled a detailed examination of the aircrew's cue-processing and the captain's decision making with respect to his plan for dealing with the weather cell. Second, the modeling process forced a reexamination of the capabilities of the model. For the model to faithfully represent individual human performance, it had to capture *why* actions were taken and *how* they were carried out. The model provided a framework for understanding and implementing the complete sequence of events leading to the accident with the specific exception of the formulation of the captain's briefing. Addressing the particulars of the formulation of the briefing, an instance of an error in otherwise expert performance, led to a proposed role for *narratives* (see Bruner, 1996) and what we will describe as *narrative synthesis* in decision making and action selection. The role for narratives was derived from insights into possible sources for the briefing as a factor contributing to the sequence of events leading to the accident—they are extensions that will be a significant challenge to implement in the models.

We begin by describing the microburst-induced commercial aviation accident at the Charlotte/Douglas Airport in 1994. We then provide background material on the modeling environment that was used to emulate the accident with particular emphasis on the human performance models for the accident aircraft's captain and first officer. With the background in place, we examine the weather situation that the crew encountered on their approach to the Charlotte/Douglas Airport and the captain's plan for addressing the threat posed by the weather cell at the threshold of the runway. We then follow the accident scenario as the captain communicated a plan for addressing the threat presented by the weather cell—a plan that was subsequently put into action and then revised, but was unsuccessful in avoiding the accident. The analysis of the accident focuses on the captain's plan for addressing the microburst threat and how it might have been generated.

The 1994 Commercial Aviation Accident at the Charlotte/Douglas Airport

The issues at hand in the accident scenario at Charlotte/Douglas Airport are representative of complex, real-world problems that expert practitioners are called on to address. Approach and landing is a flight phase in which an aircrew is sharing its attention among several demanding, concurrent tasks. The flight phase is further complicated when the crew must additionally deal with the microburst threat posed by a weather cell at the threshold of the runway.

Weather cells can be the source of large, rapid thunderstorm-induced changes in wind speed—a microburst that is one form of windshear. As described in the NTSB report on the accident at Charlotte/Douglas Airport, "A microburst is defined as a precipitation induced downdraft, which in turn, produces an outflow when the downdraft reaches the earth's surface" (NTSB, 1995, p. 80). An aircraft entering the microburst experiences a rapid increase in airspeed. The microburst-induced increase in airspeed is not a threat to the aircraft. However, on transiting the core of the microburst, the aircraft is then chasing the outflow and hence experiences a rapid decrease in airspeed. When late in the approach and landing process with airspeed reduced in preparation for the landing, the encounter with the downdraft and the subsequent rapid decrease in airspeed with the resultant loss of lift presents a very significant threat to the aircraft.

On 2 July 1994, about 18:43 Eastern Daylight Time, a Douglas DC-9-31, N954VJ, operated by USAir, Inc. as Flight 1016, collided with trees and a private residence near the Charlotte/Douglas International Airport, Charlotte, North Carolina, shortly after the flightcrew executed a missed approach from the instrument landing system approach to Runway 18R (NTSB, 1995, p. 1).

The aircraft had entered a weather cell at the threshold of the runway subsequent to the onset of a microburst. The tactic for coping with the potential microburst that the captain had preplanned and communicated to the first officer, as the pilot flying, was not sufficient to address the situation as it evolved.

Prior to selecting the USAir 1016 accident for further investigation, we had reviewed a series of accidents, culled from a set identified by Orasanu as involving "plan continuation errors" (Orasanu, Martin, and Davison, 2001), that is, accidents preceded by a series of cues that suggested another course of action, but for which the aircrew held fast to its original plan. In fact, the NTSB (1995) report focused on the captain's decision to continue the approach and landing in the face of a series of cues that strongly suggested aborting the landing at an earlier point in the approach.

The USAir 1016 captain was attending and acting on the cues presenting themselves. As the aircraft was on the final approach to Runway 18 Right, in response to the weather cell's threat, the captain briefed the tactic that he wanted executed should they have to abort the landing: "if we have to bail out...it looks like we bail out to the right" (NTSB, 1995, p. 157), to which the first officer responded "amen" (NTSB, 1995, p. 157), indicating agreement with the plan. When the situation deteriorated to the point where the captain called for acting on the plan, the first officer as pilot-flying executed the plan as prearranged, providing further evidence for agreement on the suitability of the plan as announced by the captain. Several seconds later, it became clear that the planned actions were not adequate to address the evolving situation. The captain's new orders and the first officer's execution of them were not successful in attempting to escape the microburst.

With the initial review of the accident report completed, we then developed a model of the scenario as it unfolded. Having developed previous models for approach and landing, the modeling effort went forward as expected, but with a singular exception: producing the seemingly simple plan developed by the captain for addressing the microburst threat. In fact, the plan developed by the captain and discussed with the first officer was never successfully reproduced in its exact detail in the model—the first indication that the plan did not call for the simple substitution of one procedure for another. Had the captain mistakenly invoked a missed approach procedure as suggested in Item 14 of the Conclusions of the Accident Report (NTSB, 1995), it would have been readily reproduced in the model.

Alerted to the subtlety in the captain's briefing through the modeling effort, we then attempted to identify the factors contributing to the plan as expressed in the briefing. Our sense was that further attention to the captain's briefing would contribute to a more complete understanding of the causes of the accident and provide better insight into how error can intrude on otherwise expert performance.

Model Building: From Theory to the Emulation of Human Performance

At its best, creating a theoretical foundation for human performance modeling is an iterative process. The literatures of a broad range of disciplines—cognitive, experimental, and social psychology, neuroscience and cognitive neuroscience, and anthropology—provide a wealth of *empirical* data to support model architecture development and component design. More recently, these traditional fields for the study of perception, cognition, and motor control have been supplemented by the work in macrocognition (Klein et al., 2003), providing new insights into cognition in the field that is directly relevant to model building. Having begun our work before the emergence of the discipline of macrocognition, we relied on the traditional literature in forming the foundation for model architecture design and model building.

The basic elements of the modeling framework, the Distributed Operator Model Architecture (D-OMAR[1]), that we will describe are its goal-driven behaviors, the coordination of behavioral processing elements through a publish-subscribe protocol, and the interleaving of multiple tasks, none of which requires an explicit executive (Clark, 1997). Rather than employing a production-rule system, decisions, usually in the form of simple predicates are embedded in procedures that when combined with the pattern matching established by the publish-subscribe protocol implements a process much like recognition-primed decision making (Klein, 1998).

Modeling the Response to the Demands of Concurrent Tasks

Human operators in natural environments must frequently address multiple demands, hence a principal area of research in the development of D-OMAR has been that

---

1   Source code and documentation for D-OMAR is available at <http://omar.bbn.com>.

of human multi-task performance. In developing D-OMAR, we have provided a computational framework in which to assemble a functional mix of perceptual, cognitive, and motor capabilities that operate in parallel, subject to appropriate constraints that enable the model to exhibit the multiple task behaviors of skilled human operators, that is, aircrews and air traffic controllers. The desired behaviors have a combination of proactive and reactive components. That is, the operators have an agenda that they are pursuing, but must also respond to events as they occur. It is the proactive agenda and established reactive responses implemented as goals and their procedures rather than production rules that drive a model's behaviors. Within this framework, there are limits imposed by the capabilities of their functional resources. Consequently, in pursuing the proactive agenda, there may be newly motivated tasks for which ongoing tasks must be deferred.

The bounds on what can be accomplished concurrently take several forms. A typical behavior on the flight deck will be to set aside an in-person conversation in order to respond to an air traffic controller communication. At a more basic level, two competing tasks may each require the use of the pilot's eyes—one to guide an immediately required manual operation that may preempt the use of the eyes for a background instrument scan operation. In the first instance, it is a matter of operating procedures or protocols that must be established; in the second, it is contention for perceptual and motor resources necessary to complete a task that must be arbitrated. The arbitration with respect to protocol and resources is based on task priority. While priorities can vary dynamically, we have found very few instances in which static priorities did not yield realistic performance. Unlike EPIC (Meyer and Kieras, 1997), the model does not rely on an explicit executive process governing the execution of contenting procedures. In D-OMAR, the contention is resolved among the contending procedures themselves based on priority and resource requirements.

The Model as Operator Goals that Establish a Dynamic Network of Procedures

The core of a D-OMAR model, established almost twenty years ago, is a network of procedures activated by the operator's overall strategy for addressing his or her tasks. The network's activation varies in response to events in the world that are then channeled as necessary to achieve the operator's goals. From a top-down perspective, currently activated goals represent the operator's proactive agenda for managing his or her tasks. Each goal activates a plan that typically includes a series of sub-goals and procedures. The goals and sub-goals express the operator's desired outcomes; the procedures are the implementation of the actions to achieve those outcomes. The operator's current agenda is carried out by the set of active procedures within the network. Most procedures are in a wait-state—they represent the operator's potential to address expected events consistent with the proactive agenda and the ability to cope with a changing world. The network represents the operator's long-term procedural memory for how to carry out the actions necessary to manage and respond to the changing series of events with which an expert operator is likely to be confronted. Working memory consists of the currently activated goals and procedures and the long-term and newly acquired elements of declarative memory that are maintained as values on the slots of these goals and procedures. Declarative memory items are thus bound in the context established by the procedures through which they were acquired.

From a bottom-up perspective, there is an assembly of distinct perceptual, cognitive, and motor capabilities that are recruited as procedures to address current goals and sub-goals. The functional components supporting human task execution exist at particular local centers with the coordinated operation of several such centers required to accomplish any given task; it is this operational framework that we emulate through the procedure networks and communication protocol in our models.

With specific functional capabilities modeled as concurrently operating procedures that represent the work of the several distinct perceptual, cognitive, and motor centers acting in support of the completion of a task, the form that coordination of computational elements might take is of particular importance for developing model behaviors. The architecture supporting the model must provide for coordination among the procedures at distinct functional centers necessary to accomplish a task; it must also provide for the movement of information among the remote centers. A *publish-subscribe* protocol is used to address both of these requirements. The unit of information transfer is a message identified by a name specifying the message *type* and containing a set of related information elements. As a procedure representing a functional center develops informational elements, they are assembled into an appropriately type-named message that is then *published*. Procedures requiring informational elements of a particular message, *subscribe* to the message by message type. Having subscribed to a message type, the receiving procedures are alerted to publication of the message and the message content may then be processed by the receiving procedures. There can be more than one subscription to any given message type. The publish-subscribe protocol at once provides for the movement of information among procedures representing the centers while its triggering mechanism supports the coordination of procedure execution establishing a *data flow* architecture (Arvind and Culler, 1983). The arrival of one or more messages, as required, will trigger the processing at the receiving center, thereby establishing the necessary coordination among concurrently operating procedures.

A subscribing procedure may also prescreen the message content to determine whether or not to process a message. The screening of messages further enriches the *pattern-matching* capability of the network. Having received and processed a signal, a procedure must renew its subscription to the message type if it is to act on the next message of that type. It is the requirement for the renewal of the subscription to a message type that leads to a dynamically structured network. The publish-subscribe protocol thereby creates dynamic links in the network of computational centers—the procedures that represent the operator's perceptual, cognitive, and motor capabilities.

For example, the simulated verbatim representation of a spoken statement is published and subsequently received by a procedure that "generates" and then publishes the propositional content of the statement. The verbal statement's propositional content when published, triggers subscribing procedures that support downstream content-dependent decisions or actions. Within a model, information flow management, process synchronization, and pattern matching are all derivatives of the publish-subscribe protocol.

Modeling Expertise as Process, Cognitive Skills, and Pattern Matching

With goals and the procedure network as its major elements, the architecture that we created in D-OMAR is very much an architecture in which *process* has the dominant role (Edelman, 1987, 1989). It is through process, goals, and procedures, that the model's expertise leading to human-like behaviors is established. The focus on process is in keeping with Glenberg (1997) who identifies process memory—memory for what we know how to do—as memory's primary function. With the focus on process, declarative memory is relegated to a supporting role. Active procedures establish context for the short- and long-term declarative memory elements that are maintained as attribute-values on the procedures.

On another dimension, Logan (1988a, 1988b) and Bargh and Chartrand (1999) have extended the traditional view in thinking about human *skills*. Rather than skills being restricted to the perceptual and motor domains, they suggest that there are *cognitive skills* that play a significant role in human expertise. As Logan expressed it, we need to "look more broadly for automatic processes. They need not be restricted to procedural knowledge or perceptual motor skills but may permeate the most intellectual activities in the application environment" (1988b, p. 596). In constructing the models of aircrews and controllers, we have found is that much of their expertise, rather than residing in complex decision-making processes, is found in the structure of the network of procedures itself where numerous simple action selections are made, guided by the information flow through the network following the publish-subscribe protocol. In fact, automatized or skill-based cognitive capabilities are essential for coping with the very high workloads that aircrews and controllers frequently encounter. At the same time, the automatized behaviors can be a source of habit-capture error (Reason, 1990; Norman, 1988)—the immediate situation, while closely related to many encountered in the past, may in this instance have required an alternate response.

Putting the elements of the model together, we have goals and their sub-goals and plans that establish the structure for schemata instantiated as a network whose configuration adapts in relation to the operator's proactive agenda and impinging events. The dynamic network represents the things that a person knows how to do: basic person skills (for example, coordinated hand-eye actions to set a flight deck selector) and domain specific skills (for example, the captain making the decision to land). Active goals represent the operator's proactive agenda for managing his or her tasks. These top-level goals typically activate a series of sub-goals and procedures. The procedures include decision points to address variations that arise in the immediate situation.

As the schemata play out, an operator typically faces a series of anticipated, and sometimes unanticipated, events leading to necessary adjustments in the selection of actions to maintain the operator's proactive agenda. Impinging events lead to decisions that determine the particular execution path through the sub-goal and procedure network. As the procedures execute, working memory elements, implemented as attributes of the procedures, acquire particular values. The executed form of the goals and procedures act as a responsive schema with the capability to adjust to the situation at hand.

The network of procedures and the publish-subscribe protocol implement a *pattern matcher* that adapts in response to the impinging situation over time. The models perceive and interpret a sequence of events in the world. The interpretation leads to adjustments in the operator's selected actions. It is the publish-subscribe protocol that moves the products of the observations and interpretations from procedure to procedure—from procedures supporting perception to a series of cognitive procedures supporting interpretation, and finally to those motor procedures governing the operator's actions. The newly activated procedures are the ones prepared to address the particular content generated by the previous procedure's execution. In its pattern matching, the model is not necessarily comparing alternative plans, but rather more often simply making the necessary adjustments in ongoing procedure execution required by the particular situation via the activation of the appropriate procedures. As such, the procedure network with its enabling publish-subscribe protocol behaves very much like what has been termed recognition-primed decision making (Klein, 1998). While the problems that are hard for experts may require thoughtful decision making, again mindful of Logan (1988a, 1988b), we suggest that much of an operator's expertise is performed automatically requiring minimal conscious, thoughtful intervention.

In building aircrew and controller models (Deutsch and Pew, 2004, 2002; Deutsch, 1998; Deutsch and Adams, 1995), much of the work lies in capturing the procedures that the aircrew and controllers execute—domain-specific knowledge in procedural form. The domain-specific procedures are supported by a layer of domain-independent capabilities (for example, coordinated hand-eye actions, managing verbal communication) to facilitate the skilled tasks of monitoring flight deck instruments, operating flight deck controls, managing the more complex tasks of handling flight deck and ATC communications, and the interleaving of multiple tasks as situations demand.

For the research effort described here, existing aircrew and air traffic controller procedures were augmented to address an approach and landing scenario that included the weather cell threat at the threshold of the runway. The procedure networks, representing skilled aircrew behaviors, played out in often complex ways depending on the particular timing and sequence of situations presented in the scenarios. The models thus provide a representation of the capabilities of skilled aircrews. As the aircrew's heavily cross-checked procedures played out, the models generally produced error-free expert behaviors. To explain how accidents such as that described in the NTSB (1995) report came about, we then conducted a detailed examination of the trace of particular expert behaviors as represented in the models and thereby sought to better understand how aircrew errors may have emerged to become a factor contributing to the accident.

Interpreting the Captain's Missed Approach Plan

In aircrew decision making, as in decision making in general, the decision process may be simple or complex; the decisions themselves may be automatic or require much thoughtful effort. A decision can be complex because there are many interacting factors that go into it—numerous cues to be evaluated, further complicated by cue ambiguity (Orasanu and Davison, 2001). In time-pressured situations where an incorrect decision can lead to an incident or accident, risk may be difficult to accurately quantify. There may be conflicts among immediate goals and the consequences of particular actions may be difficult to accurately assess. At the same time, many, even most aircrew decisions are highly practiced—the product of highly structured and often repeated procedures in familiar situations. Extensive experience in often-repeated situations can provide immediate solutions to complex problems with little or no need for conscious, thoughtful processing. And yet, experts with considerable experience can be put in situations where their experience is more limited—situations in which they are not expert decision makers.

The captain's plan—"if we have to bail out…it looks like we bail out to the right"—was the most difficult statement to rationalize in the USAir 1016 cockpit voice recording (CVR; NTBS, 1995). Although there had been no prior CVR mention of a windshear threat, the weather cell at the threshold of the runway was clearly visible through much of the approach. Given the threat presented by the weather cell, the captain might reasonably have been expected to brief a microburst escape maneuver. Eighteen seconds after stating his plan, the captain noted the "chance of shear," yet he did not revise his plan. Even as the captain and first officer explicitly commented on the increasing airspeed, a definitive cue pointing to a microburst encounter, the captain did not modify the plan. That turbulence and airspeed were actively being tracked strongly suggests that the captain was concerned with the microburst cues that had been important elements in his USAir training experience. Having encountered a limitation in the model-based exploration of the accident scenario—the inability to reproduce the briefing in the model—we turned to a theory-based analysis as a means to understand the reported behaviors.

Rethinking Memory

Memory, or more specifically a new perspective on what memory is fundamentally about, is one starting point for the theory grounding the model architecture. The idea is a simple one: Memory's primary function is to retain key elements of today's experience so that we will be able to use that experience to perform more effectively tomorrow (Glenberg, 1997). Rather than the prominence given to declarative knowledge, that is, *what we know*, the focus is on remembrance for *what we know how to do*: procedural memory. And indeed, most of the memory in our models is in the form of goals and procedures—what the model knows how to do.

In the *modeled* Charlotte scenario, one of the low-level background activities that both the captain and the first officer execute is a scan that is sensitive to the presence of weather cells. The sighting of the weather cell first triggers a brief discussion between the captain and first officer to note the cell's presence. It also later leads to the tracking of the several cues needed to assess the threat that the cell represents as they proceed on their approach to the runway—actions also taken by the aircrew as mentioned in the CVR (NTSB, 1995). The focus is on the steps the crew executes (drawn from long-term memory) in response to the cues and the supporting knowledge needed to select and execute those actions.

However, even the mundane things that people do each day have slight twists and variations on how they were done previously. As people make their way through each day, they are always making adjustments and refinements necessary to address particular aspects of the immediate situation. In conducting their activities, they are also making slight additions, adjustments, and refinements in their *remembrance of what they know how to do*. We suggest that the conscious representation of the goals and procedures that people execute as they go about their daily business are first-person *narratives*. By adjusting the narratives, people adapt their actions to particular situations. The underlying representation of what people know how to do, consciously interpreted and manipulated as narratives, is thus constantly subject to real-world experience. The refined narratives serve as a critique of the underlying representation that extend and improve that skill based on each new experience.

Memory's ongoing function is the refinement and consolidation of the underlying representation of what we know how to do that helps us better guide future actions—to act more effectively tomorrow based on today's experience (Glenberg, 1997). We claim that this is memory's primary function and most important accomplishment. As we will see, it is the role of narratives in action selection that suggest the process leading to the captain's briefing for addressing the threat posed by the weather cell. Two of memory's more familiar functional capabilities devolve from this nucleus. The first of these, captured by the conventional view of memory, is our notably imperfect remembrance of what happened in the past. The second further capability is to thoughtfully make use of a remembrance to explore what one might do in the future and gauge the potential success of possible actions. The ability to introspect on what we know how to do, in the form of narratives, may be used to reflect on and refine the actions that we then take in a given situation. Remembrance may lead to immediate action selection (automaticity), or it may be used as the basis for thoughtful deliberation about future actions.

In building our models based on this perspective on memory, we have redefined the traditional balance in the roles of declarative and procedural memory. Virtually all human performance modeling frameworks grant equal prominence to distinctly separate declarative and procedural memory resources (for example, Anderson and Lebière, 1998; Corker and Smith, 1993; Laird, Newell, and Rosenbloom, 1987; Meyer and Kieras, 1997). Declarative memory items are selectively retrieved from a separate store to support the usually rule-based selection and retrieval of operators, essentially procedures, which then execute more or less successfully. In our models, procedural memory, what we know how to do, has precedence. Declarative memory exists as attributes of the procedures, rather than existing in a self-contained declarative memory box, where each declarative memory element is semantically linked to the context in which it is acquired and used.

Automaticity in Expert Decision Making

The theory that grounds the model also places a strong emphasis on the role of automaticity in expertise. Well-practiced skills are readily and transparently initiated (Logan, 1988a, 1988b; Bargh and Chartrand, 1999), even when a complex combination of cues must be recognized to identify the correct course of action. The caveat in this instance is that most pilots have had limited experience in encounters with a microburst. The USAir 1016 captain and first officer had *simulator-based* experience in identifying and escaping a microburst situation, but we do not know from the NTSB report whether or not either of them had ever encountered a microburst before the July 1994 approach to Charlotte 18 Right. Given their limited simulator-based experience, the USAir 1016 pilots were unlikely to have been exercising well-practiced skills. The planning for the maneuver should the landing need to be aborted was unlikely to have been grounded in often practiced experience. The first fallback in formulating a plan was the captain's training experience.

The automaticity-action link in the model that has helped us to better understand the incidence of error in other real-world situations (Deutsch and Pew, 2002) does not adequately address the formulation of the briefing. We are concerned with the difficult juncture at which the highly skilled operator is making decisions at the edges of his well-practiced expertise. The problem at hand is one for which the captain has had limited experience. And yet there was a special urgency to the situation due to the high risk to the safety of flight posed by the microburst threat at the threshold of the runway. The situation and his training demanded preparation for a microburst escape maneuver, yet while both crew members later verbally noted microburst cues (NTSB, 1995), the captain did not call for a microburst escape in his briefing. With the data at hand, we believe that working from a theory-based perspective it is possible to suggest an explanation for the captain not selecting the required plan. By extending the theory, we were able to suggest the mechanism through which the captain's plan might have been constructed.

Characterizing Thoughtful Processes in Decision Making

The plan as generated by the captain may have been produced thoughtfully or automatically. We can not assume that the plan was thoughtfully derived, but since the captain verbally communicated the plan to the first officer, it was available for thoughtful review by both parties. There are two prominent theories (Sloman, 1996), each claiming to be the foundation for such thoughtful processes. The first is that we as people ground our thought in *rules*: "if there is a windshear threat, then brief a microburst escape maneuver." The second is that we establish and reason over *mental models* to create our plans like the one briefed by the captain.

Rules such as the one just quoted, while they might potentially explain the expected call for a microburst escape maneuver, do little to suggest how the captain's alternate plan was constructed in all of its detail. The ideas that we outline are more closely aligned with mental models, but models that are active with a perception-driven component. In our interpretation, a mental model is not as a purely objective, static representation that would then require a separate process for reasoning over that representation. A mental model, as we speak of it, is *narrative* in form—rather than objective and action-independent, it is egocentric and action-oriented (Clark, 1997). In being action-oriented, it is readily represented in one or more D-OMAR procedures. It is an evolving model that adapts in response to the multi-sensory interactions with a structured environment that drives decision making. We will provide further detail on this interpretation of mental models as first-person narratives as we discuss the processes leading to the captain's briefing for addressing the threat posed by the weather cell.

Understanding the Captain's Briefing: Narratives and their Role in Action Selection

In the *model*, the captain having viewed the weather cell on the approach path briefed a microburst escape maneuver. Based on the USAir training, this is just what he would have been expected to do. Given that the model did not predict the real-world captain's briefing, we were faced with the problem of how the briefing as given was constructed with its particular content.

The problem that we as modelers were confronted with usually comes in a slightly different form: A working model exists with capabilities to successfully address a range of situations, but within a variation of one of those situations, the cues are slightly different such that the behavior the model produces fails to be human-like in some sense. As humans in the same situation, we might readily detect the variance in the cues and adapt an almost correct retrieved response to meet the requirements of the particular situation.

The problem here was a variation on adapting to slightly different cues in an otherwise familiar situation. During the approach, the cues were consistent with past training experiences, but the captain in interpreting the cues appeared to feel the need to amend the trained response. A variation on a standard response to the situation was determined to be required. One potential source for that response was the set of *narratives* that described responses to closely related situations. In such circumstances, a narrative may be selected and slightly amended or two or more narratives may contribute segments that are interleaved in a new way to provide the response. This is a process leading to what Orasanu and Fischer (1997; Orasanu and Martin, 1998) have termed a *creative* problem solution. Rather than adopting an alternate automatized or otherwise straightforward procedure, multiple procedural components were composed in a *unique* manner to determine the response to the situation. In this instance, it was unfortunate that "creativity" was employed where a straightforward and more appropriate solution was available.

For the particular USAir 1016 captain's briefing, potential contributing narratives included: (1) a microburst escape maneuver; (2) a standard missed approach procedure, and (3) an en route maneuver to skirt a weather cell. The situation the captain appeared to be planning for was the possibility that they would act to avoid the weather cell; he planned to proceed with a standard missed approach, but he would ignore the approach plate directive that they climb to 3600 feet and instead immediately divert to the right to literally go around the cell, following the pattern of circumventing the en route cells as he had done earlier in the approach. Elements of each of the three related first-person narratives with the noted adjustments readily converge through a subjective process to produce "if we have to bail out...it looks like we bail out to the right." They will execute a standard missed approach; they will avoid the cell by going around to the right as they had successfully done to avoid weather cells earlier in their approach; and they will monitor the windshear cues as taught in the microburst escape training. This last element of the plan, while not explicitly briefed, is evidenced by the aircrew's explicit processing and discussion (NTSB, 1995) of precisely the turbulence and airspeed cues as taught in the USAir windshear training course. The cues processed by the crew that led the captain to abort the landing were drawn from the third narrative, the microburst escape narrative. Narrative synthesis suggests the process by which the plan with each of its details could have been generated.

Contention among alternate forms for the briefing suggests an approach to understanding the choice of the actual briefing. Based on Logan's (1988b) Theory of Automaticity, we suggest that there might have been several contenders to potentially drive the form of the captain's briefing. Skill-based or rule-based decision making each might have selected a missed approach or a microburst escape. It appears that the captain did not have sufficient real-world experience in microburst situations to lead to a strong automatic link from the cues at hand to the generation and briefing of the expected microburst escape plan. In like manner, the simulator-based training might not have established a strong link to the expected microburst escape plan. As another contender, we have argued that the briefing as actually delivered was the product of narrative synthesis, that is, a contending plan that did not simply align with either standard procedure. Whether as a conscious or non-conscious process, the cues presented in the evolving situation led to the construction of a compelling narrative and it was that narrative that dominated the process and drove the form of the actual plan.

We characterize the plan as "creative" because it adapted a combination of elements from multiple related standard procedures in a unique manner. Given the proximity of the aircraft to the ground, the need to abort the landing was of primary importance and was clearly present in the plan as discussed. "Take it around" serves as a variation on "go around," a vernacular for executing a missed approach procedure. It is a "go around" in the specific sense of aborting the landing and circling back to reenter the approach pattern rather than going around an obstacle such as a weather cell.

The standard procedure for a missed approach is spelled out on the approach plate for the targeted runway that should have been reviewed by the crew prior to the approach—the CVR (NTSB, 1995) did not include evidence of the review. The approach plate calls for a "Climb to 3600' outbound" explicitly denying a turn left or right. In calling for a turn to the right, the captain was amending the standard missed-approach procedure. We have an instance in which recent related decisions and actions can influence the selection of the plan for future actions (Holyoak and Simon, 1999). Successful decisions and actions taken with respect to circumventing the weather cells encountered earlier on the approach path suggest themselves as an influence in amending the planning process to avoiding the weather cell at the threshold to Charlotte 18 Right. Recent outcomes (the ready avoidance of the weather cells earlier in the approach) could have played a role in establishing the briefing decision to amend the missed approach to include the turn to the right.

The standard criteria for electing a missed approach are the failure to visually acquire the runway or the failure to stabilize the aircraft on the approach. The criteria used by the captain in electing to abort the landing were windspeed cues directly associated with a microburst as taught in the windshear training exercises. Rather than being simply the selection of the incorrect procedure, the plan was creative in the sense that it entailed an impromptu combination of a standard missed approach adapted to attempt to skirt the weather cell with triggering cues for a microburst escape procedure. The plan as briefed was accepted by the first officer as the pilot-flying—a point at which the plan might have been questioned and revised.

Three minutes later as they continued their approach, the aircrew observed and commented on the sequence of increased airspeed cues affirming an encounter with a microburst. The captain directed the first officer as pilot-flying to initiate the plan as called for in the briefing. It was almost four seconds after initiating the planned maneuver when the plan was explicitly revised, when the captain noting further increased airspeed cues called for "max power"—a key element in a microburst escape maneuver. The plan as briefed was not working and unfortunately, the belated call for a microburst escape maneuver did not lead to a successful outcome.

Discussion: Adaptive Behavior through Narrative Synthesis

Orasanu and Martin (1998, p. 101) described *creative* action selection as the response to "situations in which no suitable options are readily available and the decision maker must invent at least one to meet the demands of the situation." In the Charlotte windshear threat situation, there was a straightforward plan for a microburst escape to be put in place, yet a creative action selection process was employed. Had the error been grounded in a skill- or rule-based decision, the exact execution of an alternate procedure would have been expected, yet the plan and its later execution were found to be a composite that drew on elements from three procedures.

The narrative-based creative action selection process is suggested as central to adaptive behaviors that people regularly exhibit. It is something that we as people do many times a day as we extend what we know how to do by adjusting experience-based action selection to meet the demands of the immediate situation. Our new experience is then consolidated as an extension to what we know how to do. The consolidation need not be in propositional form. It is more likely a skill-level representation for guiding a similar mix of perceptual, cognitive, and motor actions when again encountering similar cues. Our hypothesis is that our experience base— what we know how to do in combination with what we have observed ourselves doing in the past—can then be recovered as first-person narratives derived from the skill-level representation and past experience as retained in autobiographical or episodic memory. Access to that experience cache as narratives, both at the conscious and non-conscious levels, then makes possible adaptive behaviors through narrative synthesis—a process that draws upon one or more narratives and adjusts and merges them to meet the needs of the immediate situation.

Instantiating the narrative perspective on creative action selection in a model presents difficult challenges. The procedure-based representation of skilled behaviors will be an important element in the revised model. The model will need to provide for the consolidation of recent experience that refines the existing procedural representations. When addressing a novel situation, there will need to be an extension of the current process of action selection. Selection methods for procedures with the potential to provide a solution path for the novel situation will need to be defined, the narrative form of the procedures derived, and the process for narrative synthesis developed.

References

Anderson, J.R. and Lebière, C.J. (1998). *The Atomic Components of Thought.* Mahwah, NJ: Lawrence Erlbaum Associates.

Arvind and Culler, D.E. (1983). *Why Dataflow Architectures*, Computational Structures Group Memo 229–1, Laboratory for Computer Science, Massachusetts Institute of Technology.

Bargh, J.A. and Chartrand, T.L. (1999). "The unbearable automaticity of being." *American Psychologist*, 54: 462–79.

Bruner, J. (1996). *The Culture of Education.* Cambridge, MA: Harvard University Press.

Cacciabue, P.C. and Hollnagel, E. (1995). "Simulation of cognition: Applications." In J.M. Hoc, P.C. Cacciabue, and E. Hollnagel (eds), *Expertise and Technology: Cognition and Human–Computer Cooperation.* Mahwah, NJ: Lawrence Erlbaum Associates (pp. 55–73).

Clark, A. (1997). *Being There: Putting Brain, Body and World Together Again.* Cambridge, MA: The MIT Press.

Corker, K.M. and Smith, B.R. (1993). "An architecture and model for cognitive engineering simulation analysis: Application to advanced aviation automation." In *Proceedings of the AIAA Computing in Aerospace 9 Conference.* San Diego, CA (pp. 1079–88).

Deutsch, S.E. (1998). "Interdisciplinary foundations for multiple-task human performance modeling in OMAR." In M.A. Gernsbacher and S.J. Derry (eds), *Proceedings of the 20th Annual Meeting of the Cognitive Science Society.* Mahwah, NJ: Lawrence Erlbaum Associates (pp. 303–308).

Deutsch, S.E. and Adams, M.J. (1995). "The operator model architecture and its psychological framework." In *Preprints of the 6th IFAC/IFIP/IFOR/IEA Symposium on Analysis, Design, and Evaluation of Man-Machine Systems.* Cambridge, MA: The MIT Press (pp. 41–6).

Deutsch, S.E. and Pew, R.W. (2002). "Modeling human error in a real-world teamwork environment." In *Proceedings of the 24th Annual Meeting of the Cognitive Science Society.* Fairfax, VA (pp. 274–9).

—— (2004). "Examining new flight deck technology using human performance modeling." In *Proceedings of the 48th Annual Meeting of the Human Factors and Ergonomics Society.* Santa Monica, CA: HFES (pp. 108–12).

Edelman, G.M. (1987). *Neural Darwinism: The Theory of Neuronal Group Selection.* New York: Basic Books.

—— (1989). *The Remembered Present: A Biological Theory of Consciousness.* New York: Basic Books.

Glenberg, A.M. (1997). "What memory is for." *Behavioral and Brain Sciences*, 20: 1–55.

Holyoak, K.J. and Simon, D. (1999). "Bidirectional reasoning in decision making by constraint satisfaction." *Journal of Experimental Psychology: General*, 128: 3–31.

Klein, G. (1998). *Sources of Power: How People Make Decisions.* Cambridge, MA: The MIT Press.

Klein, G., Ross, K.G., Moon, B.M., Klein, D.E., Hoffman, R.R., and Hollnagel, E. (2003). "Macrocognition." *IEEE Intelligent Systems*, 18(3): 81–5.

Klinect, J.R., Wilhelm, J.A., and Helmreich, R.L. (1999). "Threat and error management: Data from line operations safety audits." In *Proceedings of the 10th International Symposium on Aviation Psychology*. Columbus, OH: The Ohio State University (pp. 683–8).

Laird, J.E., Newell, A., and Rosenbloom, P.S. (1987). "SOAR: An architecture for general intelligence." *Artificial Intelligence*, 33: 1–64.

Logan, G.D. (1988a). "Automaticity, resources, and memory: Theoretical controversies and practical implications." *Human Factors*, 30: 583–98.

——(1988b). "Toward an instance theory of automatization." *Psychological Review*, 95: 492–527.

Meyer, D.E. and Kieras, D.E. (1997). "A computational theory of executive cognitive processes and multiple-task performance: Part 1. Basic mechanisms." *Psychological Review*, 104: 3–65.

National Transportation Safety Board (1995), *Aircraft Accident Report: Flight into Terrain during Missed Approach, USAir Flight 1016, DC–9–31, N954VJ, Charlotte/Douglas International Airport, Charlotte, North Carolina, 2 July 1994* (NTSB/AAR–95/03). Springfield, VA: National Technical Information Service.

Norman, D.A. (1988). *The Psychology of Everyday Things*. New York: Basic Books.

Orasanu, J. and Davison, J. (2001). "The role of risk assessment in flight safety: Strategies for enhancing pilot decision making." In *Proceedings of the 4th International Workshop on Human Error, Safety, and System Development*. Linköping, Sweden: Linköping University (pp. 83–94).

Orasanu, J. and Fischer, U. (1997). "Finding decisions in natural environments: The view from the cockpit." In C.E. Zsambok and G. Klein (eds), *Naturalistic Decision Making*. Mahwah, NJ: Lawrence Erlbaum Associates (pp. 343–58).

Orasanu, J. and Martin, L. (1998). "Errors in aviation decision making: A factor in accidents and incidents." In *Proceedings of the Workshop on Human Error, Safety, and Systems Development*. Seattle, WA: Battelle (pp. 100–107).

Orasanu, J., Martin, L. and Davison, J. (2001). "Context and contextual factors in aviation accidents: Decision errors." In A. Klein and E. Salas (eds), *Linking Expertise and Naturalistic Decision Making*. Mahwah, NJ: Lawrence Erlbaum Associates (pp. 209–25).

Reason, J. (1990). *Human Error*. Cambridge: Cambridge University Press.

Roth, E.M., Patterson, E.S., and Mumaw, R.J. (2002). "Cognitive engineering: Issues in user-centered system design." In J.J. Marciniak (ed.), *Encyclopedia of Software Engineering*, 2nd edn. New York, NY: John Wiley and Sons (pp. 163–79).

Sloman, S.A. (1996). "The empirical case for two systems of reasoning." *Psychological Bulletin*, 119: 3–22.

Chapter 16

Cognitive Engineering Based on Expert Skill: Notes on Success and Surprises

James J. Staszewski

The literature on scientific discovery views surprise as a significant experience, because surprises frequently foreshadow important advances (Simon, 1989). This paper describes the numerous surprises encountered in two projects that applied an approach called Cognitive Engineering Based on Expert Skill (CEBES) to the problem of landmine detection. CEBES is a form of cognitive engineering that uses models of human expertise as its knowledge base. In this case, models of expert operation of two handheld landmine detection systems served as foundations for designing training programs intended to develop the capabilities of novice operators.

Both projects succeeded beyond any reasonable predictions that could have been generated at their outset. They created entirely novel programs for training operators of the handheld landmine detection equipment used since the early 1990s by the US Army, the Army/Navy Portable Special Search–12 (PSS-12) and the technologically advanced, dual-sensor Army/Navy Portable Special Search–14 (PSS-14). The latter has been successfully deployed recently, and will gradually replace the PSS-12. Both systems are shown in Figures 16.1 and 16.2. Results from multiple tests of each of the new training programs showed that they produced manifold improvements in detection rates against the most difficult-to-find targets, landmines with plastic bodies and the smallest possible metal components in their firing mechanisms (Staszewski and Davison, 2000; Staszewski, 2004). These results had an impact beyond the research community; the US Army has adopted both training programs and uses them to prepare troops for countermine operations. US soldiers and marines trained in the CEBES-based programs now apply the techniques and thinking they teach in countermine operations. Together, these outcomes have demonstrated the practical utility of CEBES.

More than a few pleasant surprises occurred on the path to these outcomes beyond the unforeseen large-scale institutional adoptions. These included the counterintuitive way in which the expert operators used the information their equipment provided, the ease with which the models of expertise were translated into the training designs, and the ease with which instruction was delivered. Most surprising were the magnitude and robustness of the effects each training program produced.

**Figure 16.1   The AN/PSS-12 remains as US Army standard equipment for
landmine detection**

Note: The AN/PSS-12, manufactured by Scheibel Corp., uses electromagnetic induction to
sense the metallic content of landmines.

Background

Landmines are a major military threat (LaMoe and Read, 2002; Hambric and
Schneck, 1996), and they also pose a major long-term humanitarian threat to the
indigenous peoples of lands where these long-lived weapons have been emplaced
(MacDonald et al., 2003). Detection of individual mines, a prerequisite for disposal,
is performed typically by individuals operating handheld equipment working in
close proximity to hidden ordnance. The development and deployment of relatively
inexpensive landmines containing minimal amounts of metal (designated LM)
in the 1990s escalated the detection problem. This development dramatically
increased the difficulty of detecting mines with current equipment, which, until
the deployment of the PSS-14, relied exclusively on electromagnetic induction
(EMI) sensors for detecting their metal parts (MacDonald et al., 2003). A test of
the PSS-12, which uses EMI sensing, shortly before it was fielded showed that only
3.8 percent of M14 targets—very small, low-metal anti-personnel mines—were
detected (US Army Test and Evaluation Command, 1991). This finding motivated
development of the PSS-14.

**Figure 16.2   The PSS-14 used by a US Army combat engineer in a
countermine operation**

Project Motivation

Landmine detection presented an unfamiliar, but fascinating area to investigate human
expertise and its implications for instruction. Three interrelated factors motivated
study of PSS-12 expertise, which led to investigation of PSS-14 expertise. First, it
offered an opportunity to examine expertise in a new domain and thus explore the
generality of current theory and data. Second, laboratory studies suggested that an
understanding of expert performance in a given domain offered a valuable resource
for developing the skills of others through instruction (Biederman and Shiffrar,
1987; Chase and Ericsson, 1982; Staszewski, 1988, 1990; Wenger and Payne, 1995).
If a sound model of expert skill could be developed, this project enabled testing the
instructional utility hypothesis prospectively. Third, and most importantly, successful
execution could make a contribution of considerable practical importance.

Analysis of PSS-12 Expertise

An operator with extensive experience and an impressive record of success (RR)
volunteered to support the PSS-12 study (Staszewski, 1999; Webster, 1996). Data
collection involved capturing multimodal records of RR's activities while his skills
were tested against a representative sample of landmines buried at an Army test

facility.[1] Observations included continuous viewpoint video recording of the PSS-12's search head as RR visually tracked its movements during testing, along with two synchronous and independent channels of audio recording. One recorded all output signals of the PSS-12. The other channel recorded all statements made by RR as he gave concurrent verbal reports, as per instructions, as he applied his routine techniques to find mines and mark their suspected locations.

Analysis involved examining the audio/video record and coding 204 events identified in roughly 30 hours of recording. Events included sequences that ended in successful detections, those related to missed targets, those that produced false alarms, and either aborted investigations or those resulting in correct rejections. Coding focused on movements of the detector head, response of the PSS-12, and the locations of both relative to the positions of mine targets.

Findings

Several procedural regularities appeared. First, analysis revealed that what, on initial observation, appeared complex and confusing was extremely orderly. Except for procedures RR used to check for drift in the equipment's sensitivity, the model shown in Figure 16.3 accounted almost completely for his activities within the test lanes.

RR's successful detection of mines involved three sequential stages illustrated as macro-processes in Figure 16.3: search, investigation, and decision. Search involved placing the sensor head on the ground with its center on or outside the lane[2] boundary and gliding it lightly over the ground surface at a rate of about 1 ft per sec in a path across the lane. If no signal came from the PSS-12, RR continued to glide the sensor head on a cross-lane trajectory until its center was either on or outside the lane boundary. At this point, the search head would be raised just enough to advance it about 15 cm—or about the radius of the search head—and another lane "sweep" would proceed across the lane in the opposite direction. With a trajectory parallel to the previous cross-lane sweep, the search head's path would overlap the area covered by the previous sweep by roughly half. The overlap ensured exhaustive surface coverage with the sensor head, if the sweep was executed properly. Such cross-lane sweeps in alternating directions would continue until either the detector produced an output signal or RR paused to perform a sensitivity setting check.

When an output sounded, RR typically would pause and gaze at the location of the sensor head. He would then complete his cross-lane sweep and move the sensor head back over the area where the alert occurred. With movements shorter and slower than used in the cross-lane search, he would try to reproduce the output in the same spot. If the signal could not be reproduced in the same location after several attempts, search forward would resume with successive cross-lane sweeps.

---

1   Details on targets used in testing and the testing environment can be found in Staszewski (1999) and Davison and Staszewski (2002).

2   Areas designated for clearance in testing and in live operations are referred to as lanes. Lanes are roughly the width needed to extract casualties. Army doctrine sets lane width at 1.5 m. This lane width is used at all training and testing sites described here.

Model of PSS-12 Expert's Procedures

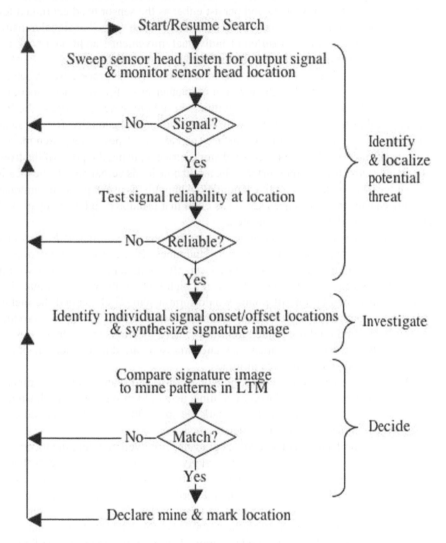

Figure 16.3   Model of the PSS-12 expert's detection procedures showing his basic macroprocesses and their relations

If a signal reliably sounded in the same location, search head movements marking the beginning of investigation would follow, immediately adjacent to the alert's location.

In the investigation phase, RR would proceed to explore the area near the alerting signal with sensor head movements like those used to confirm the alerting signal's reliability. The movements radiated progressively outward from the alerting signal's location. Movement on specific trajectories would stop when an ongoing signal stopped and remained off for a few more inches of the sensor head's movement. A backtracking movement would occur on the same trajectory to make the signal recur at the same point as previously and persist either as the sensor head continued for a few inches or remained stationary. If the locations of such "edge" or signal transition points appeared to move on different individual movements, additional back-and-forth movements were used to establish the edge points reliably.

Once an on–off transition point was located reliably, the process was repeated a few inches away to establish the location of another on–off point. These operations continued until they defined a region within whose boundaries or "edges" the PSS-12 sang and outside of which it produced no output. Occasionally, RR physically marked edge points on the ground surface (on bare soil lanes, where such marking was possible, versus lanes with crushed stone surfaces) using the point of the trowel he carried.[3] On crushed stone surfaces, he might note in his verbal reports the location of an edge if he found a distinctive landmark. Although his use of physical landmarks was infrequent, these instances served as clues that prompted spatial analysis of the auditory signals the PSS-12 generated.

The area covered with such investigatory actions varied, first as a function of whether the signals were coming from landmines. If the investigation occurred where a low-metal LM mine was buried, on–off points tended to cluster within a foot or less of one another. Alternatively, if a high-metal mine was investigated, the area circumscribed by on–off points would form a semi-circle, called the metallic (MD) halo, sometimes more than a yard in diameter. If the signals were produced by conductive material not related to mines (whose locations were obtained for the analyses), that is, clutter, the areas they encompassed varied much more in size and shape than the spatial signatures of mines.

When RR's investigation ended, he would pause and scan the area examined. If he decided the halo was produced by a mine, he would move the sensor head to its center and direct the experimenter to mark the spot. Otherwise, he resumed cross-lane sweeps searching for another alert.

The combined audio-video analysis of events involved mapping the locations at which signal onsets or offsets occurred relative to the locations of targets that RR accurately and confidently marked, a non-intuitive regularity emerged. The contours created by onset-offset or "edge" points described semi-circular halos with mines at the center, were the arcs completed to form circles. The length of the radii for these arcs covaried positively with the metallic content of the target. Although any single edge point could be obscured and moved outward by the presence of metallic clutter, in the absence of clutter the signatures showed "good" form as described by Garner (1974). Some examples of the spatial patterns produced by RR are shown in Figure 16.4.

---

3 RR used the trowel for checking his equipment for drift from the set sensitivity in the context of testing, as described in Staszewski (1999). He also used this tool to excavate suspected mines in live operations, thus providing timely and valuable feedback for detection and discrimination learning.

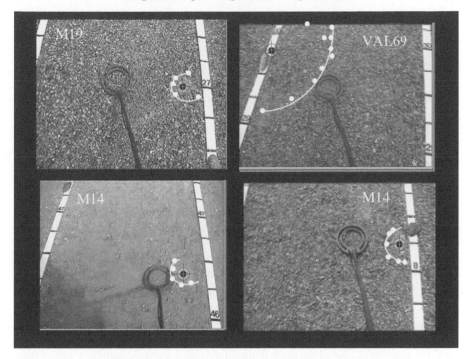

Figure 16.4   Spatial patterns inferred from analyses of the PSS-12 expert's successful detection of mines in performance testing

Crosshair marks indicate the approximate centers of the buried targets; solid white circles indicate the "edge" locations defined by either onset-offset or offset-onset of the PSS-12's auditory output signal; the white arcs shown are fitted by hand to the edge points to show pattern contours. Note the difference between the pattern sizes produced by the Valmara 69, an anti-personnel mine with high metallic content, versus the patterns of the M14, a low-metal, anti-personnel mine, and the M19, a low-metal, anti-tank mine. These mines are illustrated in Figure 16.6.

These regularities in RR's successful detection episodes led to the counterintuitive inference that he located mines by serially finding landmarks using the detector's auditory outputs, holding them in memory, and linking them to form spatial contours in his mind's eye. This process resembled that of visual pattern synthesis (Finke, 1990; Klatzky and Thompson, 1975; Palmer, 1977; Pearson and Logie, 2000; Thompson and Klatzky, 1978). RR presumably then matched these patterns, or parts thereof, against signatures held in his knowledge base of previously detected mines. Thus, spatial pattern recognition appeared to support expertise in landmine detection, consistent with findings from laboratory studies of expertise in a variety of visuo-spatial task-domains (Chase, 1986).

Analysis of PSS-14 Expertise

The PSS-12 expertise project led to the analysis of expertise on the newer piece of detection equipment. While the PSS-12 effort was underway, the intended successor of the PSS-12, the Handheld Standoff Mine Detection System (HSTAMIDS) prototype (later renamed the PSS-14) was undergoing operational testing. After nine years and $38 million had been invested in its development, the results, shown in Figure 16.14, revealed substandard performance and cast doubt upon the system's capability. The government project sponsor therefore convened a "Red Team" to review the development program and the results it produced. The Red Team included the author, due to the relevance of the PSS-12 effort. The review identified the training operators received as a contributing factor to the system's performance (Guckert, 2000). It also identified a single high-performing operator (KJ). As part of the "second chance" development effort that followed, the CEBES approach was applied to develop training for operators.

The prototype HSTAMIDS system developed by CyTerra Corporation and illustrated in Figure 16.5 consists of two main physical components. One is its search head, which houses the system's EMI and ground-penetrating radar (GPR) antennae in a plastic case. The head is attached to an adjustable-length wand that has a set of controls at its opposite end. Here, on a panel just above the pistol grip used by the operator to hold the unit, are switches that power the system up or down, adjust sensitivity levels of the GPR subsystem, and adjust the volume of the unit's auditory outputs. Cabling runs along the shaft, linking the sensors to the controls. Another cable connects the wand's components to a rectangular metal-cased unit weighing about 35 lbs and worn like a backpack by the operator. This unit houses electronic circuitry for the two independent sensor channels that process the raw signals taken in by each sensor, using separate detection algorithms, and generating auditory outputs based on the results.

An operator controls movements of the search by grasping a pistol grip attached to the wand and cinching his forearm to the wand, thus making it an extension of his arm. This arrangement enables the operator to move the search head laterally above the ground about a meter forward of his feet, providing a buffer from threats ahead.

Because the HSTAMIDS, like the PSS-12, used EMI sensing, the expert's technique was expected to resemble RR's. Learning how information afforded by GPR sensing was acquired and used would be a matter of exploration, but at least the preceding study of PSS-12 expertise provided some guidance in modeling the HSTAMIDS expert's skill.

The benefits of this prospective approach to analysis of HSTAMIDS expertise were magnified by factors limiting data collection opportunities. First, with only one prototype system available, observations of the expert's use of the system were limited to test sessions interleaved between lengthy intervals of hardware and software development. The development included continuous changes to detection algorithms, with system software undergoing modifications up to the month before the first training test. Some enhancements proved less successful than intended. The system changes and the less-than-optimal reliability, not unexpected with any complex technology under development, reduced observation time and complicated interpretation of the observations made.

Figure 16.5   E-PDRR HSTAMIDS

Note: The HSTAMIDS prototype shown in the configuration used for training development and testing in its extended product development risk reduction or "second-chance" phase.

Instrumentation used for data collection added further challenges to the analysis. A computer-controlled camera system being developed to give trainees feedback on sweep coverage (Herman and Inglesias, 1999; Herman, McMahill, and Kantor, 2000) was also used to capture the expert's activities. Reliability issues limited the corpus of recorded observations available for off-line analysis. Thus, analysis of the expert's skill relied to a worrisome degree on direct, real-time observations of system operation as the empirical basis for abstracting regularities. These observations covered an estimated 90 to 120 hours of testing spread over five developmental test sessions at three different test locations. Roughly 850 encounters with mine targets like those shown in Figure 16.6 were observed, but a much smaller, unsystematic sample of these observations, totaling roughly ten hours of audio/video recording, was captured and available for detailed examination. These records were used to test tentative findings about the expert's techniques and thought processes based on the live observations and hand-recorded notations.

Landmines

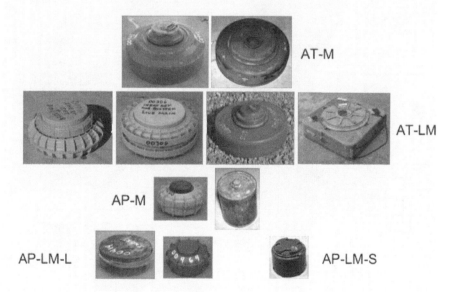

Figure 16.6   Mine targets used in testing

Notes: Top row: left to right, shows anti-tank mines with metal bodies and firing mechanisms, the TM62M and M15. Second row: low metallic, anti-tank mines, VS1.6, VS2.2, TM62P3, and M16. Third row: anti-personnel mines with high metallic content, VS-50, Valmara 69. Bottom row: PMA3, TS-50, M14. The latter are divided into two subgroups on the basis of physical size and are distinguished in results on HSTAMIDS test results. For size reference, the larger circular AT mines are roughly 12 in. in diameter, whereas the M14, which contains but 0.6 grams of metal, is 2.2 in. in diameter.

Findings

Observations of the HSTAMIDS expert's detection procedures showed that the model of RR's procedures generalized surprisingly well.[4] The model shown in Figure 16.7 with its procedural sequence of search, investigation, and decision operations provided an excellent characterization of his detection activities. An alerting signal encountered during search sweep, received as an output from either the MD or GPR subsystem, would trigger investigation in the area where the alert occurred. KJ then typically started investigation with the metal detection system, using it in a way that was functionally similar to RR's procedures for investigating alerts. That is, he used the MD to locate MD edges and build in a piecemeal fashion the contours of the patterns that constitute the metallic halos of mines. His MD investigation was characterized by relatively slow continuous movements of the sensor head (roughly 2–8 cm/sec) in scallop-shaped trajectories producing consistent changes in the frequency of the MD output signal. These changes allowed him to continuously trace the edges of an MD halo, first working from the "six o'clock" position to the "three o'clock" and then back around to the "nine o'clock" position.

He would then use the established MD pattern, particularly if it described the semi-circular pattern produced by mines, along with his knowledge of the sizes and shapes of mines to guide investigation with GPR. If MD investigation produced a large semi-circular contour .75 to 1.5 m at its widest—characteristic of an AT-M or AP-M target—GPR investigation would be conducted inside the metal halo. If the MD halo was small (~ 2.5–8 cm), suggesting a low-metal target, investigation would start outside the halo, especially if the alerting signal came via GPR followed by an MD signal as detector head sweeps advanced beyond the GPR alert. If the alert occurred first via MD, GPR investigation proceeded either inside or over the edges of the MD halo, depending upon its size. GPR investigation involved movements that were typically performed at a speed 1–5 cm per sec faster than head velocities in MD. The sensor head trajectories for GPR investigation involved back-and-forth movements on a path anywhere from 10–20 cm in length running perpendicular to the six o'clock–twelve o'clock axis of the MD signature. For AP targets, the initial trajectories would carry the head back and forth over the three o'clock–nine o'clock axis with GPR outputs starting as the head passed over the edge of the mine and ceasing after the head passed over the far edge. Pulling back to the six o'clock position, KJ applied similar back-and-forth movements in a forward direction, producing GPR outputs as the head passed over the near edge of a mine target and continued toward the three o'clock–nine o'clock axis. Further "GPR short sweeps" that advanced forward would then produce GPR outputs that diminished in frequency (for mines with circular shape) until the sensor head passed over the far edge of the target and GPR outputs ceased. For the larger AT mines, GPR investigation would be performed first at the three o'clock position, then at the nine, then at the six, and finally at the twelve. Essentially, when a mine was present, GPR output signals occurred whenever the search head passed from off the body of the mine to over it,

---

4   RR and his techniques were unknown to KJ until similarities were pointed out to him by the author, at which point this knowledge had no influence on his technique.

Model of HSTAMIDS Expert's Procedures

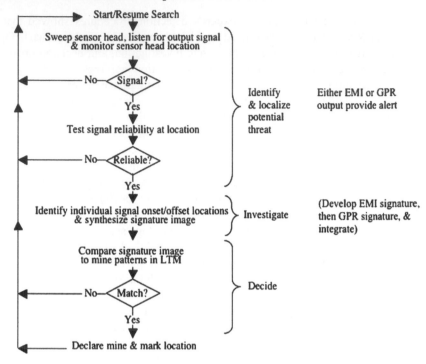

Figure 16.7   Model of the HSTAMIDS expert's basic procedures

Note the similarity to that of the PSS-12 expert model in Figure 16.3.

and continued until the head's sweep carried it off the body. Mapping where GPR outputs occurred made it possible to infer the shape of the buried target. The patterns and the spacing of GPR onset and offset points for the major categories on mines are shown in Figures 16.8–16.11.

Although KJ's decision process, like RR's, was based on pattern matching, the additional information provided by GPR increased its complexity. First, the contours produced by GPR investigation were matched to the expert's familiarity with the various sizes and shapes of mines. This was inferred from statements like "OK, we've got a M19. It's square," or "This looks small, maybe an M14." Comparison of the spatial relations between the MD and GPR patterns added to the representational load. The outcome of such comparisons predicted his "mine/no mine" decisions. This was suggested by a statement KJ commonly made prior to successful mine declarations: "Got MD. Got solid GPR. MD and GPR are correlated." His explanation of "correlated" was that MD arcs and GPR shapes produced by investigating mines shared common axes of symmetry, whose intersections were spatially coincident. Thus, for mines with circular bodies, the MD halo and the GPR outline in his mind's eye would have to share the same center point for him to make a mine declaration. His declaration mark would also be placed precisely at that point.

Figure 16.8   HSTAMIDS sensor output pattern for an M16 (AP-M)

Figure 16.9   HSTAMIDS sensor output pattern for an M14 (AP-LM-S)

Figure 16.10 HSTAMIDS sensor output pattern for an M15 (AT-M)

Completion of site preparations for the initial HSTAMIDS training test provided an opportunity to prospectively test the validity of the inference character of the expert's representation of target patterns. Targets buried for training drills were marked at their centerpoints, but the expert was kept blind to their identities. His task was to locate the MD and GPR pattern edges on designated vectors radiating from each target's center mark for each of 81 targets. Exemplars of the mines in the target set are shown in Figure 16.6. Each edge point was marked and its distance from the center point was measured. Examples of the resulting patterns for specific targets are shown in Figure 16.8–16.11.

Analysis procedures adapted the approach used by Rosch et al. (1976) to identify prototypes of basic-level perceptual categories. Radial distances from the target center to the same edge points of targets within five broad target groupings were averaged. The resulting spatial patterns shown in Figure 16.12 emerged. Despite local, mine-to-mine variability, stable signature prototypes emerged for each category, as seen by the relatively small standard errors associated with each mean distance. These results confirmed inferences drawn about experts' mine patterns derived from observations taken during developmental tests.

To sum up, analyses of expert operators of the PSS-12 and PSS-14 suggest that like experts in other areas, both expert operators rely on recognition of visuo-spatial patterns created and held in memory to locate buried mines successfully. Both employ a search process, that is, search sweep, to identify the general area where

Figure 16.11 HSTAMIDS sensor output pattern for a TM62P3 (AT-LM)

Notes to figures 16.8–16.11: Target centers are marked with crosshairs. White markers indicate edges of metallic halo. Black markers show onset/offset locations of GPR signals. Signal processing algorithms for GPR return signals detect changes in the returned radar signal produced by the differential dielectric constant values of a mine's body and the surrounding soil.

mine patterns are located. The location of the sensor head at the time of an alerting signal marks the general area. The patterns that identify mines are not found whole at these locations. Rather, the experts appear to build them using a mental construction process similar to that described in the literature on mental synthesis (Finke, 1990; Klatzky and Thompson, 1975; Palmer, 1977; Pearson and Logie, 2000; Thompson and Klatzky, 1978). The pattern parts are acquired by iterative investigation routines that serially identify a series of locations held in the mind's eye where each system's auditory outputs define pattern boundary or edge points. Imaginary connection of adjacent edge points constructs contours (for both MD and GPR, in the case of the PSS-14), whose sizes and shapes are matched against those produced by the operators' previous investigations of mines. If the pattern produced by investigation is sufficiently similar to the patterns of previously encountered mines, the operator declares a mine and marks its location.

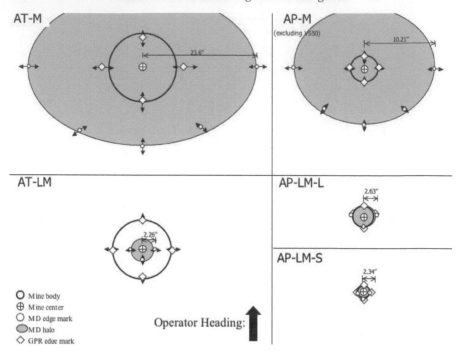

Figure 16.12 Prototypical HSTAMIDS footprints for mine categories

Note: Prototypical HSTAMIDS footprints for mine categories produced by averaging the distances from target center point to edge markers for multiple exemplars of each of five mine categories. Sensor signal onset and offset points (white circles for EMI signals, white diamonds for GPR signals) are illustrated relative to the target mine body (outlined in black with crosshairs marking the target center points). The positions of the edge markers represent the mean distances produced by aggregating the measurements taken from patterns like those illustrated in Figures 16.8–16.11.

Training Based on Expert Skill

Supplied with an understanding of how expert PSS-12 and PSS-14 operators achieved exemplary performance with their equipment, attention turned to designing training for new operators of each system and testing its effectiveness. Four principles guided the design process for each system's training program. They were:

1. Teach what works,
2. Identify how learning will be assessed,
3. Provide practice with feedback, and
4. Minimize working memory load to avoid overloading trainees with information, communicate how procedural steps are organized, including their linkages, or, conditions for executing one action after another.

Implementing the first principle involved using the process models outlined above to provide content and organization for the training. The performance of the experts showed that these procedures worked.

Second, detection rates were identified as the primary performance measure for training assessment over either target-clutter discrimination or area coverage speed. It was reasoned that the latter would both improve with task practice aimed at maximizing detection rates. Moreover, discrimination and coverage speed are related and both can be categorized as measures of efficiency. In live operations, however, a missed mine that resulted in a casualty, would end up costing efficiency and obviously much more. Thus, maximizing detection rates was prioritized.

Third, practice with feedback is essential for skill development. Feedback in these programs involved performance feedback on exercises (for example, detection rate after a search exercise) as well as diagnostic feedback on technique that failed to achieve desired results.

Several means were devised to minimize working memory load in the portions of training that focused on cognitive and associative stages of skill training. The first involved parsing the task of detection into component units and incorporating training units that covered each component. The macro-processes of the models in Figures 16.3 and 16.7 identified the units, which were further subdivided into still smaller units (for example, developing a pattern with the EMI sensor, or finding a single location of a pattern with the EMI sensor—or the GPR sensor, in the case of the PSS-14). Conveying the organization of these units, which involved communicating the organization of the skills to be acquired, exploited chunking to reduce memory load. The hierarchical structure of the experts' skills supported this strategy. For instance, instructors told trainees that successful mine detection involved learning to do just three things, after learning how to properly prepare the detector for operation:

1. Locate alerting signals;
2. Search for patterns where alerting signals occurred;
3. Match the patterns created in a given locations to the patterns they would learn are produced by mines.

Following instruction and demonstration, trainees proceeded to practice each skill component, receiving feedback and any needed remediation from trainers. Thus, trainees first learned component skills before combining them in "whole task" practice exercises. Another approach used to minimize memory load in pattern learning involved using physical marks to locate metal (or GPR, for the PSS-14) pattern edge points. Thus, after all edge points were marked for a particular pattern, a visible pattern of marking points was available to both trainees and trainers.

Communicating skill organization also exploited a robust finding from research modeling problem solving and skilled performance showing that processing formalisms known as productions can organize and control sequences of actions— either mental or behavioral. Productions are conditional if-then rules with two parts. The first part specifies a condition, which describes the state of affairs under which a particular action is appropriate. The second part, the action component describes the activity to be executed when specified features related to either the external state of the world or the problem-solver's knowledge (including current goal) are present.

One example of such a rule is,

> IF <an alerting signal is heard>,
>
> THEN <move the detector's sensor head back to where it was when the signal was heard, to reproduce it in the same place>.

Another, related to investigation with the metal sensor of both systems is,

> IF <the edge of the suspected target's metal halo (pattern) has been located at the six-o'clock position>,
>
> THEN <repeat the edge location operations to find the halo edge at the three o'clock position to locate the edge at that location>.

Instructions were delivered using this format where appropriate, which served to link the perceptual outcomes of trainees' actions to the action that needed to be taken next. The conditions described in such statements also served as information for either assessment by trainers or self-assessment of performance by trainees.

A comprehensive description of the training, even for just one detector, is beyond the scope of this chapter, but it should be noted that implementing the above principles in a context in which multiple operators could be training simultaneously required a novel design for the physical training environment, as well as creating new training aids for performance assessment. Details about the training programs can be found in (Davison and Staszewski, 2002; Headquarters, Department of the Army, 2002, 2006; Staszewski and Davison, 2000).

Training Tests

Tests of the two training programs produced multiple surprises. The most important was the strength of the training effects obtained in initial tests of prototype PSS-12 and PSS-14 training programs. In the initial test of PSS-12 training (see Staszewski and Davison, 2000) conducted at the US Army Engineer School at Ft Leonard Wood, MO, combat engineers who had just completed advanced individual training participated as trainees. All had received standard training on use of the PSS-12. After randomly assigning soldiers to treatment (N=10) and control groups (N=10), all were pre-tested. The two groups' performances were indistinguishable, and detection rates for low-metal targets were dangerously low, especially for the targets with the least metal content, M14s. After the soldiers in the experimental group received instruction and 12–15 hours of practice per man with feedback, both groups were again tested. The control group, which did not have the opportunity to practice, essentially replicated its pre-test performance. The experimental group's detection rates rose, as shown in Figure 16.13, to unexpected levels, given the relatively brief training.

Because these performance gains seemed too good to be true, a second test was performed to test the reliability of the experimental group's gains. To justify this effort, six of its soldiers were suited up in cumbersome body armor, introducing the issue of generality to the test. Results replicated the performance observed in the initial test (Staszewski and Davison, 2000).

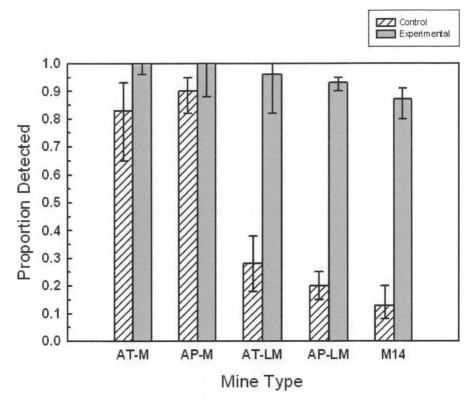

Figure 16.13 PSS-12 test 1 results

Note: Initial PSS-12 test results produced by combat engineers who had either received experimental training and controls, who had only then-standard Army training, showing proportion of target detected as a function of target category. Error bars represent 95 percent confidence intervals. Note that pre-testing of both groups showed them performing at equivalent levels.

Because the targets used for testing thus far had been simulated mines instead of actual mines, a more important test of the experimental group's skills involved transporting the group to Aberdeen Proving Ground, where actual demilitarized landmines were emplaced. Against these targets, the group achieved an aggregate detection rate of 0.97, which included a detection rate of 1.00 against real M14s (Staszewski and Davison, 2000).

The limited availability of control-group soldiers complicated definitive interpretation of the experimental group's gains relative to the control soldiers, but the pre-post treatment performance gains and the consistently high detection performance for all soldiers receiving the experimental treatment were nonetheless impressive.

The first PSS-14 test, the results of which are shown in Figure 16.14, involved five operators; two Army non-commissioned officers (NCOs) serving as project liaisons, and three civilians. Equipment reliability issues substantially added challenges with potentially deleterious consequences and problems encountered during training with the new systems as well as with instrumentation used for scoring and providing performance feedback made the gains achieved in the first test all the more surprising.

Because the trainees in the first test all had prior exposure to the system before the training, which might have inflated their performance, trainees chosen for the second test were completely unfamiliar with the HSTAMIDS. This test tapped the military population most likely to use the system and thus need training—if the system showed sufficient potential to merit further development and eventual deployment. Four combat engineers, two from the US Army and two from the US Marine Corps served as trainees. They received the same training content and amount of practice, however, improved system reliability diminished the frustrating interruptions and delays encountered in the previous test. Detection rates, also shown in Figure 16.14, were statistically indistinguishable from that of the first group.

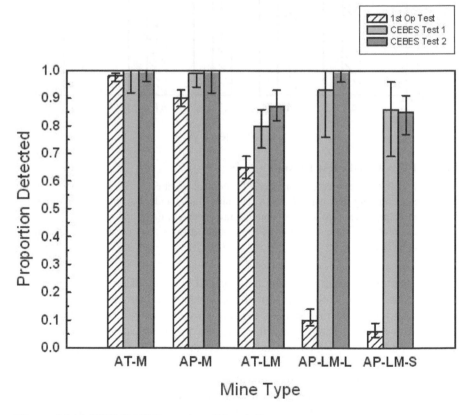

Figure 16.14 HSTAMIDS test 1 and 2 results

Note: Proportion of targets detected as a function of target category for each of two tests of the HSTAMIDS training program. Results from original operational testing of the HSTAMIDS prototype are shown for reference. Error bars represent 95 percent confidence intervals. Note the asymptotic performance shown for AT-LM and AP-LM-S targets that served to diagnose system limitations that were remedied by subsequent hardware/software modifications.

Taken together, the PSS-14 test results had three important effects: (1) the gains in detection rates relative to the initial test performance were sufficient to justify further program funding for developing the HSTAMIDS; (2) the results were sufficiently stable across trainees to diagnose system shortcomings and thus guide subsequent and successful hardware refinements; and (3) the results showed the substantial contribution made by properly trained operators to system performance. This latter realization led to the adoption of a policy of only distributing the system (renamed the AN/PSS-14 for deployment) to units who also received required CEBES training.

Two further tests of the PSS-12 training—prior to its official adoption by US Army—showed its effects to be robust. The first was motivated by an invitation from combat engineer and military police units at the Joint Readiness Training Center

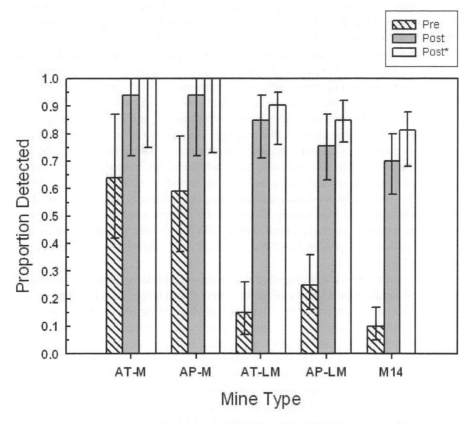

Figure 16.15 Pre-test and post-test performance as a function of mine type for non-commissioned officers trained at the Joint Readiness Training Center, Ft Polk, LA, in August 2001

Note: Results for group labeled Post* eliminate the data of two soldiers whose post-test detection rates were five and seven standard errors below that which both had shown in earlier blind search drills. Both showed an apparent loss of motivation upon learning that a long holiday weekend leave, which had been scheduled to begin upon conclusion of the mine detection training, was cancelled and that their unit would start two weeks of field training immediately.

(JRTC), Ft Polk, LA. These units were scheduled for overseas deployment to areas where landmines were a major threat. The invitation requested that the author and collaborator Alan Davison build a permanent mine detection training site at JRTC and administer the still-experimental PSS-12 training. Doing so presented an opportunity to explore the generality of the training, since NCOs would participate as trainees, and the training (and training site design) could be extended to include training and testing in a forested environment. Severe weather and unexpected last-minute reductions in the availability of troops for training reduced the time for drill and practice to the range of 3–4 hours per soldier. Other extraneous events that occurred in the midst of post-testing appeared to have negatively affected the motivation of some of the trainees (Soldiers were informed that long holiday weekend leaves which would begin upon conclusion of their mine detection training were cancelled and that they would begin two weeks of field training immediately). Results showed detection rates lower than those observed in training drills, but the pre-test–post-test gains, shown in Figure 16.15, were greater than expected.

Another unexpected outcome of the training program came soon after the JRTC exercise. It involved the delivery of the PSS-12 training by JRTC personnel in charge of combat engineer training and assessment. JRTC personnel who had participated as trainees in the training described above served as instructors. Neither the training designer/instructor nor Davison were present, and did not participate in or oversee the administration of this training. The minimally experienced trainers trained six platoons of combat engineers (approximately 180 soldiers) from the US Army's 10th Mountain Division who were scheduled for immediate deployment. They used materials and training aids left from previous exercise and followed its procedures—including pre-testing trainees using the previously taught then-standard Army techniques and post-testing following the experimental training—with one exception: schedule constraints limited drill and practice time to approximately one hour per trainee. The pre- and post-test detection rates achieved by the trainees are shown in Figure 16.16.

In short, the PSS-12 training effects have remained relatively robust in spite of contextual variation on a variety of factors, more than a few of which seemed likely to effect learning adversely. These factors include training time, ground surfaces, weather conditions, soil moisture and humidity levels, mine targets, equipment condition, trainee rank and military experience, military specialization, instructor experience, to name a few. The PSS-14 training has also proven pleasantly robust to variables such as variations in sensor head height and velocity.

Three more features were surprising about the CEBES training effects common to both training programs. The first was the speed with which these effects appeared.

The PSS-12 program provided between 12–15 hours of drill and practice per trainee. The HSTAMIDS program provided roughly 28–32 hours, due to the system's operational complexity and the increased cognitive and perceptual-motor demands it places upon its operators. Note that students of basic algebra and American football players (in one season) typically receive more than an order of magnitude more practice time. Despite the novelty and the comparatively brief training periods for both of the CEBES mine detection training programs, both produced the large improvements in detection capability.

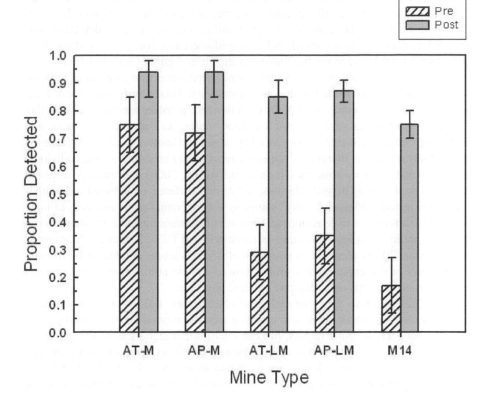

Figure 16.16 Pre-test and post-test detection rates

Note: Pre-test and post-test performance as a function of target type for combat engineers from the 10th Mountain Division. The training was administered by JRTC NCOs trained in the August training exercise and allowed for only one hour for drill and practice of the new expert detection techniques. Data collected by JRTC training personnel and provided courtesy of LTC Tommy Mize.

Even more surprising were the effects observed well before operators had completed training. In each training session, operators first received instruction and practice on part tasks before engaging in blind search drills, which require integration of the individual skill components that the part-task exercises were intended to develop. The initial blind search drills in each program involved each trainee searching for mines in lanes 7.5 m long and 1.5 m wide. (Areas designated for search in testing and clearance in live operations are referred to as lanes. Lanes are roughly the width needed to extract casualties. Army doctrine sets lane width at 1.5 m. This lane width is used at all training and testing sites described here.) Later drills doubled lane length and final drills used lane lengths equivalent to those used for exit testing, namely, 5 m for the PSS-12 and 25 m for the HSTAMIDS. The surprise was that the trainee groups in each of the four initial training sessions (initial PSS-12 training, JRTC training, and both HSTAMIDS tests) achieved near-asymptotic detection rates on the first set of blind search exercises. That is, the first time trainees applied expert techniques in the full detection task, they achieved detection rates that were statistically equal to those produced in exit testing. In all four cases, detection rates of 0.92 or better were achieved and manifold increases were observed for M14 detection rates relative to the baseline rates shown in Figures 16.13 and 16.14.

Two additional surprises are worth noting. First, detection rates were uniform across trainees in all of the training administrations. This is surprising because the pattern development component of the detection task involves mentally constructing spatial patterns using each detector's auditory output to define pattern parts, and matching them against patterns produced by mines and learned in drills. Because the individual differences literature documents substantial variability among individual in spatial and imagery abilities (Hegarty and Waller, 2006) and trainees were not screened on the basis of such abilities, substantial differences between individuals were anticipated. Although "lucky" sampling may account for the absences of large individual differences in detection rates in three of the training administrations, it can be ruled out in the original training administration of PSS-12 training (Staszewski and Davison, 2000). Here, an examination of ASVAB scores obtained after training for soldiers in the experimental group showed a wide ability range; however, all trainees achieved final detection rates of 0.92 or greater.

The final surprise was trainees' false alarm rates. False alarms result from encounters either in training or testing with materials on or under the ground surface that are not parts of mine targets, but can elicit output signals from the detectors. Despite no explicit instruction or drills to support operators' learning to discriminate targets from clutter, false alarm rates were remarkably low. In all PSS-12 training tests, aggregate false alarm rates (0.33 FA/m2) nearly achieved the extremely severe goal of 0.30 false alarms per square meter set for the HSTAMIDS. The HSTAMIDS rates bettered this criterion with .08 false alarms per square meter in training drills and 0.25 in subsequent testing. Figure 16.17 shows the receiver operating characteristic obtained for the soldiers who received the initial PSS-12 CEBES training and the military operators participating in the second HSTAMIDS training exercise. The data points for PSS-12 and HSTAMIDS training exercises show detection and false alarm rates for a common test environment at Aberdeen Proving Ground.

Definitive evaluation of discrimination requires a controlled manipulation of clutter, which was impossible given the available resources. Moreover, the amount of clutter in the training and testing areas was reduced relative to the amount originally found, further complicating interpretation and generalization of the false alarm rates shown here. Nonetheless, observations of both programs' trainees during testing showed numerous instances in which operators encountered clutter, but did not make mine declaration. The discrimination capabilities that both programs produced were unexpected in the absence of any explicit discrimination training or practice.

To conclude this section, the literature on human expertise often generalizes Hayes' (1981) conclusion that ten years of practice are required to attain expert-level performance, although the relatively few laboratory studies of the development of expertise show that the amounts of time and practice needed can be much less (Biederman and Shiffrar, 1987; Chase and Ericsson, 1982; Staszewski, 1988, 1990; Wenger and Payne, 1995). To put the training times in these studies in perspective,

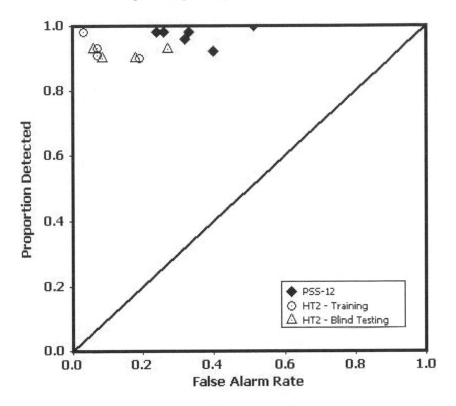

Figure 16.17 Relations between detection and false alarm rates for CEBES-trained PSS-12 operators and HSTAMIDS operators achieved in the same test environment

Note: HSTAMIDS measures came from the second test of training using military operators previously unfamiliar with the HSTAMIDS. Detection rates and false alarm rates for the HSTAMIDS operators were collected both in tests built into the training exercises for assessment purposes and in final, summative testing at the conclusion of training.

VanLehn (1990, pp. 561) defines a novice as "someone with several hundred hours of training (approximately a college course's worth)." Taking these estimates of the practice requirements for skill acquisition into consideration and considering (a) the baseline performance levels that preceded these applications of CEBES training, (b) the absolute levels of performance achieved by CEBES trainees on a task on which a single error has dire consequences, and (c) the relatively tiny amounts of task practice trainees received, the effects described here are quite surprising.

Conclusions and Caveats

The PSS-12 and PSS-14 CEBES training development projects represent successful cognitive engineering efforts. Both substantially improved performance on a practical task of considerable difficulty, danger, and importance. The improvements in operators' detection rates were compelling enough for military leaders to adopt, disseminate, and implement training programs and policies that involved neither trivial changes in past practice nor expense.

What is the significance of these CEBES projects? Together, the results of the two projects link practical applications to basic science on expertise. They corroborate theoretical claims about the role of pattern recognition as a key general element of human expertise (Chase, 1986; Simon, 1979, 1990) and also show that the domain specificity of expert knowledge and skill is not intractably adapted to particular task environments. They also confirm Glaser's (1984, 1989) claims about the potential of basic research on human expertise to inform instructional design, which have been questioned (Alexander, 2003; Patel et al., 1996), not unjustifiably. They show that the non-trivial costs of cognitive task analysis (Schraagen, Chipman, and Shalin, 2000) can be justified by the significant gains that CEBES can deliver. Finally, they show the practical value of CEBES as an approach capable of supporting instructional design and assessment.

It would be premature, however, to conclude that CEBES is a powerfully effective approach to instructional design, at least on the basis of these studies alone. Although their results are consistent with findings from laboratory studies of human expertise and Glaser's instructional theory, limitations imposed on the execution of these projects and fundamental design issues prevent drawing any definitive, general conclusions. The robust effects that these CEBES programs have produced may be the kind that Gilbert, Mosteller, and Tukey (1976) call "interocular,"[5] but the current studies fall short of the design criteria for rigorous program evaluation. Their findings should not be easily dismissed though.

The outcomes of these projects suggest that the CEBES approach has considerable potential to make instructional design more scientifically principled and effective. Thus, despite the limitations that should be heeded in evaluating the CEBES approach to designing instruction, the results described here invite systematic investigation and evaluation of the approach that produced them. Such efforts will likely require a large-scale and expensive effort. This work suggests that sufficient potential is there to justify the investments needed to carefully evaluate the CEBES approach to instructional design, examine its generality, refine its practice, and realize its benefits.

CEBES and Macrocognition

Several features of the projects described here, particularly those related to their analyses of expertise, are consistent with the tenets of macrocognition as they are outlined in Klein et al. (2003), Crandall, Klein, and Hoffman (2006), and throughout the chapters of this volume.

---

5   Gilbert et al. (1976, p. 298) describe interocular effects as "large and easily replicated effects…that hit us between the eyes."

First, and foremost, these projects examined landmine detection expertise *in situ*, under the most naturalistic possible conditions. Although instrumenting the experts and asking them to give concurrent verbal protocols as they demonstrated their skills is not pure naturalistic observation, data collection was made as unobtrusive as possible to minimize impact upon ecological validity. Consistent with both the tenets of macrocognition and information-processing psychology, as it has been applied to investigating cognitively complex tasks such as problem solving (Newell and Simon, 1972), expert performance (Chi, Glaser, and Farr, 1988; Ericsson, 1996) and skill acquisition (Chase and Ericsson, 1982; Staszewski, 1988), natural "whole tasks" (Newell, 1973) were the object of study using a general approach Simon (1989) labeled "experimenting without an independent variable." Maintaining the organizational integrity of the thinking and behavior (Miller, 1986) used to perform well on tasks posed in the natural environment has two important advantages. Theoretically, it permits inferences about the mechanisms that people use to perceive their task environment and adapt effectively to proficiently achieve their goals in this environment. Practically, it increases the chances that the theoretical conclusions cognitive analyses yield will be practically applicable.

The grain-size of the mechanisms used to describe the thought processes of the PSS-12 and PSS-14 experts is compatible with the macro level of functional description advocated by macrocognition proponents. Also, the mechanisms inferred are consistent with those found in studies of naturalistic decision making, as well as studies of expertise conducted in the information-processing tradition. The core of the landmine experts' skills is their ability to use their instruments to infer patterns from the sequential, spatially distributed auditory outputs acquired by systematic search in regions where they acquire an alerting signal. Their mine declaration decisions and their marking of mine locations appear to be mediated by their recognition and categorization of patterns which they had learned from previous episodes of successful mine detection.

The processes postulated to precede pattern recognition and decision in the detection model—search sweep, and edge detection—are similarly high-level functions; nowhere do the observations collected support inferences at a finer temporal resolution, one that conceivably might include moment-to-moment changes in attentional focus. It should be emphasized that the grain-size of neither the models' mechanisms nor the analyses that produced them was theoretically or methodologically predetermined. Rather, it was determined by the empirical regularities uncovered through exploration of the observations collected. Indeed, this is also the case for inferences drawn about the experts' goal structures and the use of the production formalism to explain how the mechanisms identified were serially related.

Additional compatibilities between the methods and findings of this work and the postulates of macrocognition could be listed. They suggest compatibility as well as a complementarity between macrocognitive and information-processing approaches to the study of complex cognition in naturalistic settings.

This convergence is significant and encouraging. Thinking and knowledge are inherently abstract phenomena, and thus not directly observable. This, of course, adds to the challenge of drawing valid and useful inferences from empirical observation. As physicists confronted with this problem in the 1920s have shown and as cognitive psychology has known and shown since its inception (Garner, Hake, and Erikson, 1956; Miller, 1986), the greatest confidence can be placed in convergent findings produced by different approaches. The convergence of findings between the two studies of expert landmine detection speaks to their validity. Their convergence with generalizations about the core role pattern recognition plays in expert performance produced by laboratory studies of expertise (Chase, 1986; Simon, 1990) and naturalistic studies (Klein et al., 2003) strengthens confidence in this conclusion.

The successful outcomes of the tests of training programs, whose instructional content was based on the idea that recognition and interpretation of patterns of a specified character were a key to effective landmine detection, further validate the inferences drawn. The outcomes also confirm the assertion that the better the understanding of cognition in natural settings, "the better we should be able to design better ways for using information technology, better interfaces, and better training programs" (Klein et al., 2003, p. 84).

The implications are that cognitive psychology, with diverse (and sometimes contentious) theoretical and methodological viewpoints, has accumulated principles, theories, and methods sound and comprehensive enough to support engineering solutions to important practical problems.

References

Alexander, P.A. (2003). "Can we get there from here?" *Educational Researcher*, 32 (8): 3–4.

Biederman, I. and Shiffrar, M.M. (1987). "Sexing day-old chicks: A case study and expert systems analysis of a difficult perceptual-learning task." *Journal of Experimental Psychology Learning, Memory, and Cognition*, 13: 640–45.

Chase, W.G. (1986). "Visual information processing." In K.R. Boff, L. Kaufman, and J.P. Thomas (eds), *Handbook of Perception and Human Performance* Vol. 2. New York, NY, Wiley (pp. 28–71).

Chase, W.G. and Ericsson, K.A. (1982). "Skill and working memory." In G.H. Bower (ed.), *The Psychology of Learning and Motivation* Vol. 16. New York, NY: Academic Press (pp. 1–58).

Chi, M.T.H., Glaser, R., and Farr, M.J. (eds), (1988). *The Nature of Expertise*. Hillsdale, NJ: Erlbaum.

Crandall, B., Klein, G., and Hoffman, R.R. (2006). *Working Minds: A Practitioner's Guide to Cognitive Task Analysis*. Cambridge, MA: The MIT Press.

Davison, A. and Staszewski, J. (2002). *Handheld Mine Detection Based on Expert Skill: Reference Guide for Training Plan Design and Training Site Development*. Aberdeen Proving Ground, MD: Army Research Laboratory Human Research and Engineering Directorate.

Ericsson, K.A. (ed.) (1996). *The Road to Excellence: The Acquisition of Expert Performance in the Arts and Sciences, Sports and Games*. Mahwah, NJ: Erlbaum.

Finke, R.A. (1990). *Creative Imagery: Discoveries and Inventions in Visualization*. Hillsdale, NJ: Erlbaum Associates.

Garner, W.R. (1974). *The Processing of Information and Structure*. Potomac, MD: Erlbaum Associates.

Garner, W.R., Hake, H.W., and Eriksen, C.W. (1956). "Operationism and the concept of perception." *Psychological Review*, 63: 149–59.

Gilbert, J.P., Mosteller, F., and Tukey, J.W. (1976). "Steady social progress requires quantitative evaluation to be searching." In C.C. Abt (ed.), *The Evaluation of Social Programs*. Beverly Hills, CA: Sage Publications (pp. 295–312).

Glaser, R. (1984). "Education and thinking: The role of knowledge." *American Psychologist*, 39: 93–104.

—— (1989). "Expertise and learning: How do we think about instructional processes now that we have discovered knowledge structures?" In D. Klahr and K. Kotovsky (eds), *Complex Information Processing: The Impact of Herbert A. Simon*. Hillsdale, NJ: Erlbaum (pp. 269–82).

Guckert, R. (2000). "HSTAMIDS Red Team experiences/Lessons learned." *Proceedings of the UXO/Countermine Forum*, 2–4 May 2000, Anaheim, CA.

Hall, G.S. (1898). "Some aspects of the early sense of self." *American Journal of Psychology*, 9: 351–95.

Hambric, H.H. and Schneck, W.C. (1996). "The antipersonnel mine threat." *Proceedings of the Technology and the Mine Problem Symposium*, Vol. 1, Naval Postgraduate School, 18–21 November 1996, pp. 3–45.

Hayes, J.R. (1981). *The Complete Problem Solver*. Philadelphia, PA: Franklin Institute Press.

Headquarters, Department of the Army (2002). *Mine/Countermine Operations: FM20–32 C3*. Washington, DC: Department of the Army.

—— (2006). *Operator's and Unit Maintenance Manual for Detecting set, Mine AN/PSS-14: TM 5–6665–373–12&P*. Washington, DC: Department of the Army.

Hegarty, M. and Waller, D. (2006). "Individual differences in spatial abilities." In P. Shah and A. Miyake (eds), *Handbook of Higher-level Spatial Cognition*. Oxford: Oxford University Press (pp. 121–69).

Herman, H. and Inglesias, D. (1999). "Human-in-the-loop issues for demining." In T. Broach, A.C. Dubey, R.E. Dugan, and J. Harvey (eds), *Detection and Remediation Technologies for Mines and Minelike Targets IV, Proceedings of the Society for Photo-Optical Instrumentation Engineers 13th Annual Meeting*, 3710: 797–805.

Herman, H., McMahill, J., and Kantor, G. (2000). "Training and performance assessment of landmine detector operator using motion tracking and virtual mine lane." In A.C. Dubey, J.F. Harvey, J.T. Broach, and R.E. Dugan (eds), *Detection and Remediation Technologies for Mines and Minelike Targets V*, Proceedings of the Society for Photo-Optical Instrumentation Engineers 13th Annual Meeting, Vol. 4038: 110–21.

Klatzky. R.L. and Thompson, A. (1975). "Integration of features in comparing multifeature stimuli." *Perception and Psychophysics*, 18: 428–32.

Klein, G., Ross, K.G., Moon, B., Klein, D.E., Hoffman, R.R., and Hollnagel, E. (2003). "Macrocognition." *IEEE Intelligent Systems*, 18: 81–5.

LaMoe, J.P. and Read, T. (2002). "Countermine operations in the contemporary operational environment." *Engineer*, 32 (April): 21–3.

MacDonald, J., Lockwood, J.R., McFee, J., Altshuler, T., Broach, T., Carin, L., Harmon, R., Rappaport, C., Scott, W., and Weaver, R. (2003). *Alternative for Landmine Detection*. Santa Monica, CA: Rand Science and Technology Policy Institute.

Miller, G.A. (1986). "Dismembering cognition." In S.E. Hulse and B.F. Green (eds), *One Hundred Years of Psychological Research in America: G. Stanley Hall and the Johns Hopkins Tradition*. Baltimore, MD: The Johns Hopkins University Press (pp. 277–98).

Newell, A.N. (1973). "You can't play twenty questions with nature and win." In W.G. Chase (ed.), *Visual Information Processing*. New York: Academic Press (pp. 283–308).

Newell, A.N. and Simon, H.A. (1972). *Human Problem Solving*. New York: Prentice-Hall.

Palmer, S.E. (1977). "Hierarchical structure in perceptual representation." *Cognitive Psychology*, 9: 441–74.

Patel, V.L., Kaufman, D.R., and Magder, S.A. (1996). "The acquisition of medical expertise in complex dynamic environments." In K.A. Ericsson (ed.), *The Road to Excellence*. Hillsdale, NJ: Lawrence Erlbaum Associates.

Pearson, D.G. and Logie, R.H. (2000). "Working memory and mental synthesis: A dual task approach." In S. Nualláin (ed.), *Spatial Cognition: Foundations and Applications*. Amsterdam: John Benjamins Publishing pp 347–59.

Rosch, E., Mervis, C.B., Gray, W., Johnson, D., and Boyes-Braem, P. (1976). "Basic objects in natural categories." *Cognitive Psychology*, 8: 382–439.

Schraagen, J.M., Chipman, S., and Shalin, V.L. (2000). "Introduction to cognitive task analysis." In S. Chipman, J.M. Schraagen, and V.L. Shalin (eds), *Cognitive Task Analysis*. Mahwah, NJ: Lawrence Erlbaum Associates (pp. 3–23).

Simon, H.A. (1979). "Human information processing models of cognition." *Annual Review of Psychology*, 30 (January): 363–96.

—— (1989). "The scientist as problem solver." In D. Klahr and K. Kotovsky (eds), *Complex Information Processing: The Impact of Herbert A. Simon*. Hillsdale, NJ: Lawrence Erlbaum Associates (pp. 375–98).

—— (1990). "Invariants of human behavior." *Annual Review of Psychology*, 14: 1–19.

Staszewski, J. (1988). "Skilled memory in expert mental calculation." In M.T.H. Chi, R. Glaser, and M.J. Farr (eds), *The Nature of Expertise*. Hillsdale, NJ: Lawrence Erlbaum Associates (pp. 71–128).

—— (1990). "Exceptional memory: The influence of practice and knowledge on the development of elaborative encoding strategies." In W. Schneider and F.E. Weinert (eds), *Interactions Among Aptitudes, Strategies, and Knowledge in Cognitive Performance*. New York: Springer-Verlag (pp. 252–85).

—— (1999). "Information processing analysis of human land mine detection skill." In T. Broach, A.C. Dubey, R.E. Dugan, and J. Harvey (eds), *Detection and Remediation Technologies for Mines and Minelike Targets IV, Proceedings of the Society for Photo-Optical Instrumentation Engineers 13th Annual Meeting*, Vol. 3710: 766–77.

—— (2004). "Models of expertise as blueprints cognitive engineering: Applications to landmine detection." *Proceedings of the 48th Annual Meeting of the Human Factors and Ergonomics Society*, New Orleans, LA, 20–24 September 2004.

Staszewski, J. and Davison, A. (2000). "Mine detection training based on expert skill." In A.C. Dubey, J.F. Harvey, J.T. Broach, and R.E. Dugan (eds), *Detection and Remediation Technologies for Mines and Mine-like Targets V, Proceedings of Society of Photo-Optical Instrumentation Engineers 14th Annual Meeting*, 4038: 90–101.

Thompson, A.L. and Klatzky, R.L. (1978). "Studies of visual synthesis: Integration of fragments into forms." *Journal of Experimental Psychology: Human Perception and Performance*, 4: 244–63.

US Army Test and Experimentation Command (1991). *Expanded Test Report: AN/ PSS12 Metallic-mine Detector* (Report EU–E/CS–1600) Ft Hood, TX: US Army Test and Experimentation Command.

VanLehn, K. (1990). "Problem solving and cognitive skill acquisition." In M.I. Posner (ed.), *Foundations of Cognitive Science*. Cambridge, MA: MIT Press (pp. 527–79).

Webster, D. (1996). *Aftermath: The Remnants of War*. New York: Pantheon Books.

Wenger, M. and Payne, D. (1995). "On the acquisition of mnemonic skill: Application of skilled memory theory." *Journal of Experimental Psychology: Applied*, 1: 195–215.

PART IV
Alternative Approaches

Chapter 17

Representation of Spatio-Temporal Resource Constraints in Network-Based Command and Control

Rogier Woltjer, Kip Smith, and Erik Hollnagel

Introduction

Command and control is in a state of change. Both civilian emergency managers and military commanders seek to implement network-based organizations. The drive for network-based command and control stems from coordination problems in complex environments that cannot be solved by traditional hierarchic command structures (Alberts et al., 1999; Cebrowski and Garstka, 1998). For example, civilian (emergency management) agencies want to improve inter-agency coordination (Smith and Dowell, 2000) and the military wants to overcome problems with combat power and speed. Moreover, joint operations by a growing number of civilian agencies (for example, rescue services, local and national governments, and private-sector agencies) and military services are often required to resolve emergency or conflict situations. The international nature of these emergencies, disasters, and conflicts compels civilian and military services to cooperate across traditional organizational and cultural boundaries. The prospect of having to coordinate activities across agencies, nations, and cultures suggests that a whole new range of technical and social problems may emerge that can only be addressed by a network-based command-and-control structure.

The concept of network-based command and control is proposed to enable forces (civilian or military) to organize in a bottom-up fashion and to self-synchronize in order to meet a commander's intent (Brehmer and Sundin, 2000; Cebrowski and Garstka, 1998). This concept steps away from the traditional "platform-centric" hierarchical command structure where power is partially lost in top-down command-directed synchronization. Network-based command and control envisions a high-performance information grid that makes sensor information available to the entire network. Part of this vision is that filtering, aggregation, and presentation of data will be done by computer systems. Nodes in the network are foreseen to be able to communicate with all other nodes, so that informed and appropriate action can be taken locally. The vision relies heavily on emerging information technologies.

The unit of analysis for studies of network-based command and control is much larger than for studies of traditional hierarchical control. In traditional organizations, the chain of command is orderly. The network-based vision entails a much greater flexibility in the implementation of communication and action. Large parts of the network necessarily become the unit of analysis. This situation presents a challenge to studies of network-based command and control. Two of the challenges are (1) how to describe and model joint behavior, and (2) how to design information technology that supports network-based command and control.

In this chapter, we address these challenges by outlining a method for analysis of joint behavior using constraints and opportunities for action represented in state spaces. The method is meant to inform the design of information technology to be used in network-based command-and-control centers. Our research goal is to develop and verify a methodology to identify constraints generated by the actions of distributed collaborative decision makers. By addressing this goal, we aim to provide insight in distributed collaborative command and control in dynamic network-based settings. Understanding the reciprocal relationship between constraints and actions of joint systems contributes to the science of network-based command and control and to the design of systems that support command and control. In this chapter, we restrict our analysis to the management of emergency situations in distributed collaborative command and control.

Constraints and Their Representation

The concept "constraint" has been addressed by scientists from a variety of disciplines and perspectives and has implications for cognitive systems engineering (CSE) (Ashby, 1956; Checkland, 1981; Gibson, 1986; Hollnagel and Woods, 2005; Leveson, 2004; Norman, 1998; Reitman, 1964; Vicente, 1999; Vicente and Rasmussen, 1992; Woods, 1995). Combining these perspectives we define constraint as either a limit on goal-directed behavior, or an opportunity for goal-directed behavior, or both. Constraints are invariably described as essential factors in the functioning of (cognitive) systems. This work addresses constraints explicitly in the analysis of behavior of joint cognitive systems. In CSE, constraints are said to shape the selection of appropriate action (Hollnagel and Woods, 2005). Similarly, the related disciplines of cybernetics and systems theory consider the concept of constraint to be of major importance, because constraints facilitate control.

In cybernetics, constraints play a significant role in the control of processes. One of the fundamental principles of control and regulation in cybernetics is the Law of Requisite Variety (Ashby, 1956). This law states that a controller of a process needs to have at least as much variety (behavioral diversity) as the controlled process. Constraints in cybernetics are described as limits on variety. Constraints can refer to both the variety of the process to be controlled and the variety of the controller. This means that if a specific constraint limits the variety of a process, less variety is required of the controller. Similarly, if a specific constraint limits the variety of the controller, less variety of the process can be met when trying to exercise control. Regarding the prediction of future behavior, Ashby notes that if a system is predictable, then there are constraints that the system adheres to. Knowing about constraints on variety makes it possible to anticipate future behavior of the controlled process, and thereby facilitates control.

The recognition of constraints on action has been observed in field studies as a strategy to cope with complexity in sociotechnical joint cognitive systems. Example domains include refinery process control (De Jong and Köster, 1974), ship navigation (Hutchins, 1991a, 1991b), air traffic control (Chapman et al., 2001; Smith, 2001), and trading in the spot currency markets (Smith, 1997). Similarities of system complexity and coupling in these domains and command and control suggest that the recognition of constraints is likely a useful approach in command and control as well.

Woods (1986) emphasizes the importance of spatial representations when providing support to controllers. In this chapter, we develop the visualization of constraints as the support strategy. Representation design (Woods, 1995) and ecological interface design (Vicente and Rasmussen, 1992) offer design guidelines concerned with constraint. Decision support systems should facilitate the discovery of constraints, represent constraints in a way that makes the possibilities for action and resolution evident, and highlight the time-dependency of constraints. One representation scheme for such discovery is the state space.

State space representation is a graphic method for representing the change in state of process variables over time (Ashby, 1960; Knecht and Smith, 2001; Port and Van Gelder, 1995; Stappers and Flach, 2004). The variables are represented by the axes. States are defined by points in the space. Time is represented implicitly by the traces of states through the space. Figure 17.1 is an example state space in which the variables are fuel range and distance. The regions with different shades represent alternative opportunities for action. The lines between these regions are constraints on those actions. The arrows represent traces of states of a pair of processes through the space as their states change over time. The opportunities for action have yet to change for the first process, but they have changed for the second.

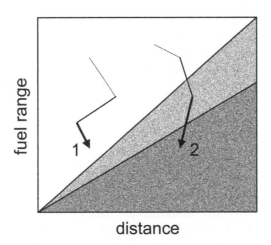

Figure 17.1 An example of a state space

Note: Constraints are represented as thresholds between regions associated with alternative opportunities for action. The arrows represent a pair of processes as their states change over time.

Recent work in cognitive systems engineering, human factors, and ergonomics reflects a renewed interest in the state space representation. Stappers and Flach (2004) describe state space representation as a promising and rich visualization method of the behavior of cognitive systems. They state that it steps away from the typical block diagrams of perception, mental processing, and motor control that hinder appropriate thinking about dynamics. They describe the metaphor of state spaces as "that of the cartographer's map, showing a landscape with roads and pitfalls, mountains and rivers, i.e., opportunities and threats, but leaving the reader's mind free to roam and imagine developing states and possible routes through the territory" (p. 825).

Command and Control and Decision Making

The design of any support system is necessarily based on the model that the designers have of the task that is to be supported. Sometimes this model is only a figment of the designers' minds. It would be distinctly preferable if designers were to analyze the task they mean to support in detail, so that their design of tools could be based on an explicit and well-informed model. Moreover, their design would benefit if they were to envision how their design would change the task and whether it would serve the actual purpose of the joint system that it is designed for (Hollnagel and Woods, 2005). The modeling of command and control and decision making is therefore a critical precursor of the design of support systems.

Diverse research traditions including, but not limited to, cognitive systems engineering, and decision making address the modeling of command and control. Here we offer a selective review of several models linking control, command and control, and decision making.

Joint Command and Control Systems

We adhere to the paradigm of cognitive systems engineering (Hollnagel and Woods, 1983, 2005). CSE addresses questions such as how to make use of the power of contemporary technology to facilitate human cognition, how to understand the interactions between humans and technologies, and how to aid design and evaluation of digital artifacts. Related to distributed cognition (Hollan et al., 2000) and macrocognition (Klein et al., 2003), CSE takes an ecological view regarding the importance of the context when addressing cognition. It focuses on constraints that are external to the cognitive system (present in the environment) rather than on the internal constraints (memory limitations, and so on) that are foci of the information processing perspective. For CSE, it is these external constraints that determine behavior.

The unit of analysis in cognitive systems engineering is the "joint cognitive system." A cognitive system (Hollnagel and Woods, 2005) is a system that can control its goal-directed behavior on the basis of experience. The term "joint cognitive system" (Hollnagel and Woods, 2005) means that control is accomplished by an ensemble of cognitive systems and (physical and social) artifacts that exhibit goal-directed behavior. In the areas of interest to cognitive systems engineering, typically one or several persons (controllers) and one or several support systems constitute a joint cognitive system which is typically engaged in some sort of process control in a complex environment.

An early account of modern command-and-control centers is provided by Adelson (1961), who describes decisions in what he called "command control centers." He states that "command centers are typically nodes in networks constituting command control systems" (Adelson, 1961, p. 726), where the word "center" means "a place where decisions are made," and does not imply centralization of decision making. This suggests that the network-based environments envisioned today are not as conceptually revolutionary as one may think. Adelson emphasizes that both the complexity and urgency of command in 1961 had increased compared to earlier times. Two outcomes of this were a need for a higher rate of information handling and more severe consequences of decisions. Adelson goes on to enumerate a large number of important factors that the "decider" must know about, including the recent history of relevant variables' behavior, the present state of the world, predicted future states, including the opponents' behavior, their own system's and environment's characteristics, alternatives for action, predicted outcomes, and objectives. Adelson even mentions war games and business games as tools to simulate and predict processes. Furthermore, he demonstrates problem-driven joint-systems thinking in that he emphasizes the importance of the objectives of technological improvement of command centers based on the real needs of "the assemblage of men and machines that constitute the system" (Adelson, 1961, p. 729).

The Dynamics of Decision Making and Control

Researchers often associate command and control with decision making. The term "decision making" refers to a mental process that precedes the execution of action. The classical prescriptive model of the decision-making process is normative decision theory. According to this theory, decision making includes four sequential steps: the generation of all possible options for action, independent assessments of the probability and utility of each option, and the selection of the option with the highest expected utility (see, for example, Von Neumann and Morgenstern, 1953).

Although normative decision theory has proven to be difficult to apply to dynamic, complex environments like command and control (Brehmer, 1992; Hollnagel, 1986, 1993, 1999; Klein and Calderwood, 1991), it has dominated the "expert system" paradigm for decision support systems. Like normative decision theory generally, decision support systems based on the option generation and evaluation paradigm can perform well only in limited, well-defined, and predictable domains. Predictably, they are typically inappropriate for network-based command and control, because the underlying model of "decision making" does not fit the distributed nature of the task. The literature is replete with reports about the pervasively negative consequences of support systems based on decision theoretic designs including, but not limited to, lack of user acceptance, brittle performance when faced with unanticipated novelty, users' over-reliance on the machine's "expertise," and biasing users' cognitive and decision processes (Bainbridge, 1983; Dekker and Woods, 1999, 2002; Hollnagel, 1999; Hollnagel and Woods, 2005; Lee and See, 2004; Roth et al., 1997).

Fortunately, alternative models have emerged that are more appropriate for command-and-control applications. Numerous studies of decision makers in so-called "naturalistic," high-stakes, time-critical, and complex settings have led Gary Klein and colleagues to propose a model of naturalistic decision making (NDM) called "recognition-primed decision making" (for example, Klein and Calderwood, 1991). Their critical insight is that "decision makers" in naturalistic settings actually do not make conscious analytical decisions and only rarely consider more than one option. The recognition-primed decision-making model states that decision makers generate one, and only one, option based on recognition of familiarity, evaluate whether that option will work (using "mental simulation"), and implement it if it will. If it will not work, they generate another option and evaluate it. The process cycles until a satisfactory option is found. Adequate recognition and simulation of the situation are thus the keys to successful "decision making" according to this model.

A second approach is called "dynamic decision making" (DDM) (Brehmer, 1992). This approach focuses on the functions served by decision making in order to gain control, or to achieve some desired state of affairs, rather than on the decisions themselves. As its name implies, dynamic decision making involves tasks with a dynamic character. This implies that a series of interdependent decisions in real time is required to reach the goal, and that the state of the decision problem changes both autonomously and as a consequence of the decision-maker's actions. This approach describes decision making as the intersection of two control processes: (1) the system that the decision maker aims to control and (2) the means for achieving control. The second process is used to control the first (Brehmer and Allard, 1991). The DOODA loop (Brehmer, 2006) is a contextualized model of this control loop approach, integrating a loop of (1) sensemaking, decision making, action, and information collection, with (2) more military-specific concepts such as command, order, mission, and sensors, into a comprehensive model of command and control.

The third approach to modeling that we review are the models that form the foundations of cognitive systems engineering (Hollnagel and Woods, 2005): the description of joint cognitive systems, that is, human-machine ensembles that jointly perform their functions, the cyclical model of action as it is performed by joint cognitive systems, and contextual control models that emphasize the CSE principle that gaining and maintaining control in dynamic situations is paramount. The cyclical model of human action is the heart of cognitive systems engineering, and is based on Neisser's (1976) perceptual cycle. This model, also called the "CAEC loop," basically states that an operator's current understanding (or construct, C) directs the operator's next action. This action (A) produces some information or feedback. This information (or event, E) modifies the operator's current understanding, and so on. This does not mean that these concepts always occur in this order: event evaluation and action often occur in an intermingled or iterative fashion. The emphasis on control is based on the idea that behavior towards goals, and thereby control, is a combination of feedback (compensatory) and feedforward (anticipatory) control. CSE models control as simultaneous control loops that range in their character on a gradual scale from short-term compensatory control to long-term anticipatory control.

Control Support Through the Recognition of Constraints

These alternative models of cognition and decision making are congruent with the cognitive systems engineering view that states: "From the joint cognitive system perspective, the intelligence in decision support lies not in the machine itself (the tool), but in the tool builder and in the tool user" (Woods, 1986, p. 173). In terms of the three models discussed above, people are involved in the recognition of the situation and generating appropriate action (NDM), or sensemaking and decision making (DDM), or constructing the current understanding and selecting action (CSE). The CSE view on intelligent decision support for process control is based on the principle that a joint cognitive system controls its behavior to achieve its goals. Rather than supporting decision making, cognitive systems engineering aims to support actions directed at gaining and maintaining control. In this view, interaction between the various parts of the joint cognitive system must be designed to enable the joint cognitive system to cope with complexity and to handle disturbances on the way to retaining control. This view recognizes that process-control environments are often fundamentally complex and that oversimplifying the interaction would likely reduce the joint cognitive system's ability to maintain control. Supporting control means supporting the evaluation of the situation and anticipation of future events, the selection of action, and the performance of action, as well as ensuring that the time available for actions is sufficient (Hollnagel and Woods, 2005).

In the many tasks where the option generation and evaluation paradigm of intelligent decision support systems seems inappropriate, other ways of supporting the joint cognitive system to gain and maintain control may be found. For the domain of air traffic control, Dekker and Woods (1999) mention an alternative form of supporting controllers: "In one situation, controllers suggested that telling aircraft in general where *not* to go was an easier (and sufficient) intervention than telling each individual where to go" (p. 94). This statement suggests that operators may necessarily and sufficiently be informed about the constraints on their actions, rather than all possible actions. Fundamental similarities between the air traffic control and emergency management command-and-control tasks indicate that this idea may transfer well to the emergency management command-and-control domain.

Accordingly, the approach to decision support that we are currently investigating is to support control through recognition of constraints. The theoretical basis for this has been outlined in the sections on the rationale for looking at constraints to analyze behavior, and to support goal-directed behavior. Particular models of command and control also encourage the investigation of a constraint-based approach to the analysis and support of joint systems' behavior in command and control.

Dynamic decision-making research in Brehmer's view "is concerned with people's formulation of goals and models as a function of the observability and action possibilities of the system to be controlled" (Brehmer, 1992, p. 218). Thus the question what constitutes observability and action possibilities is a research problem. The discussion of constraints above is exactly concerned with this problem. Constraints influence the action possibilities. Hence, it is important that people's model of the system to be controlled is adapted to the possibilities for action and therefore the constraints on action to be able to control. To know the possibilities for action, they must be discernible. As mentioned above, exhaustive option generation is not meaningful in many complex environments with conflicting goals and multiple ways to achieve these goals. Therefore, to be able to adapt one's model to the possibilities for action, it may be more useful to be able to clearly and more directly observe the constraints on action. This is the perspective on command-and-control support investigated in this paper, in the form of visualization of constraints.

The constraint concept has recently been discussed in a command-and-control context. As a result of a study employing grounded theory, Persson (1997) defines command and control as "the enduring management of internal and external constraints by actors in an organization in order to achieve imposed and internal goals" (p. 131). He argues that the core of command and control is "constraint management," and defines constraint as something that is an obstacle to action or goal-achievement. Persson indicates the importance of the recognition, resolution, and thereby management of constraints in command and control.

A Method for the Recognition of Constraints

From the discussion above, we can conclude that constraints are acknowledged in work practice and described in the literature to shape action and enable control, and that being aware of constraints is essential for process controllers. However, in complex and dynamic environments where constraints continually change, where people must work collaboratively and where information and people are distributed, keeping abreast of all relevant constraints can be difficult. To address this difficulty, this chapter investigates a method for recognizing constraints in process control tasks. The method is intended to become part of the support systems that Brehmer and Sundin (2000) and Cebrowski and Garstka (1998) envision. The method has three steps, which are discussed in turn.

(1) The first step in the method is a functional resonance analysis of the joint cognitive system's process control task. The analysis is conducted by (or in consultation with) a domain expert. The modeling of the joint cognitive system's process control task is done by connecting modules that represent the functions that need to be performed. All functions are described through and may be linked via six aspects; input, output, preconditions, resources, time, and control. This functional modeling scheme was recently developed by Hollnagel (2004) for accident modeling and complex system analysis under the acronym FRAM (functional resonance accident model). The modules of functions and their characteristics are therefore called FRAM-modules. Because the modeling method is developed for complex joint human-machine system analysis, it is also suitable for modeling the tasks and functions that are performed in network-based command and control, which typically involves such complex systems.

Figure 17.2 describes the six aspects that a FRAM-module addresses for each function that is identified. To find the FRAM-modules, one may start with the top-level goal, which may translate into the top-level function, or one may start with any function and move on to related functions. Typically, the goal in command and control is to get some sort of process under control, such as a progressing fire, a

Function Aspect	Description
Input (I)	That which the function uses or transforms
Output (O)	That which the function produces
Preconditions (P)	Conditions that must be fulfilled before the function can be carried out
Resources (R)	That which the function needs or consumes when it is carried out (for example, matter, energy, information, manpower)
Time (T)	Time available, as a resource or a constraint
Control (C)	That which supervises or adjusts the function (for example, controller, guideline, plan, procedure)

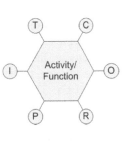

Figure 17.2   FRAM-module description of function aspects and hexagonal function representation

Source: Hollnagel, 2004.

crowd of panicking people, a spreading oil spill, or a moving adversarial military force. This translates into the top-level function of, for example, fighting the fire, directing the flow of people, blocking and cleaning the oil spill, or defeating the adversary. From there on, the functions that are necessary or otherwise related to the input of the function—its timing, control, required resources, and preconditions—determine the other functions in the model of a task.

FRAM has possibilities and advantages similar to goals-means task analysis (GMTA) (Hollnagel, 1993), which is a highly related method. Moreover, the tasks identified through a GMTA may aid in the definition of functions in FRAM. Both FRAM and GMTA have a recursive structure, focus on the goals of the joint human-machine system rather than strict function allocation, and can be applied to operational as well as hypothetical systems. The Test-Operate-Test-Exit unit (Miller et al., 1960) and the means-ends analysis in the General Problem Solver (Newell and Simon, 1961) are early examples of goals-means task analysis. More recently, the principles of GMTA have successfully been applied to the modeling, design, and evaluation of process control tasks (Hollnagel, 1993; Hollnagel and Bye, 2000; Hollnagel and Woods, 2005; Lind and Larsen, 1995).

(2) The second step in the method is to identify the essential variables, the subset of process variables that the joint cognitive system can both observe and (indirectly) affect. This step in the method identifies the variables that change during the performance of the tasks identified in the FRAM analysis. These are the variables that

are affected by the joint cognitive system's goal-directed actions, and the variables upon which the joint cognitive system bases its understanding of the situation. These variables stem directly from the components of the FRAM module, that is, the constraints associated with time, control, preconditions, and resources.

Following Neisser (1976) and Hollnagel and Woods (2005), support for people doing their work should focus on assuring a satisfactory understanding of the situation and an effective execution of action toward the goals of the joint cognitive system. (Support in the recognition-primed and dynamic decision-making models similarly would focus on facilitating recognition and sensemaking.) The act of controlling steers the process to be controlled toward the fulfillment of the joint cognitive system's goals. The act of controlling is modeled as an assembly of feedback (compensatory) and feedforward (anticipatory) loops (Hollnagel and Woods, 2005). Feedback and feedforward control, and the understanding of the situation, are based on observations of the state that the system is in.

In the process control domains that are of interest to cognitive systems engineering, the units of observation are typically states, defined by the current instantiation of a large number of variables that can change rapidly over time. Following Ashby (1956), who used the term "essential variables" for the variables that must be kept within assigned limits for an organism to survive, we use the term "essential variables" for the subset of process variables that the joint cognitive system can both observe and (indirectly) affect. We imply that essential variables must be known for the joint cognitive system to function adequately.

(3) The third step of the method has three parts: (a) sampling the values of the essential variables, (b) juxtaposing these variables to form a set of multidimensional state spaces, and (c) comparing the location of the data in the state spaces to the thresholds between regions specifying differing opportunities for action. Step 1, the FRAM analysis, identifies the functions for and constraints on the execution of a task, and Step 2 identifies the variables that change during its execution and the constraints on these variables. Step 3 juxtaposes variables associated with related tasks to form state spaces that make these preconditions explicit by specifying regions where the constraints are (not) met. Boundaries between regions form thresholds that represent constraints on action. In the example of Figure 17.1, there are two such thresholds that divide the space into three regions.

In the remainder of this chapter, we discuss an experiment that provides an existence proof that the representation of data and constraints in state spaces can form the basis for the design of information systems to support operators in command-and-control environments.

Application of the Constraint Recognition Method

This section describes the microworld study that we have conducted, and more importantly the insights we have gained regarding the application of our method for the recognition of constraints.

Microworlds and C3Fire

Microworlds are simulated task environments that (a) provide a task that can be made more complex, challenging, and realistic than traditional laboratory studies but that (b) generalize to interesting parts of real world problem solving while remaining (c) more controllable, tractable, reproducible, and flexibly designable research environments than a field study (Brehmer and Dörner, 1993; Funke, 2001; Gray, 2002).

C3Fire (Granlund, 2002) is a firefighting microworld in which a group of people work together to gain control of a computer-simulated fire. Their task is to collaborate in an experimentally controlled configuration for command-and-control interaction, under observation of a researcher who manages the experiment. Figure 17.3 presents an overview of the C3Fire microworld. Its elements include a map showing vegetation, buildings, vehicles, and the fire, and the computer network for controlling the vehicles and communicating by e-mail.

In the C3Fire setting used in this experiment, various mobile units and stationary units are located on the map. Mobile units (trucks) need fuel to move across the map. Stationary units have a fixed position on the map and cannot move. Fire trucks are mobile units that can fight fires. To do so, they need water. Water-providing units can provide other units with water. This class is formed by water trucks and water stations. Similarly, fuel-providing units are units that can supply all moving units with

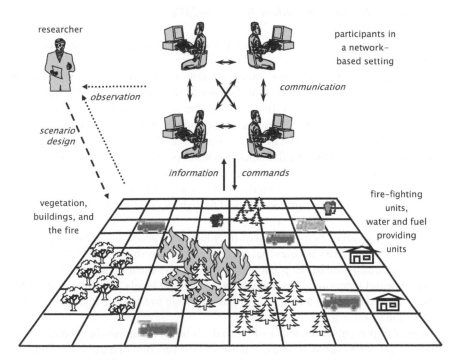

Figure 17.3 An overview of the C3Fire microworld setting used in the study

Source: adapted from Granlund, 2002.

fuel. These can be divided into fuel trucks and fuel stations. The class of stationary units includes water and fuel stations as well as vegetation and buildings.

The rate of burning and spreading of the fire can be set by the researcher, and is typically made dependent on vegetation, terrain, the presence of buildings, and wind direction and speed. Thus, these properties of the environment constrain the development of the fire.

Participants direct the units where to drive by interacting with an interface showing a map with the dynamic simulation, an e-mail tool, and data describing the state and characteristics of the trucks.

The C3Fire microworld captures many of the characteristics of complex environments: it implements high degrees of connectivity, complexity, uncertainty, time pressure, and polytely (the existence of multiple goals). It provides a test bed for research on collaborative and distributed command and control in a dynamic, volatile, uncertain environment. The various units in C3Fire are mutually enabling and constraining because of interdependencies in the consumption and provision of water and fuel. Interdependencies among decision makers arise whenever different classes of units are assigned to different participants in the simulated command-and-control center. For example, if fire trucks need water and fuel, water trucks and fuel trucks constrain the actions of fire trucks. If different people have control over these different units, their actions are mutually facilitating and limiting.

The Constraint Recognition Method Applied to C3Fire

This section presents the application of the constraint recognition method to the C3Fire microworld.

A small part of the functional modeling task analysis of C3Fire is shown in Figure 17.4. The analysis reveals how functions are linked and thereby how units are mutually facilitating and limiting. To take the example of the "fire trucks fight fire" function, the input of this function is burning vegetation and buildings, which may be saved from or lost to the fire (the output of the function). The preconditions of fighting a fire are (1) that the fire truck has water (which is the output of the water refilling function) and (2) that the fire truck is collocated with that fire (which is the output of the moving-to-fire function). The time and pace of fire trucks fighting fire is influenced by the fire spread rate and firefighting rate for example. The main resource used for firefighting is water.

The essential variables were subsequently identified according to the second step in the method. These are variables that are both affected by the actions of the network of participants, and used by the participants to assess the situation that unfolds during the experimental scenarios. They include fuel and water levels, distances to intended positions and other trucks and stations, time available to fight fire and move across the map, and so on.

Table 17.1 presents an example of an analysis of essential variables linked to the various aspects of a FRAM-module. It describes the function of a truck *T* refilling its resource *R* from a source *S* as an example. This abstract description of the function may then be instantiated with, for example, "fire truck F4" for truck *T*, "water" for resource *R*, and "water truck W12" for source *S*. Table 17.1 also illustrates how

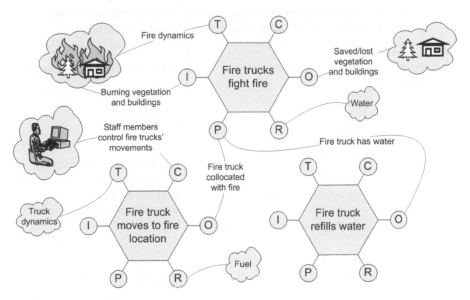

Figure 17.4   Functional modeling (FRAM) analysis of a small part of the task in the C3Fire microworld

operators such as "increase," "level," "same," "time," "between," and "rate," may be used to express actions on or properties of the various quantities in the simulation. These operators enable the linking of aspects of one function to the other function. For example, changing the position of truck T or source S in order to fulfill the precondition "same(pos(truck T), pos(source S))" is the output of the truck moving function, which in turn specifies the indirectly related (and therefore marked with an asterisk) variable of "time(between(truck T, source S))*."

Experimentation with C3Fire consisted of multiple trials in which different scenarios of a simulated firefighting task were presented to a multi-person team. Thirty-two Swedish men with a mean age of 24.6 years participated in the study for a monetary compensation. The participants were not trained in command and control. The participants filled out a form stating their informed consent to anonymous participation in the study. There were a total of eight scenarios with changing geographical layout and task difficulty. The purpose of these manipulations was to keep the scenarios challenging and the participants engaged. The 32 participants engaged in the C3Fire scenarios in eight four-person teams, each playing eight scenarios independently so that 64 trials of empirical data were collected. C3Fire captures and logs a large number of variables. Performance variables include the number of squares that were still burning, closed out, or burnt out at the end of each scenario, which will not be discussed further here. For the purpose of the analysis of constraints, all variables associated with the fire truck, water truck, and fuel trucks' behavior were logged.

The third step of the method for recognizing constraints entails the sampling of the values of the essential variables, juxtaposing these variables in multidimensional

Table 17.1   Description of the function module "truck *T* refills resource *R* from source *S*" with examples of essential variables

Truck T refills resource R from source S	Description	Essential variables, examples
Input	truck *T* source *S* resource *R*	pos(truck *T*) pos(source *S*)
Output	increase(level(truck *T*, resource *R*))	new level(truck *T*, resource *R*)
Preconditions	same(pos(truck *T*), pos(source *S*)) level(source *S*, resource *R*) > 0	pos(truck *T*), pos(source *S*) time(between(truck *T*, source *S*))* level(source *S*, resource *R*)
Resources		
Time	refill dynamics	rate(increase(level(truck *T*, resource *R*, source *S*)))
Control		

state spaces, and comparing the location of the data in the state spaces to the thresholds between regions specifying differing opportunities for action. Our discussion emphasizes the state space representation of behavior relative to constraints, tells the stories the state spaces contain, and illustrates the utility of state space representations in a command-and-control environment.

An example of a state space for constraints is shown in Figure 17.5. This figure juxtaposes the variables "distance to the nearest fuel providing unit" and "fuel range" and plots data of all fire trucks (F1–6) and all water trucks (W7–9) during one entire trial. The distance to the nearest fuel-providing unit is the distance of a unit to the nearest fuel truck or fuel station. The fuel range indicates how many squares the truck can drive with the current fuel level. The higher black line indicates the states where the truck's fuel is exactly sufficient to reach the nearest fuel-providing unit. In the region above this line, the distance to the nearest fuel-providing unit is smaller than the distance that the current fuel level allows the truck to drive. In other words, when a truck is above the line, it can make it on its own to a fuel-providing unit. In the region below the lower constraint line, the truck does not have sufficient fuel to reach the nearest fuel-providing unit, and a fuel truck will have to come to it instead to refuel. It may be a strategy at a particular point in time that the fire trucks in a specific area with a small fire have to refuel themselves because fuel trucks are occupied elsewhere, and that fuel trucks in another area have to actively supply fire trucks with fuel because they are too occupied fighting extensive fires. This state space representation thus links the goal structure of the task, the constraints of the environment, strategies for performing the task, and the actual states of the control process in relation to the three former concepts.

Between these constraint lines, there is a gray zone where it is not certain whether the truck can reach its intended position. The existence of the gray zone in the state space reflects the uncertainty that is often encountered when traveling in the real world, where one often does not know the exact rate of fuel consumption. Often however one can estimate the upper and lower limits on fuel use. The exact fuel consumption is difficult to predict exactly because of local influences such as road condition, weather, and so on. In this way, C3Fire captures some of the inherent uncertainty in the command-and-control domain.

Strictly horizontal lines mean that the truck is not moving (fuel range stays the same) and the nearest fuel-providing unit is moving either towards or away from the truck. In cases when the fuel-providing unit moves away, another fuel-providing unit may become the nearest.

The vertical line in these state spaces at $X = 1$ reflects the fact that when trucks refuel, they are one square removed from the nearest fuel-providing unit. As they refill, their fuel range increases to a maximum of 20, tracing the vertical line at $X = 1$.

In Figure 17.5, we can again see that trucks generally had sufficient fuel ranges to make it to the nearest fuel-providing unit. Examination of the trace of truck W7 through the state space is informative. The oval in Figure 17.5 shows W7 at states (6.5, 8) and (6.5, 7), where it had enough fuel to reach the nearest fuel-providing unit. Between these states, W7 kept moving (decreasing fuel level) but the line goes

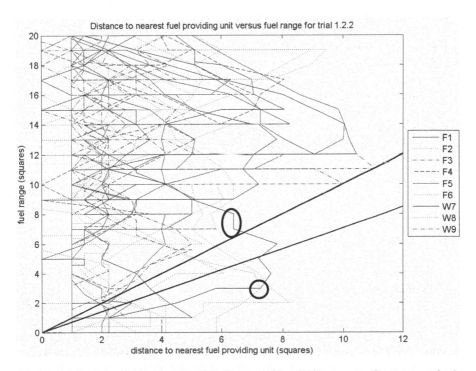

Figure 17.5 **Sample state space for the variables "distance to the nearest fuel-providing unit" and "fuel range" for fire trucks and water trucks**

straight down, indicating that the distance to the nearest fuel-providing unit stayed constant. It can be inferred that the nearest fuel-providing unit was moving in the same direction as W7. Thereafter the distance to the fuel-providing unit grew, and truck W7 then made a long excursion into the region of the state space that indicates that it could not reach a fuel-providing unit. It crossed both of the black constraint lines, eventually reaching a state with approximate coordinates (7.2, 3), highlighted by the circle in Figure 17.5. At this time, the horizontal line indicates that a fuel-providing unit approached truck W7. Thereafter, the diagonal line from (5.8, 3) towards the origin indicates that W7 started to burn fuel as it moved towards the fuel-providing unit.

The vertical lines at the bottom of the figure that intersect the horizontal axis reveal that trucks F2, F4, F5, F6, W7, and W8 all ran out of fuel one or more times. They were not positioned next to a fuel providing unit while their fuel ranges were zero. In these cases a fuel truck would have to come to them. Knowing this information in advance or being able to predict when this situation may happen could have been a major advantage to participants trying to engage in anticipatory constraint control (De Jong and Köster, 1974).

Discussion

A large pool of state spaces was obtained during the experiment, juxtaposing many different pairs (and even triples) of variables. The recognition of constraints and their illustration in state spaces has fulfilled two goals of this study. First, the state spaces *illustrate behavior* in terms of movement of vehicles, resource usage, and strategy and tactics in handling constraints. The participants' coping with constraints is illustrated in the state space by the movement of states towards or away from constraint lines. Review of the state spaces has identified trials with many crossings of constraint lines. In these trials, trucks frequently ran into difficulties and the fires burned out of control. In other state spaces, there were few or no crossings of constraint lines. This contrast indicates a large variety in the handling of constraints across teams and scenarios. Second, the study suggests that *representation design* in decision support systems and interfaces may usefully be based on the state space representation. In this study, information made available post-hoc in the state spaces suggests that the spaces give useful insight through the combination of essential variables and their behavior over time. Figure 17.6 recapitulates the state spaces in an abstract conceptual form.

The state space on the left-hand side of Figure 17.6 illustrates regions specifying differing opportunities for action. For states in the white region, fuel range is sufficient to bridge the distance. In the light gray zone, one cannot be certain if fuel range is sufficient, and the dark gray region indicates that fuel is insufficient to bridge the distance. The first example of a state space (Figure 17.5) is a concrete example of this kind of state space. The state space on the right of Figure 17.6 illustrates a more gradual specification of constraint, and more interlaced regions with opportunities for action. An example of such a state space may juxtapose the time available for an activity with the time needed for a related activity. In such cases, the combination and juxtaposition of constraints may not be as rigid, but can nevertheless be relevant in a goal-directed context. Other shapes of regions of opportunities for action with more constraint lines combined may be identified in the future, as they have been reported in the literature in other domains. The combination of essential variables into one visualization can be highly meaningful and useful.

State spaces form a very powerful representation. They can combine essential variables to form meaningful ensembles that are relevant to the goals of command and control. A value of an essential variable may not be informative in itself, but in the context of other variables and a commander's goals, meaning is added. Thus, the visually lucid nature of state spaces lets their observer interpret the values of essential variables in a goal-directed context. State spaces rearrange data in a way that facilitates (the macrocognitive concepts of) sensemaking and situation assessment and extraction of meaning, leaving the cognitive work and interpretation of data to the human controller, and thus avoiding the pitfalls of automation. The detection of problems in the form of "violations" of constraints and direction of attention at these problems are also likely to benefit from methods and representations that recognize constraints. The fact that a constraint is not adhered to may be part of people's strategy to cope with the unfolding situation. It may however also indicate undetected problems that require attention in order to adapt to unexpected circumstances and/or replan. Future research on the use of representations of constraint in displays will have to show how to train operators in their use, how these representations may be best fitted to actual work practice, and how the practice of work changes with the introduction of these displays.

To control a process means to know the history of the process, its current situation or state, and to be able to anticipate and affect future states (Hollnagel and Woods, 2005). State spaces address the behavior of essential variables in the process to be controlled through the recent history of the process, the current state of the process. They enable extrapolation of the past and present to anticipate future behavior.

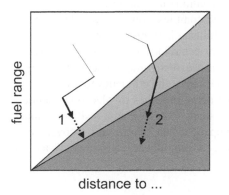 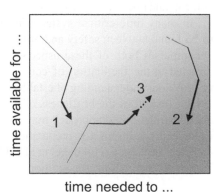

Figure 17.6   Conceptual idea of illustrating behavior and representing constraints in state spaces

Note: Arrows represent state transitions over time. Thickness of the solid lines illustrates recency of state transition. Dotted arrows illustrate possible projections of future behavior.

It is important to discuss the assumptions inherited from the experimental platform, C3Fire. Although C3Fire captures many essential factors and properties of network-based command-and-control systems, there are some obvious differences between C3Fire and command-and-control environments. The C3Fire environment is clearly defined in terms of states of variables and their values. It is entirely deterministic, and can be called a closed-loop system. In command-and-control settings, the variables that play a role in the control of complex systems may not always be clearly identifiable. Information about the values of these variables may be uncertain, unreliable, or incomplete. The utility of the method of recognizing constraints is not necessarily affected by these admittedly large differences. The state space representation does not interpret data; it merely enables human operators to interpret data and find meaningful patterns and relationships. The method does not assume that data are certain, reliable, or complete; it merely rearranges data in a manner that makes the constraints on opportunities for action obvious. It is important to inform operators whether and when information may be uncertain, unreliable, or incomplete, so that they can judge for themselves if the display is useful of not, just as this is the case with raw sensor data.

The representations developed here are meant to be incorporated in the network-based visions of distributed collaborative command and control. It therefore assumes the availability of data by means of a broadcasting structure of communication, where each information-providing entity sends information to a network, and entities connected to this network have access to this information, listening in on the information they need to fulfill their tasks. Although the method surely can be applied to a certain extent to situations where operators have to actively seek information, the information considered here is assumed to be readily available. The set of essential variables should therefore be identified either at the design stage of support system development, or in real time by the commanders, given a flexible representation of data. Issues of network bandwidth, data filtering, and communication technology should therefore be ascribed to the idea of the network structure of command and control, and not to the constraint recognition method outlined here.

In many time-critical systems in dynamic complex domains, a balance needs to be found between safety and efficiency (Hollnagel, 1999). The notion of control encompasses both of these aspects of system behavior, and research approaches directed at the facilitation of control such as the recognition of constraints may improve our understanding of establishing both safety and efficiency to a satisfactory extent. Instead of the brittleness of many earlier approaches to "decision support," the constraint-based approach may enable systems to behave in a resilient manner. Enhanced conditions for preparation, anticipation, adaptation, and recovery may facilitate resilience (Hollnagel et al., 2006), a concept that may also deserve a place in the literature on macrocognition.

Acknowledgement

The Swedish Defense Material Administration is gratefully acknowledged for supporting this research.

References

Adelson, M. (1961). "Human decisions in command control centers." *Annals of the New York Academy of Sciences*, 89: 726–31.

Alberts, D.S., Garstka, J.J., and Stein, F.P. (1999). *Network-centric Warfare: Developing and Leveraging Information Superiority* (2nd (revised) edn). Washington, DC: National Defense University Press.

Ashby, W.R. (1956). *An Introduction to Cybernetics* (Internet (1999) edn). London: Chapman and Hall. Available at <http://pcp.vub.ac.be/books/IntroCyb.pdf>.

—— (1960). *Design for a Brain: The Origin of Adaptive Behavior* (2nd (revised) edn). London: Chapman and Hall Ltd.

Bainbridge, L. (1983). "Ironies of automation." *Automatica*, 19 (6): 775–9.

Brehmer, B. (1992). "Dynamic decision making: Human control of complex systems." *Acta Psychologica*, 81: 211–41.

—— (2006). "One loop to rule them all." Paper presented at the 11th International Command and Control Research and Technology Symposium, Cambridge, UK.

Brehmer, B. and Allard, R. (1991). "Real-time dynamic decision making: Effects of task complexity and feedback delays." In J. Rasmussen, B. Brehmer, and J. Leplat (eds), *Distributed Decision Making: Cognitive Models for Cooperative Work*. Chichester: Wiley (pp. 319–47).

Brehmer, B. and Dörner, D. (1993). "Experiments with computer-simulated microworlds: Escaping both the narrow straits of the laboratory and the deep blue sea of the field study." *Computers in Human Behavior*, 9 (2–3): 171–84.

Brehmer, B. and Sundin, C. (2000). "Command and control in network-centric warfare." In C. Sundin and H. Friman (eds), *ROLF 2010 – The Way Ahead and the First Step*. Stockholm: The Swedish National Defence College (pp. 45–54).

Cebrowski, A.K. and Garstka, J.J. (1998). "Network-centric warfare: Its origin and future." *US Naval Institute Proceedings*, January: 28–35.

Chapman, R., Smith, P.J., Billings, C.E., McCoy, C.E., and Heintz Obradovich, J. (2001). "Collaborative constraint propagation as a planning strategy in the national airspace system." Paper presented at the 2001 Annual Meeting of the IEEE Society on Systems, Man and Cybernetics, Tucson, AZ.

Checkland, P. (1981). *Systems Thinking, Systems Practice*. Chichester: John Wiley and Sons.

De Jong, J.J. and Köster, E.P. (1974). "The human operator in the computer-controlled refinery." In E. Edwards and F.P. Lees (eds), *The Human Operator in Process Control*. London: Taylor and Francis Ltd (pp. 196–205).

Dekker, S.W.A. and Woods, D.D. (1999). "To intervene or not to intervene: The dilemma of management by exception." *Cognition, Technology and Work*, 1(2): 86–96.

—— (2002). "Maba-maba or abracadabra? Progress on human–automation coordination." *Cognition, Technology and Work*, 4(4): 240–44.

Funke, J. (2001). "Dynamic systems as tools for analysing human judgement." *Thinking and Reasoning*, 7(1): 69–89.

Gibson, J.J. (1986). *The Ecological Approach to Visual Perception*. Hillsdale, NJ: Lawrence Erlbaum Associates.

Granlund, R. (2002). "Monitoring distributed teamwork training." Unpublished PhD thesis, Linköpings universitet, Linköping.

Gray, W.D. (2002). "Simulated task environments: The role of high-fidelity simulations, scaled worlds, synthetic environments, and laboratory tasks in basic and applied cognitive research." *Cognitive Science Quarterly*, 2: 205–27.

Hollan, J., Hutchins, E., and Kirsch, D. (2000). "Distributed cognition: Toward a new foundation for human-computer interaction research." *ACM Transactions on Computer–Human Interaction*, 7(2): 174–96.

Hollnagel, E. (1986). "Cognitive systems performance analysis." In E. Hollnagel, G. Mancini, and D.D. Woods (eds), *Intelligent Decision Support in Process Environments*. Berlin: Springer-Verlag (pp. 211–26).

—— (1993). *Human Reliability Analysis: Context and Control*. London; San Diego, CA: Academic Press.

—— (1999). "From function allocation to function congruence." In S. Dekker and E. Hollnagel (eds), *Coping with Computers in the Cockpit*. Aldershot: Ashgate (pp. 29–53).

—— (2004). *Barriers and Accident Prevention*. Mahwah, NJ: Lawrence Erlbaum Associates.

Hollnagel, E. and Bye, A. (2000). "Principles for modelling function allocation." *International Journal of Human–Computer Studies*, 52(2): 253–65.

Hollnagel, E. and Woods, D.D. (1983). "Cognitive systems engineering: New wine in new bottles." *International Journal of Man–Machine Studies*, 18: 583–600.

—— (2005). *Joint cognitive systems: Foundations of cognitive systems engineering*. Boca Raton, FL: CRC Press.

Hollnagel, E., Woods, D.D., and Leveson, N. (eds) (2006). *Resilience Engineering: Concepts and Precepts*. Aldershot: Ashgate.

Hutchins, E. (1991a). "Organizing work by adaptation." *Organization Science*, 2(1): 14–39.

—— (1991b). "The social organization of distributed cognition." In L. Resnick, J. Levine, and S. Teasley (eds), *Perspectives on Socially Shared Cognition*. Washington, DC: APA Press (pp. 283–307).

Klein, G.A. and Calderwood, R. (1991). "Decision models: Some lessons from the field." *IEEE Transactions on Systems, Man and Cybernetics*, 21(5): 1018–26.

Klein, G.A., Ross, K.G., Moon, B.M., Klein, D.E., Hoffman, R.R., and Hollnagel, E. (2003). "Macrocognition." *IEEE Intelligent Systems*, 18(3): 81–5.

Knecht, W.R. and Smith, K. (2001). "The manoeuvre space: A new aid to aircraft tactical separation." In D. Harris (ed.), *Engineering Psychology and Cognitive Ergonomics*, Vol. 5. Aldershot: Ashgate (pp. 197–202).

Lee, J.D. and See, K.A. (2004). "Trust in automation: Designing for appropriate reliance." *Human Factors*, 46(1): 50–80.

Leveson, N.G. (2004). "A new accident model for engineering safer systems." *Safety Science*, 42: 237–70.

Lind, M. and Larsen, M.N. (1995). "Planning support and the intentionality of dynamic environments." In J.-M. Hoc, P.C. Cacciabue, and E. Hollnagel (eds), *Expertise and Technology: Cognition and Human–Computer Cooperation*. Hillsdale, NJ: Lawrence Erlbaum Associates (pp. 255–78).

Miller, G.A., Galanter, E., and Pribram, K.H. (1960). *Plans and the Structure of Behavior*. New York: Holt, Rinehart and Winston, Inc.

Neisser, U. (1976). *Cognition and Reality: Principles and Implications of Cognitive Psychology*. San Francisco, CA: W.H. Freeman and Company.

Newell, A. and Simon, H.A. (1961). "GPS: A program that simulates human thought." In H. Billings (ed.), *Lernende Automaten*. München: R. Oldenbourg (pp. 109–24).

Norman, D.A. (1998). *The Design of Everyday Things*. London: The MIT Press.

Persson, P.-A. (1997). "Toward a grounded theory for support of command and control in military coalitions." Unpublished Licentiate thesis, Linköpings universitet, Linköping, Sweden.

Port, R.F. and Van Gelder, T. (eds) (1995). *Mind as Motion: Explorations in the Dynamics of Cognition*. Cambridge, MA: The MIT Press.

Reitman, W.R. (1964). "Heuristic decision procedures, open constraints, and the structure of ill-defined problems." In M.W. Shelly II and G.L. Bryan (eds), *Human Judgments and Optimality*. New York: John Wiley and Sons, Inc. (pp. 282–315).

Roth, E.M., Malin, J.T., and Schreckenghost, D.L. (1997). "Paradigms for intelligent interface design." In M. Helander, T.K. Landauer, and P. Prabhu (eds), *Handbook of Human–Computer Interaction* (2nd edn). Amsterdam: Elsevier Science (pp. 1177–201).

Smith, K. (1997). "How currency traders think about the spot market's thinking." Paper presented at the Cognitive Science Society 19th Annual Conference.

—— (2001). "Incompatible goals, uncertain information, and conflicting incentives: The dispatch dilemma." *Human Factors and Aerospace Safety*, 1(4): 361–81.

Smith, W. and Dowell, J. (2000). "A case study of coordinative decision-making in disaster management." *Ergonomics*, 43(8): 1153–66.

Stappers, P.J. and Flach, J.M. (2004). "Visualizing cognitive systems: Getting past block diagrams." Paper presented at the IEEE International Conference on Systems, Man and Cybernetics, The Hague, The Netherlands.

Vicente, K.J. (1999). *Cognitive Work Analysis: Towards Safe, Productive, and Healthy Computer-based Work*. Mahwah, NJ: Lawrence Erlbaum Associates.

Vicente, K.J. and Rasmussen, J. (1992). "Ecological interface design: Theoretical foundations." *IEEE Transactions on Systems, Man and Cybernetics*, 22(4): 589–606.

Von Neumann, J. and Morgenstern, O. (1953). *Theory of Games and Economic Behavior* (3rd edn). Princeton, NJ: Princeton University Press.

Woods, D.D. (1986). "Paradigms for intelligent decision support." In E. Hollnagel, G. Mancini, and D.D. Woods (eds), *Intelligent Decision Support in Process Environments*. Berlin: Springer-Verlag (pp. 153–73).

—— (1995). "Toward a theoretical base for representation design in the computer medium: Ecological perception and aiding human cognition." In J. Flach, P. Hancock, J. Caird, and K. Vicente (eds), *Global Perspectives on the Ecology of Human–Machine Systems*. Hillsdale, NJ: Lawrence Erlbaum Associates (pp. 157–88).

Chapter 18

Situated Cognitive Engineering for Complex Task Environments

Mark A. Neerincx and Jasper Lindenberg

Introduction

Effective, efficient, and easy-to-learn operation support is crucial for human-system performance in complex task environments, such as space laboratories, ship control centers and process control rooms. Progress in Information and Communication Technology and Artificial Intelligence enables the development of Human-Machine Collaboration (HMC), that is, systems in which human and machine actors perform their tasks jointly (Hoc, 2001). Such machines will be designed to cooperate with humans, and to assess and adapt to human goals (Castelfranchi, 1998). The general aim is to accommodate user characteristics, tasks, and contexts in order to provide the "right" information, services, and functions at the "right" time and in the "right" way (Fischer, 2001). Due to the adaptive nature of both the human and machine behavior, it is difficult to provide generic and detailed predictions on the overall human-machine performance. There is a lack of theories and empirical evidence to derive such predictions. An important question is how to improve and apply the knowledge base on the most relevant human and machine factors, and their effects on the concerning operations for the envisioned adaptive cognitive systems.

Cognitive engineering (CE) approaches originated in the 1980s to improve computer-supported task performance by increasing insight in the cognitive factors of human-computer interaction (for example, Hollnagel and Woods, 1983; Norman, 1986; Rasmussen, 1986). These approaches guide the iterative process of development in which an artifact is specified in more and more detail and specifications are assessed more or less regularly to refine the specification, to test it, and to adjust or extend it. For cooperative systems, the "classical" methods must be extended with an explicit technology input for two reasons. First, the technological design space sets a focus in the process of specification and generation of ideas. Second, the reciprocal effects of technology and human factors are made explicit and are integrated in the development process. The human factors knowledge provides relevant theories, guidelines, support concepts, and methods for the specification and assessment of HMC. In the specification, both the guidelines and the technological design space must be addressed concurrently. In the assessment, it is checked whether the specifications agree with these guidelines and the technological design space. An assessment will provide qualitative or quantitative results in terms of effectiveness, efficiency, satisfaction, learnability, and trust which are used to refine, adjust, or extend the specification. Eventually, the process of iteration stops when the assessment shows that the HMC satisfies the requirements (as far as the resources and completion date allow). So, we propose a CE+ method, adding a technology perspective into "common" human factors engineering practices (Maanen et al., 2005).

In addition to the added focus on the technology that can be applied in the concerning situation, we propose to develop practical theories and methods that are *situated* in the domain. In this approach, above-mentioned assessments are being used to improve the situated cognitive engineering methods, to establish a common design knowledge base, and to develop a kind of design guide. For realizing adequate HMC performance, generic human factors knowledge and HMC solutions are refined, contextualized, and tested within the domain, as part of the HMC development process. The resulting situated cognitive engineering method comprises an increasing knowledge base on the most relevant human and machine factors for the overall task performance.

The situated CE+ framework has been developed and applied in the defense and space domain for the design of cognitive support that enhances the capacities of teams and team members during critical and complex operations (for example, to improve task load management, trouble-shooting, and situation awareness). It is based on experiences with previous and current (space, naval) missions and based on practical theories on the object of support (for example, cognitive task load; Neerincx, 2003). Such a theory comprises accepted features of human cognition, to be "contextualized or instantiated" for the application domain. Corresponding to the concept of *macrocognition* (Klein et al., 2003), these features reflect cognitive functions at the level of "natural" decision-making settings and work demands. This chapter presents two case studies that show the development and application of the situated CE+ framework, respectively for naval ships and space stations.

Task Support in Naval Ships

Design of information and communication technology for naval ships is characterized by the involvement of diverse stakeholders and the implementation of diverse applications (platform and weapon systems). There was a need for a concise and coherent design approach for the development processes for future naval ships, in order to realize adequate deployment of human and technical resources for the new set of naval missions (for example, for littoral operations). This approach should incorporate state-of-the-art human factors knowledge, facilitate the application of new enabling technology, and fit the specific defense context.

Figure 18.1 shows the design approach and guidance that we are developing for the Royal Netherlands Navy SEAMATE (Situated Engineering for Adaptive Maritime Aiding Task Environments). First, we developed a model of cognitive task load (CTL) that could be used to harmonize the task demands to the human capacities via task allocation and design of cognitive support, for example, to prevent overload and cognitive lock-up during damage control on the naval ship. This theory is being parametrized for the naval domain (for example, the specific regions of overload and cognitive lock-up were defined for naval operators; Grootjen et al., 2006c). Second,

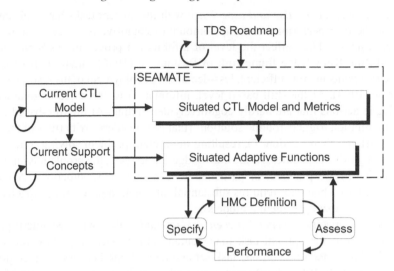

Figure 18.1 Design guidance for intelligent user interfaces of naval ships

Notes: CTL = Cognitive Task Load; TDS = Technological Design Space; HMC = Human-Machine Collaboration.

we distinguished current support concepts for the harmonization and, based on these concepts, defined a specific set of support functions and support modes as "building blocks" for the adaptive functions in future naval ships. Third, we are specifying a Technological Design Space (TDS) roadmap with a scope of five to ten years in which we describe the expected developments of the enabling technology for the adaptive functions. These functions are currently being developed in an iterative process. When system development processes for future naval ships start, SEAMATE can be used to guide the development of the intelligent user interfaces. The next three sections will provide an overview of the cognitive ingredients of SEAMATE: the CTL model, the support concepts, and the adaptive functions.

Cognitive Task Load (CTL)

For attuning the user interface to the individual cognitive capacities, we developed a practical theory on cognitive task load and support for supervision and damage control (Neerincx, 2003). This theory distinguishes three cognitive factors.

First, *cognitive processing speed* is the speed of executing relatively over-learned or automatized elementary cognitive processes, especially when high mental efficiency (that is, attention and focused concentration) is required (see Carroll, 1993). Cognitive processing speed is usually measured by tasks that require rapid cognitive processing, but little thinking. It contains elements such as perceptual speed (the ability to rapidly search for and compare known visual symbols or patterns), rate-of-test-taking (the ability to rapidly perform tests which are relatively easy or that require very simple decisions), and number facility (the ability to rapidly and accurately manipulate and deal with numbers).

Second, user's *expertise and experience* with the current tasks have substantial effect on the user performance and the amount of cognitive resources required for this performance. They affect the level of information processing as indicated by the Skill-Rule-Knowledge framework of Rasmussen (1986): higher expertise and experience results in more efficient, less-demanding deployment of the resources (see Anderson, 1993). At the skill-based level, information is processed automatically, resulting in actions that are hardly cognitively demanding. At the rule-based level, input information triggers routine solutions (that is, procedures with rules of the type "f <event/state> then <action>"), resulting in efficient problem solving in terms of required cognitive capacities. At the knowledge-based level, the problem is analyzed and solution(s) are planned, in particular to deal with new situations. Accurate knowledge-based behavior requires substantial information processing capacity and knowledge (for example, mental models).

Third, the capacity to *switch between task-sets* and dealing with task interruptions is often required for optimal task management and interleaving. However, users are inclined to concentrate on one task-set and neglect another task-set ("cognitive lockup," Kerstholt, 1997). Furthermore, switching can be a major mental load factor in itself and the elderly seem to have generally less capacity for switching (Sit and Fisk, 1999). Kramer et al. (1999) found large age-related differences in switch-costs early in practice (that is, the costs in reaction time and errors due to switching between two tasks). After relatively modest amounts of practice, the switch costs for old and young adults became equivalent and maintained equivalent across a two-month retention period. However, under high memory loads, older adults were unable to capitalize on practice and the switch costs remained higher for them. To address task demands and individual capacities, user interfaces must be designed to accommodate people's memory limitations relative to task-set switching and to being interrupted (McFarlane, 1999).

Cognitive Support

In order to address operator limitations for processing speed, expertise and experience, and task-set switching capability, four functions of cognitive support were developed to improve supervision and damage control (Neerincx, 2003). An overview of the cognitive factors, support concepts and support functions is given in Table 18.1.

First, the *information handler* filters and combines information to improve task performance time and situation awareness, that is, knowledge of the state of the system and its environment. Due to the increasing availability of information and speed of information changes, situation awareness can deteriorate without support. Sensor information should therefore be combined into alarms that are structured according to their function, such as fire control, propulsion, and energy supply. Furthermore, information handling can support the operators in keeping overview by making the structure of the complete system explicit at a global level and by indicating functional relationships between system components. Relevant information should be presented at the right time, at the right abstraction level, and be compatible with the human cognitive processing capacity.

Table 18.1 Cognitive factors with accompanying support concepts and functions for supervision and damage control

Cognitive factor	Support concept	Support function
Processing speed	Combining and structuring information	Information Handler
Expertise and experience	Providing normative procedures	Rule Provider
	Guidance of diagnostic processes	Diagnosis Guide
Task-set switching	Providing an overall work plan	Task Scheduler

Second, the *rule provider* gives the normative procedure for solving (a part of) the current problem, complementing the user's procedural knowledge. Due to training and experience, people develop and retain procedures for efficient task performance. Performance deficiencies may arise when the task is performed rarely so that procedures will not be learned or will be forgotten, or when the information does not trigger the corresponding procedure in human memory.

Third, the *diagnosis guide* complements user expertise and experience. The level of information processing increases when no complete (executable) procedure is available to deal with the current alarms and situation. This support function guides the operator during the diagnosis, resulting in an adequate problem-solving strategy for a specific task.

Fourth, the *task scheduler* affects the task-set switches by providing an overall continuously updated work plan for emergency handling. Task priorities are dynamically set and shown in a task-overview to the operator resulting in effective and efficient switches.

Prototypes that contain such functions have been developed and tested in specific application domains; in general, the support proves to have a beneficial effect on task performance (for example, Neerincx and De Greef, 1998; Neerincx, 2003). However, this support is still static and neither attuned to the dynamics of the task and context demands, nor to the specific (momentary) individual capacities of an operator. To establish the envisioned human-machine collaboration, we will personalize the support functions in so-called adaptive functions: first, to determine which support functions a specific user needs at a specific moment, and second, to establish the specific support-mode and mode-control conditions.

Adaptive Functions

Personalization concerns the use of information about general and momentary user needs, to deliver appropriate services and content, in a format tailor-made to the user and to the context-of-use (Cremers and Neerincx, 2004). Our aim is to provide personalized support by accommodating individual user characteristics, tasks, and contexts in order to establish human-computer collaboration in which the computer complements the momentary cognitive resources of the user (see Fischer, 2001).

A first requisite for providing cognitive support functions tuned to the momentary or predicted future capacities of the user is the ability to make real-time assessments and/or predictions of these capacities (Grootjen et al., 2006a). In the previous section, we described different cognitive capacities influencing performance in human-computer tasks. The basic capacities of a user can be measured and contained in a user model (experience, expertise, spatial ability, and so on). As we argued in the introduction, context can be of major influence on cognitive capacities. Therefore, information about the current context should be placed in a context model. Earlier, a number of support concepts were described; if available, such support concepts can be put in the system model accompanied by other information about the overall system (that is, component status). Further, users are trying to accomplish certain goals by performing tasks and interacting with the system. The task model should therefore contain the relevant user tasks and specify which tasks are active currently. Based on the information contained in the four models, personalized support can be provided. A personalized support module determines the momentary capacities of a specific user, based on his or her "default capacities" from the user model and the information from the context model. Combining the user's momentary capacity with information about the tasks that are to be performed and the available support modules, this personalization module can select and activate the appropriate support function.

The four functions of cognitive support (Table 18.1) were developed for static support and used as a basis for the development of dynamic, personalized support mechanisms (see Figure 18.2). Support modules can be switched on or off, based on the momentary operator capacities. Reducing support at certain times can also be useful to prevent underload and reduction of skills. In addition, alarms (or clusters of alarms) can be set in certain modes (determined by operator, system, or both). These modes determine the behavior of the support functions and thus provide an extra opportunity to fine-tune the support to the available capacities. There are five modes for each alarm affecting the behavior of the procedure support function with regard to that specific alarm:

- *Manual.* All procedure steps must be performed individually.
- *Critiqued.* All procedure steps must be performed by the operator. However, the system looks over the shoulder of the operator and critiques the operator when necessary (for example to point to specific risks, conventions or rules that apply to the situation and actions).
- *Supervised.* All procedure steps are performed by the operator. However, another human operator (or manager) is supervising and might be asked to confirm specific critical assessments or action plans.
- *Confirm.* The system performs all procedure steps it can. The operator steps in to supervise difficult (or sensitive) steps or to perform steps in the procedure which cannot be performed by the system. This mode can be viewed as a specific type of supervisory control in which the technical system performs most functions autonomously, and the operator confirms critical assessments or action plans.
- *Automatic.* All steps are performed by the system autonomously.

Not all support levels are available all the time for each alarm (type). For example, there are procedures that cannot be (*or* are not allowed to be) performed automatically. In that case "confirm" would be the highest level of automation.

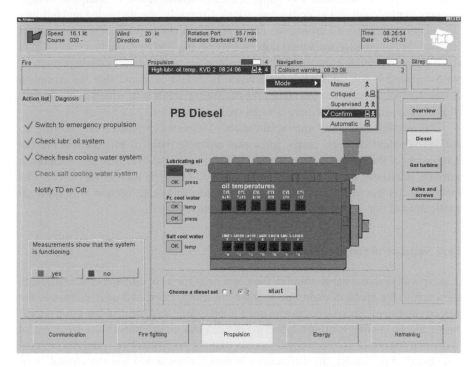

Figure 18.2 **Adaptive support interface in the confirm mode: "joint machine-operator" procedure execution to handle high oil-pressure alarm**

Iterations

The two components of SEAMATE (that is, model and metrics, and functions) have been specified and assessed in numerous cycles in case studies with the Royal Netherlands Navy (see Figure 18.1). First, we established and refined the situated CTL model and metrics in empirical studies on board of naval frigates and in a high-fidelity simulator (Grootjen et al., 2006a, 2006c). Second, we transformed general cognitive support functions into situated "building blocks" of support functions for naval ship operations, and evaluated these "building blocks" in several experiments to refine the support functions (Grootjen et al., 2002; Neerincx, 2003; Neerincx et al., 2004). Results of these evaluations provided (situated) requirements for adaptive mechanisms to deal with the dynamics of the operator capacities, operational demands, and environmental constraints. Examples are requirements to support operators' understanding of the adaptation mechanism (such as automatic reconfiguration of energy supply in critical conditions) and to allow operators to access the models that affect this mechanism (such as the definition of ship's readiness state). Subsequently, we designed situated adaptive functions and integrated these functions into the overall support interface. For example, to add platform supervision tasks to the bridge, it was determined which combinations of crew settings, platform and navigation tasks, contexts, support functions, and support modes would result in adequate human resource deployment and effective operations. Currently, we are testing and refining this adaptive interface to attune the momentary support provision (that is, support functions and modes) to the actual state of the operator and environment (Grootjen et al., 2006b). Figure 18.2 shows such a user interface that integrates damage control, platform supervision, and navigation tasks. It should be noted that the complete design comprises adaptive interfaces for multi-user settings in which dynamic task allocation among crew members is supported.

This section discussed our application of situated cognitive engineering to the development of naval ships. Important characteristics of this domain are the complexity of the technical systems, the risks of operations in critical and sometimes extreme situations, and the design for envisioned missions that seldom occur. The space domain has similar characteristics, and the following section will show an application of our design approach for space stations.

Task Support in Space Stations

Corresponding to the naval domain, there is a need for a concise and coherent design approach in the space domain due to the complexity, risks, and uncertainties in the task environment. Space laboratories are a good example of the problems that appear during the design of operation support in complex task environments: the involvement of diverse stakeholders, the implementation of diverse applications (platform systems and so-called payloads), the differences in design approaches and the separation of a task and a user-interface design community. In previous manned space missions, procedural support, the mapping of task procedures on the user interfaces, the usability of the individual systems (including the fit to "context of use"), and the consistency between interfaces showed serious shortcomings (Neerincx et al., 2004). This resulted in extensive training and preparation efforts, and non-optimal task performance of the astronauts and cosmonauts. A common cognitive engineering design approach for both the tasks and interfaces is required to structurally enhance the effectiveness and efficiency of the crew operations. Such an approach should be tailored to the specific requirements of the application domain with respect to the development process (for example, for adequate contributions of the different stakeholders' expertise) and end-user operations (for example, for fault detection, isolation, and recovery).

For supporting the diverse stakeholders of the overall design process, we developed a cognitive engineering toolkit, called SUITE (Situated Usability engineering for Interactive Task Environments), similar to the SEAMATE structure (Figure 18.3). Whereas SEAMATE aims at systems over five to ten years, SUITE aims at the current systems that are being developed for space missions. First, based on current cognitive engineering approaches, we formulated an initial design method and user-interface framework. According to this method, procedure and interface design is an iterative process, in which high-level task specifications and user requirements are being refined systematically into detailed action schemes and displays and, finally, the actual task documentation and user interface. The method was provided as a usability engineering handbook that contains context- and user-tailored views on the recommended Human Factors method, guidelines, and best practices (that is, the development of procedures and user interfaces for three different payloads). Furthermore, SUITE provides a task support and dialogue framework, called Supporting Crew OPErations (SCOPE), as both an implementation of these methods and guidelines, and an instance of current interaction and AI technology for human-machine collaboration. This framework defines a common multi-modal interaction with a system, including multimedia information access, virtual control panels, alarm management services, and the integrated provision of context-specific task support for nominal and off-nominal situations. In addition to the support of supervision and damage control, it provides support to access and process multimedia information, for instance, for preparation of actions, training on the job, or maintenance tasks.

The next two sections will present some cognitive ingredients of SUITE: the hypermedia processing model and corresponding support functions.

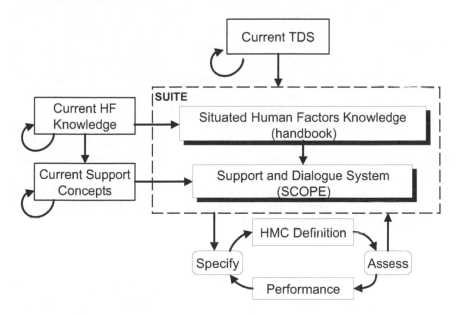

Figure 18.3   Design guidance for intelligent user interfaces of space stations

Notes: HF = Human Factors; TDS = Technological Design Space; HMC = Human-Machine Collaboration.

Hypermedia Processing

The four cognitive factors of supervision in the Navy (Table 18.1) also have a substantial effect on an astronaut's task performance. For the space mission, however, we had a broader support focus and, consequently, addressed other relevant factors as well to be included in a practical theory. Astronauts must process a lot of information during space operations. For searching and navigating in hypermedia environments via manual and speech commands, we therefore developed a practical theory on user and context effects, distinguishing two cognitive factors (Neerincx et al., 1999, 2000, 2001).

The *spatial ability* of users proves to affect the Web-navigation performance. Users with poor spatial ability have more problems with navigation in websites, requiring extra search and navigation support (Czaja, 1997). Höök et al. (1996) found that spatial ability is related to the time spent in completing a set of tasks in a large, hypermedia, information structure. Particularly, certain aspects of spatial ability were related to the ability to navigate in hypermedia, namely those related to solving spatial problems mentally rather than solving spatial problems in the physical world. An experiment by Vicente et al. (1987) demonstrated that psychometric tests of vocabulary and spatial visualization are the best predictors of task performance (searching in a hierarchical file system), accounting for 45 percent of the variance. The spatial predictor was found to be most influential.

A second cognitive factor that plays an important role in computer-supported tasks is *memory capacity*. Especially working memory can influence task performance. The working memory resources determine our ability to retain and manipulate the limited amount of material that falls within our focus of attention. Differences are small or nonexistent for tasks that simply require people to retain a small amount of information for short periods (digit-span task; see Kausler, 1994), but are larger when people must simultaneously process, or manipulate, the material (reading span task). For language, the elderly experience relatively large limitations in overall working memory capacity at times of high demand (Just and Carpenter, 1992).

It is interesting to note that contextual factors, such as background noise or distracting events, can provide similar constraints to the interaction as individual capacities, such as specific auditory limitations or attention reductions. Furthermore, context and capacities can interact in such a way that specific tasks may become difficult to perform for some persons. For example, some people may be hindered severely by distracting phone calls when fatigued (for example, at night), whereas others can proceed with their work easily.

Cognitive Support

Corresponding to the Navy example, we developed support concepts to deal with limitations in processing speed and expertise and experience (that is, combining and structuring of information, providing normative procedures and guidance of diagnostic processes; see Table 18.1). In order to address the contextual and individual affected limitations on spatial ability and memory for searching and navigating in hypermedia environments, we first developed three support concepts (Table 18.2).

The *Categorizing Landmarks* are cues that are added to the interface to support the users in recognizing their presence in a certain part of the multimedia content (that is, it arranges information into categories that are meaningful for the user's task). This should help the users to perceive the information in meaningful clusters and prevents the user from getting lost. For example, specific categories of information have a dedicated background color; hyperlinks that refer to this content have the same color.

Table 18.2   Cognitive factors with accompanying support concepts and functions for searching and navigating in hypermedia environments

Cognitive factor	Support concept	Support function
Spatial ability	Support the users in recognizing their presence in a certain part of a hypermedia environment.	Categorizing Landmarks
Memory	Provide an overview of the overall structure of a hypermedia environment including the user's current and past views.	History Map
	Aid users' comprehension of current speech commands that can be used.	Speech Command View

The basis for the *History Map* is a "sitemap": a representation of the structure of the multimedia content. It shows a hierarchical overview or tree to indicate the location of the content currently being viewed and may include a separate presentation of the current "leaf" ("breadcrumb"). History information is annotated in the overview. This memory aid should improve users' comprehension of the content structure in relation to their task and provide information about the status of their various sub-goals.

A *Speech Command View* presents the specific commands that a specific user is allowed to use for controlling the current active part of the application. Current and most recent commands can be indicated in the view.

The *Rule Provider* for astronauts is rather similar to the Rule Provider for naval operators (Table 18.1). The *Diagnosis Guide* for the astronauts has been developed in more detail than for naval operators, and is an important support function of SCOPE. It detects system failures, guides the isolation of the root causes of failures, and presents the relevant repair procedures in textual, graphical and multimedia formats (Bos et al., 2004). The diagnosis is a joint astronaut-SCOPE activity: when needed, SCOPE asks the astronaut to perform additional measurements in order to help resolve uncertainties, ambiguities, or conflicts in the current machine status model. SCOPE will ask the user to supply values to input variables it has no sensors for measuring by itself. Each new question is chosen on the basis of an evaluation function that can incorporate both a cost factor (choose the variable with the lowest cost) and a usefulness factor (choose the variable that will provide the largest amount of new information to the diagnosis engine). After each answer, the diagnosis reevaluates the possible fault modes of the system on the basis of the additional values (and new samples for the ones that can be measured). As soon as SCOPE has determined the most likely health state(s) of the system with sufficient probability, it presents these states to the user, possibly with suggestions for appropriate repair procedures that can be added to the to-do list and executed. As soon as the machine has been repaired, SCOPE will detect and reflect this.

Iterations

According to our general cognitive engineering approach, we apply and refine SUITE during "real" development processes. In internal and external reviews, interim versions of the cognitive engineering method were refined and the final versions were evaluated. The first application focused on the Cardiopres payload (a system for continuous physiological measurement, such as blood pressure and ECG, which will be used in several space missions). The user interface of the prototype was running on a Tablet PC with direct manipulation and speech dialogue (see Figure 18.4). Via a domain analysis and technology assessment, we defined the scope of this prototype (that is, the scenarios and technology to be implemented). The prototype design (that is, the user interface framework) was tested via expert reviews, user walkthroughs with astronauts and a usability test with ten participants measuring effectiveness, efficiency, satisfaction, and learnability. In the evaluations of the SCOPE system for the Cardiopres, the user interface and cooperative task support functions proved to be effective, efficient, and easy to learn, and astronauts were very satisfied with the system (Neerincx et al., 2004).

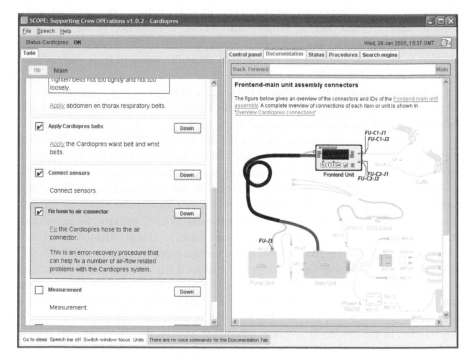

Figure 18.4 SCOPE showing a successful completion of a cooperative diagnosis process (status bar at top), procedure generation (left), and reference documentation (right)

The development of the SUITE tool kit is an iterative process in itself, and new experiences with its application will improve it. Recently, the SCOPE framework was applied for the development of an intelligent user interface for the Pulmonary Function System (PFS) payload (Figure 18.5; Bos et al., 2006). Its task support functions will be improved to deal with dependencies of actions with each other and the usage context. Assessments will help to establish adequate performance and user experience of this component (see Figure 18.3). In general, the SUITE toolkit reduces the time and cost of development efforts, whereas it improves the usability of intelligent interfaces. Embedded in a cognitive engineering process, user interfaces and the underlying AI methods are systematically and coherently specified, implemented, and assessed from early development phases on, which is in itself efficient and prevents the need for late harmonization efforts between user requirements and technological constraints.

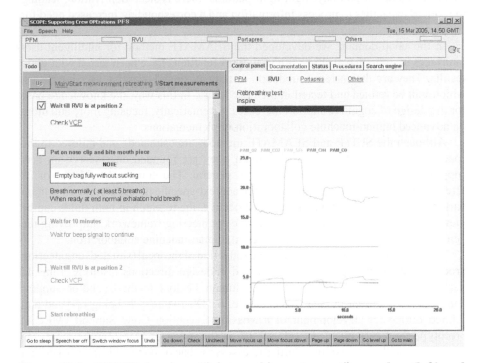

Figure 18.5   SCOPE showing a "joint machine-astronaut" procedure (left) and control panel (right)

Conclusion

This chapter presented a generic cognitive engineering approach for the design of human-machine collaboration in which both human factors and technological constraints are being addressed in an iterative development process. Two case studies showed how this approach can be tailored to the specific needs of an application domain and refined into a toolkit for design guidance. In the space domain, the proposed SUITE method reduces the time and cost of development efforts, whereas it improves the usability of procedures and interfaces. In the project team, the SUITE design approach facilitated the contribution of different expertise types (for example, interaction design, software technology, payload) at the right time (that is, as early as possible), whereas the use scenarios facilitated the communication between the different stakeholders. The SCOPE framework proved to support nominal and off-nominal operations effectively and efficiently. Although the method and SCOPE framework can be applied independently, their combination will show the largest beneficial effects in terms of (1) better user performance, (2) earlier design changes, (3) reduced training efforts, and (4) efficient maintenance. In general, it should be stressed that the in-orbit environment and working conditions are extreme, requiring an extra need to "situate" the support. For example, due to micro-gravity, a restricted living area and unusual day-night light patterns, there is sleep deprivation, leading to reduced capacities to process information and a corresponding increased need for support.

In the naval domain, we developed the SEAMATE method for intelligent user interfaces in future naval ships. Some support functions have been tested with good results. They are the current foundation of the envisioned adaptive user interfaces, which will be refined and tested in the near future. In this way, the knowledge base for the design of cognitive support increases systematically, focusing more and more on advanced human-machine collaboration implementations.

Although the SUITE and SEAMATE methods are "situated" and tailored to the space and, respectively, the naval domain, they are founded on research on task support in complex task environments in general. The empirical evidence of its effectiveness, for instance, the usage of the context-specific procedures, is, for a main part, consistent over domains. In this perspective, the research in both fields can be viewed as exemplars of a generic cognitive engineering framework, improving the general foundation of effective and efficient human-machine collaboration.

This general framework addresses the reflection of the local (situated) engineering processes, in order to prevent being locked in design directions resulting from old Technological Design Spaces (TDSs), Human Factors methods and/or support concepts. In a current project, for example, a TDS roadmap is being updated and refined during the development of a Mission Execution Crew Support (MECA) prototype for future manned explorations on the moon or Mars. Furthermore, the project explicitly asks external parties to assess the background and (interim) design choices (Neerincx et al., 2006).

The situated cognitive engineering approach focuses on the performance of the mental activities of human actors and the cognitive functions of machine actors, to achieve the (joint) operational goals. In this way, the notion of collaboration has been extended, viewing the machine as a social actor that can take initiative to act, critique, or confirm in joint human-machine activities. The design focuses on the manifestation of these activities in "real settings," corresponding to the concept of macrocognition (Klein et al., 2003). The evaluations in the naval and space domains improved our understanding of the relevant macrocognitive functions, such as situation assessment during naval damage control and problem detection during payload operations in a space station. These insights are being incorporated in the SEAMATE and SUITE tools to improve current and future cognitive engineering practices in these domains.

Acknowledgements

The SUITE tool-kit was developed for the European Space Agency (contract C16472/02/NL/JA) in cooperation with Science and Technology (The Netherlands). The development of SEAMATE is funded by the Dutch Ministry of Defense.

References

Anderson, J.R. (1993). *Rules of the Mind*. Hillsdale, NJ: Lawrence Erlbaum Associates.

Bos, A., Breebaart, L., Neerincx, M.A., and Wolff, M. (2004). "SCOPE: An Intelligent Maintenance System for Supporting Crew Operations." *Proceedings of IEEE Autotestcon 2004*. ISBN 0-7803-8450-4 (pp. 497–503).

Bos, A., Breebaart, L., Grant, T., Neerincx, M.A., Olmedo Soler, A., Brauer, U., and Wolff, M. (2006). "Supporting Complex Astronaut Tasks: The Right Advice at the Right Time." *SMC–IT 2006: 2nd IEEE International Conference on Space Mission Challenges for Information Technology*. Los Alamitos, California: IEEE Conference Publising Services (pp. 389–96).

Carroll, J.B. (1993). *Human Cognitive Abilities: A Survey of Factor Analytic Studies*. New York: Cambridge University Press.

Castelfranchi, C. (1998). "Modelling social action for agents." *Artificial Intelligence*, 103: 157–82.

Cremers, A.H.M. and Neerincx, M.A. (2004). "Personalisation meets accessibility: Towards the design of individual user interfaces for all." In C. Stary and C. Stephanidis (eds), *User-Centered Interaction Paradigms for Universal Access in the Information Society. Lecture Notes in Computer Science*. Berlin: Springer (pp. 119–24).

Czaja, S.J. (1997). "Computer technology and the older adult." In M.G. Helander, T.K. Landauer, and P.V. Prabhu (eds), *Handbook of Human-Computer Interaction*, 2nd edn. Amsterdam: Elsevier North-Holland (pp. 797–812).

Fischer, G. (2001). "User modeling in human-computer interaction." *User Modeling and User-Adapted Interaction*, 11: 65–8.

Grootjen, M., Bierman, B., and Neerincx, M.A. (2006a). "Optimizing cognitive task load in naval ship control centres using adaptive support." *16th World Congress on Ergonomics IEA2006*. (CD–ROM). Amsterdam, The Netherlands: Elsevier (pp. 661–7).

Grootjen, M., Neerincx, M.A., and van Weert, J.C.M. (2006b). "Task-Based Interpretation of Operator State Information for Adaptive Support." *ACI2006: 2nd Augmented Cognition International Conference*, 15–20 October 2006, San Francisco, CA.

Grootjen, M., Neerincx, M.A., and Veltman, J.A. (2006c). "Cognitive task load in naval ship control centres: From identification to prediction." *Ergonomics*, 49: 1238–64.

Grootjen, M., Neerincx, M.A., and Passenier, P.O. (2002). "Cognitive task load and support on a ship's bridge: Design and evaluation of a prototype user interface." *INEC 2002 Conference Proceedings*. London: IMarEST (pp. 198–207).

Hoc, J.-M. (2001). "Towards a cognitive approach to human-machine cooperation in dynamic situations." *International Journal of Human–Computer Studies*, 54(4): 509–40.

Hollnagel, E. and Woods, D.D. (1983). "Cognitive systems engineering: New wine in new bottles." *International Journal of Man–Machine Studies*, 18: 583–600.

Höök, K., Sjölinder, M., and Dahlbäck, N. (1996). *Individual Differences and Navigation in Hypermedia*. SICS Research Report, R96:01, SICS, Sweden.

Just, M.A. and Carpenter, P.A. (1992). "A capacity theory of comprehension: individual differences in working memory." *Psychological Review*, 99(1): 122–49.

Kausler, D.H. (ed.) (1994). *Learning and Memory in Normal Aging*. San Diego, CA: Academic Press.

Kerstholt, J.H. (1997). "Dynamic decision making in non-routine situations." In R. Flin, E. Salas, M. Strub, and L. Martin (eds). *Decision Making Under Stress: Emerging Themes and Applications*. Aldershot: Ashgate Publishing Company (pp. 185–92).

Klein, G., Ross, K.G., Moon, B., Klein, D.E., Hoffman, R., and Hollnagel, E. (2003). "Macrocognition." *IEEE Intelligent Systems*, May/June 2003: 81–5.

Kramer, A.F., Hahn, S., and Gopher, D. (1999). "Task coordination and ageing: Explorations of executive control processes in the task switching paradigm." *Acta Psychologica*, 101: 339–78.

Maanen, P.P. van, Lindenberg, J., and Neerincx, M.A. (2005). "Integrating human factors and artificial intelligence in the development of human-machine cooperation." In H.R. Arabnia, and R. Joshua (eds), *Proceedings of the 2005 International Conference on Artificial Intelligence (ICAI '05)*. Las Vegas, NV: CSREA Press (pp. 10–16).

Maguire, M. (2001). "Methods to support human-centred design." *International Journal of Human–Computer Studies*, 55: 587–634.

McFarlane, D.C. (1999). "Coordinating the interruption of people in human–computer interaction." In M.A. Sasse and C. Johnson (eds), *Human-Computer Interaction – Interact '99*. Amsterdam: IOS Press (pp. 295–303).

Neerincx, M.A. (2003). "Cognitive task load design: Model, methods and examples." In E. Hollnagel (ed.), *Handbook of Cognitive Task Design*. Mahwah, NJ: Lawrence Erlbaum Associates (pp. 283–305).

Neerincx, M.A. and De Greef, H.P. (1998). "Cognitive support: Extending human knowledge and processing capacities." *Human Computer Interaction*, 13(1): 73–106.

Neerincx, M.A., Cremers, A.H.M., Bos, A., and Ruijsendaal, M. (2004). *A Tool Kit for the Design of Crew Procedures and User Interfaces in Space Stations*, Report TM–04–C026. Soesterberg, The Netherlands: TNO Human Factors.

Neerincx, M.A., Grootjen, M., and Veltman, J.A. (2004). "How to manage cognitive task load during supervision and damage control in an all-electric ship." *IASME Transactions*, 2(1): 253–8.

Neerincx, M.A., Ruijsendaal, M., and Wolff, M. (2001). "Usability engineering guide for integrated operation support in space station payloads." *International Journal of Cognitive Ergonomics*, 5(3): 187–98.

Neerincx, M.A., Lindenberg, J., Smets, N., Grant, T., Bos, A., Olmedo Soler, A., Brauer, U., and Wolff, M. (2006). "Cognitive engineering for long duration missions: Human–Machine collaboration on the Moon and Mars." *SMC–IT 2006: 2nd IEEE International Conference on Space Mission Challenges for Information Technology*. Los Alamitos, CA: IEEE Conference Publishing Services (pp. 40–46).

Neerincx, M.A. and Lindenberg, J. (1999). "Supporting individual situation awareness in web-environments." In E. Pikaar (ed.), *Ergonomie in Uitvoering: De Digitale Mens*. Utrecht: Nederlandse Vereniging voor Ergonomie (pp. 144–55).

Neerincx, M.A., Lindenberg, J., and Pemberton, S. (2001). "Support concepts for Web navigation: A cognitive engineering approach." *Proceedings of the 10th World Wide Web Conference*. New York: ACM Press (pp. 119–28).

Neerincx, M.A., Lindenberg, J., Rypkema, J., and Van Besouw, S. (2000). "A practical cognitive theory of web-navigation: Explaining age-related performance differences." In Position Paper *CHI 2000*, Workshop Basic Research Symposium, ACM.

Norman, D.A. (1986). "Cognitive engineering." In D.A. Norman, and S.W. Draper (eds), *User-Centered System Design: New Perspectives on Human–Computer Interaction*. Hillsdale, NJ: Lawrence Erlbaum Associates (pp. 31–62).

Rasmussen, J. (1986). *Information Processing and Human–Machine Interaction: An Approach to Cognitive Engineering*. Amsterdam: Elsevier.

Rogers, R.D. and Monsell, S. (1995). "Costs of predictable switch between simple cognitive tasks." *Journal of Experimental Psychology: General*, 124: 207–31.

Sit, R.A. and Fisk, A.D. (1999). "Age-related performance in a multiple task environment." *Human Factors*, 41(1): 26–34.

Vicente, K.J., Hayes, B.C., and Willeges, R.C. (1987). "Assaying and isolating individual-differences in searching a hierarchical file system." *Human Factors*, 29(3): 349–59.

Index of Names

Subject Index